Biology

for the IB Diploma

Second edition

Brenda Walpole

with
Ashby Merson-Davies
Leighton Dann

Cambridge University Press's mission is to advance learning, knowledge and research worldwide.

Our IB Diploma resources aim to:
- encourage learners to explore concepts, ideas and topics that have local and global significance
- help students develop a positive attitude to learning in preparation for higher education
- assist students in approaching complex questions, applying critical-thinking skills and forming reasoned answers.

CAMBRIDGE
UNIVERSITY PRESS

University Printing House, Cambridge CB2 8BS, United Kingdom

One Liberty Plaza, 20th Floor, New York, NY 10006, USA

477 Williamstown Road, Port Melbourne, VIC 3207, Australia

314–321, 3rd Floor, Plot 3, Splendor Forum, Jasola District Centre, New Delhi – 110025, India

103 Penang Road, #05-06/07, Visioncrest Commercial, Singapore 238467

Cambridge University Press is part of the University of Cambridge.

It furthers the University's mission by disseminating knowledge in the pursuit of education, learning and research at the highest international levels of excellence.

Information on this title:
www.cambridge.org/9781107654600 (Paperback)
www.cambridge.org/9781009331562 (Paperback + Digital Access, 2 years)
www.cambridge.org/9781107537798 (Digital Access, 2 years)

© Cambridge University Press 2011, 2014

This publication is in copyright. Subject to statutory exception and to the provisions of relevant collective licensing agreements, no reproduction of any part may take place without the written permission of Cambridge University Press.

First published 2011
Second edition 2014

20 19 18 17 16 15 14 13 12 11 10

Printed in Great Britain by CPI Group (UK) Ltd, Croydon CR0 4YY

A catalogue record for this publication is available from the British Library

ISBN 978-1-107-65460-0 Paperback
ISBN 978-1-009-33156-2 Paperback + Digital Access (2 years)
ISBN 978-1-107-53779-8 Digital Access (2 years)

Additional resources for this publication are available through Cambridge GO. Visit cambridge.org/go

Cambridge University Press has no responsibility for the persistence or accuracy of URLs for external or third-party internet websites referred to in this publication, and does not guarantee that any content on such websites is, or will remain, accurate or appropriate. Information regarding prices, travel timetables, and other factual information given in this work is correct at the time of first printing but Cambridge University Press does not guarantee the accuracy of such information thereafter.

The material has been developed independently by the publisher and the content is in no way connected with nor endorsed by the International Baccalaureate Organization.

....

NOTICE TO TEACHERS IN THE UK
It is illegal to reproduce any part of this book in material form (including photocopying and electronic storage) except under the following circumstances:
(i) where you are abiding by a licence granted to your school or institution by the Copyright Licensing Agency;
(ii) where no such licence exists, or where you wish to exceed the terms of a licence, and you have gained the written permission of Cambridge University Press;
(iii) where you are allowed to reproduce without permission under the provisions of Chapter 3 of the Copyright, Designs and Patents Act 1988, which covers, for example, the reproduction of short passages within certain types of educational anthology and reproduction for the purposes of setting examination questions.

The website accompanying this book contains further resources to support your IB Biology studies. Visit cambridge.org/go and register for access.

Separate website terms and conditions apply.

Contents

Introduction	**v**
1 Cell biology	**1**
Introduction	1
1.1 Introduction to cells	1
1.2 Ultrastructure of cells	12
1.3 Membrane structure	19
1.4 Membrane transport	23
1.5 The origin of cells	30
1.6 Cell division	33
Exam-style questions	41
2 Molecular biology	**43**
Introduction	43
2.1 Molecules to metabolism	43
2.2 Water	49
2.3 Carbohydrates and lipids	54
2.4 Proteins	59
2.5 Enzymes	62
2.6 Structure of DNA and RNA	68
2.7 DNA replication, transcription and translation	71
2.8 Cell respiration	78
2.9 Photosynthesis	82
Exam-style questions	89
3 Genetics	**93**
Introduction	93
3.1 Genes	93
3.2 Chromosomes	99
3.3 Meiosis	103
3.4 Inheritance	110
3.5 Genetic modification and biotechnology	124
Exam-style questions	133
4 Ecology	**136**
Introduction	136
4.1 Species, communities and ecosystems	136
4.2 Energy flow	141
4.3 Carbon recycling	147
4.4 Climate change	151
Exam-style questions	157
5 Evolution and biodiversity	**159**
Introduction	159
5.1 Evidence for evolution	159
5.2 Natural selection	164
5.3 Classification of biodiversity	169
5.4 Cladistics	178
Exam-style questions	185
6 Human physiology	**187**
Introduction	187
6.1 Digestion and absorption	187
6.2 The blood system	193
6.3 Defence against infectious disease	202
6.4 Gas exchange	209
6.5 Neurons and synapses	215
6.6 Hormones, homeostasis and reproduction	220
Exam-style questions	232
7 Nucleic acids (HL)	**235**
Introduction	235
7.1 DNA structure and replication	235
7.2 Transcription and gene expression	245
7.3 Translation	251
Exam-style questions	259
8 Metabolism, cell respiration and photosynthesis (HL)	**262**
Introduction	262
8.1 Metabolism	262
8.2 Cell respiration	268
8.3 Photosynthesis	276
Exam-style questions	283
9 Plant biology (HL)	**285**
Introduction	285
9.1 Transport in the xylem of plants	285
9.2 Transport in the phloem of plants	293
9.3 Growth in plants	297
9.4 Reproduction in plants	300
Exam-style questions	305

10	**Genetics and evolution (HL)**	**308**
	Introduction	308
	10.1 Meiosis	308
	10.2 Inheritance	313
	10.3 Gene pools and speciation	327
	Exam-style questions	335
11	**Animal physiology (HL)**	**337**
	Introduction	337
	11.1 Antibody production and vaccination	337
	11.2 Movement	346
	11.3 The kidney and osmoregulation	353
	11.4 Sexual reproduction	365
	Exam-style questions	377

Answers to test yourself questions	**380**
Glossary	**386**
Index	**401**
Acknowledgements	**407**

Free online material

The website accompanying this book contains further resources to support your IB Biology studies. Visit **cambridge.org/go** and register to access these resources:

Options
Option A Neurobiology and behaviour
Option B Biotechnology and bioinformatics
Option C Ecology and conservation
Option D Human physiology

Self-test questions
Assessment guidance
Model exam papers
Nature of Science
Answers to exam-style questions
Answers to Options questions

Introduction

Biology has advanced at a rapid rate over recent decades and is truly the science of the 21st century. Advances in genetics, biochemistry, medicine and cell biology have kept the subject in the forefront of international news. To keep pace with new developments, the IB Biology course is regularly updated so that IB students can understand not only the principles of modern science but also the processes and the ethical implications that go with them. The latest revision of the IB Biology syllabus will be examined in the years 2016–2022, and this second edition of *Biology for the IB Diploma* is fully updated to cover the content of that syllabus.

Biology may be studied at Standard Level (SL) or Higher Level (HL). Both share a common core, which is covered in Topics 1–6. At HL the core is extended to include Topics 7–11. In addition, at both levels, students then choose one Option to complete their studies. Each option consists of common core and additional Higher Level material. You can identify the HL content in this book by 'HL' included in the topic title (or section title in the Options), and by the red page border. The Options are included in the free online material that is accessible via **cambridge.org/go**

The structure of this book follows the structure of the IB Biology syllabus. Each topic in the book matches a syllabus topic, and the sections within each topic mirror the sections in the syllabus. Each section begins with learning objectives as starting and reference points. Test yourself questions appear throughout the text so students can check their progress and become familiar with the style and command terms used, and examination style questions appear at the end of each topic.

Theory of Knowledge (TOK) provides a cross-curricular link between different subjects. It stimulates thought about critical thinking and how we can say we know what we claim to know. Throughout this book, TOK features highlight concepts in Biology that can be considered from a TOK perspective. These are indicated by the 'TOK' logo, shown here.

Science is a truly international endeavour, being practised across all continents, frequently in international or even global partnerships. Many problems that science aims to solve are international, and will require globally implemented solutions. Throughout this book, International-Mindedness features highlight international concerns in Biology. These are indicated by the 'International-Mindedness' logo, shown here.

Nature of Science is an overarching theme of the Biology course. The theme examines the processes and concepts that are central to scientific endeavour, and how science serves and connects with the wider community. At the end of each section in this book, there is a 'Nature of Science' paragraph that discusses a particular concept or discovery from the point of view of one or more aspects of Nature of Science. A chapter giving a general introduction to the Nature of Science theme is available in the free online material.

Free online material

Additional material to support the IB Biology Diploma course is available online. Visit cambridge.org/go and register to access these resources.

Besides the Options and Nature of Science chapter, you will find a collection of resources to help with revision and exam preparation. This includes guidance on the assessments, interactive self-test questions and model exam papers. Additionally, answers to the exam-style questions in this book and to all the questions in the Options are available.

Cell biology 1

Introduction

In the middle of the seventeenth century, one of the pioneers of microscopy, Robert Hooke (1635–1703), decided to examine a piece of cork tissue with his home-built microscope. He saw numerous box-shaped structures that he thought resembled monks' cells or rooms in a monastery, so he called them 'cells'. As microscopes became more sophisticated, other scientists observed cells and found that they occurred in every organism. No organism has yet been discovered that does not have at least one cell. Living things may vary in shape and size but scientists agree that they are all composed of cells. The study of cells has enabled us to learn more about how whole organisms function.

Learning objectives

You should understand that:
- Cell theory explains that living organisms are composed of cells.
- Unicellular organisms carry out all the functions of life.
- Surface area to volume ratio is an important factor in limiting cell size.
- Interactions between their cellular components lead to new emergent properties in multicellular organisms.
- Multicellular organisms have specialised tissues, which develop as a result of cell differentiation.
- Cell differentiation results from the expression of some genes but not others.
- Stem cells are able to divide and differentiate along different pathways and are essential for embryonic development. This ability makes them suitable for therapeutic uses.

1.1 Introduction to cells

The cell theory

Today, scientists agree that the cell is the fundamental unit of all life forms. **Cell theory** proposes that all organisms are composed of one or more cells and, furthermore, that cells are the smallest units of life. An individual cell can perform all the functions of life and anything that is not made of cells, such as viruses, cannot be considered living.

One of the key life processes of all living organisms is reproduction. Therefore, one of the first principles of the cell theory is that cells can only come from pre-existing cells. Louis Pasteur (1822–1895) carried out experiments that provided evidence for this.

Extensive examination of many organisms and millions of different types of cell supports the cell theory, although a few examples have been found that do not fit the theory perfectly. One example is fungi, whose structures consist of long threads called hyphae (Figure **1.1**), which have many nuclei but are not divided into separate cells by cell walls. Another example is skeletal muscle, which is composed of muscle fibres that are much larger than a single cell and contain several hundred nuclei. Cells of some large algae are somewhat anomalous because their single cells are undifferentiated but are attached to chains of identical cells or surrounded by a matrix of extra cellular material so that they form large structures, and mammalian erythrocytes (red blood cells) do not contain nuclei once they have matured and been released into the bloodstream, which means at this stage of their life cycle they cannot carry out all the functions of life so they too depart from cell theory.

Unicellular organisms

By definition, a living organism comprising just one cell has to perform all the necessary functions for survival.

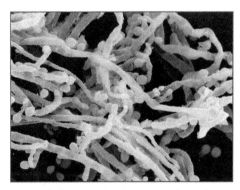

Figure 1.1 Fungal hyphae grow through material that nourishes the fungus. In this scanning electron micrograph the thread like structures are the hyphae and the pale pink spheres are the reproductive spores (×2000).

Key principles of the cell theory:

- living organisms are composed of cells
- cells are the smallest units of life
- all cells come from pre-existing cells.

Drawing cell structures

When you draw cells as they appear under a microscope, always use a sharp pencil and draw single lines to show the relative sizes and positions of the structures you can see. Do not use shading or cross-hatching on your diagram. Label each structure with a straight line so that the name of each part appears at the side of your diagram. Always include a title and the magnification of your drawing. You can see an example of how to do this in Figure **1.2**.

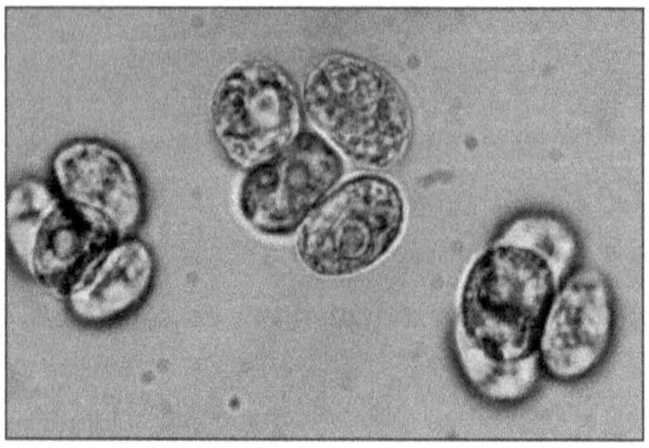

The functions of life are:
- metabolism
- growth
- response (or sensitivity)
- homeostasis
- nutrition
- reproduction
- excretion

A unicellular organism such as *Paramecium* (Figure **1.2**) needs to metabolise organic materials in order to make the chemicals needed to sustain life. It must also be able to **excrete** waste produced during metabolism and dispose of it. It must be able to detect changes in its environment, so it can **respond** to more favourable or less favourable conditions. Some unicellular organisms photosynthesise and they have a light spot that enables them to move to a brighter environment to maximise photosynthesis. A unicellular organism must also be able to control its internal environment (homeostasis), as large changes in water or salt concentrations may have a detrimental effect on metabolism and other cellular functions. It must also obtain food, whether produced from simple inorganic substances through photosynthesis (as in *Chlorella*, Figure **1.3**) or ingested as complex organic materials from outside as a source of nutrition. If the species is to survive, an organism must be able to reproduce. This could be either asexual or sexual reproduction.

Investigations of some life processes in *Paramecium* and *Chlorella*

Paramecium can be observed under a light microscope. Paramecia have cilia, which they flick in rhythmic waves to move about in water, and they also have a row of specialised cilia that waft food particles towards the

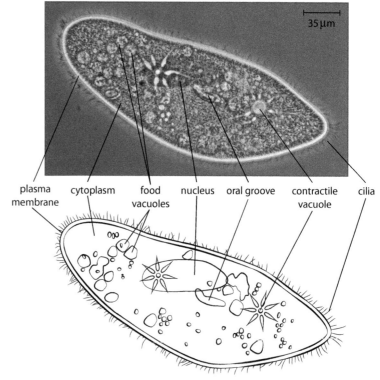

Figure 1.2 *Paramecium* carries out all the life functions within its single cell (×300).

Figure 1.3 *Chlorella* is a unicellular organism containing a chloroplast (×1200).

oral groove. If stained yeast cells are added to the culture of *Paramecium* it is possible to observe the path taken by food particles through the body of the organism and the food vacuoles that are formed. If *Paramecium* is placed in water of different salinities from distilled water – 0.1%, 0.2%, 0.4% and 0.8% sodium chloride solution, for example – the contractile vacuole, which controls the water balance of the cell, can be seen forming and emptying.

Chlorella is a photosynthetic organism with a rapid growth rate. Although its cells are small and must be viewed with a microscope, it can quickly produce large numbers of individuals, which turn water green and opaque. This is most likely to happen when *Chlorella* grows in water that is rich in nitrates or phosphates. The organism has been used in many scientific experiments; the Nobel prize winner Otto Warburg published his pioneering work on cellular metabolism following intensive experiments on *Chlorella* in 1919, and in 1961 Melvin Calvin carried out his experiments on photosynthesis using *Chlorella* (Subtopic **8.3**).

Cell size

One of the few cells large enough to be visible to the unaided eye is the mature human ovum, which has a diameter of approximately 150 μm. However, most cells are much smaller than this, and can only be seen using a microscope. Light microscopes, which can magnify up to 2000 times, reveal some internal structures such as the nucleus, but greater detail requires the use of more powerful microscopes such as the electron microscope, which magnifies cell structures up to 500 000 times. Viruses can only be seen with these microscopes, so the structure of viruses was unknown until the invention of these microscopes in the 20th century.

Electron microscopes use a beam of electrons, instead of light, to produce an image. The **resolution** of an electron microscope is much better than that of a light microscope because of the shorter wavelength of electrons. **Resolving power** is the ability of the microscope to separate objects that are close together so that more detail can be seen. Only non-living material can be observed in an electron microscope and specimens must be prepared with heavy metals or coated with carbon or gold. There are two types of electron microscope: the TEM (transmission electron microscope) and SEM (scanning electron microscope). A TEM produces clear images of thin sections of material while in an SEM electrons are bounced off objects to produce detailed images of their external appearance. Both types of microscope produce black and white images but these are often artificially coloured so that certain features can be seen more clearly. Table **1.1** compares the different types of microscope.

Even the electron microscope cannot distinguish individual molecules. Other techniques such as X-ray crystallography are needed to do this. Figure **1.4.** indicates the relative sizes of some biological structures.

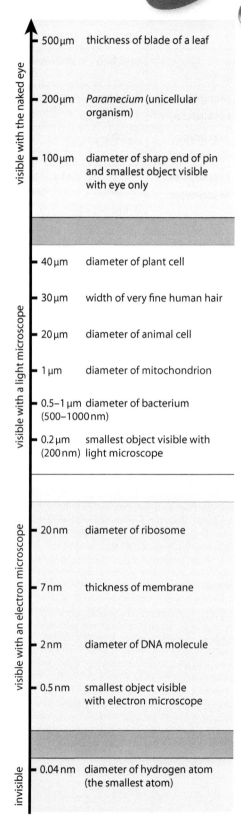

Figure 1.4 The sizes of some biological structures.

	Light microscope	**TEM**	**SEM**
	uses light to produce images	uses electron beams to produce images	uses electron beams to produce images
Maximum resolution	200 nm	1 nm	1 nm
Maximum magnification	×2000	up to ×1 000 000	×200 000
Preparation of material	thin sections of material mounted on slides living organisms can be examined	very thin sections of material supported on metal grids living organisms cannot be examined	very thin sections of material supported on metal grids living organisms cannot be examined
Stain used	coloured dyes	heavy metals	carbon or gold coating
Image	viewed directly through eyepiece lens	viewed on a screen or photographic plate	viewed on a screen or photographic plate

Table 1.1 Comparison of light microscopes with transmission electron microscope (TEM) and scanning elelctron microscope (SEM).

Magnification and scale

Knowing the sizes of objects viewed under the microscope can be very useful (Figure **1.5**). For example, a plant scientist might want to compare the relative sizes of pollen grains from plants in the same genus to help identify different species.

Magnification is defined as the ratio of the size of the image to the size of the object:

$$\text{magnification} = \frac{\text{size of image}}{\text{size of object}}$$

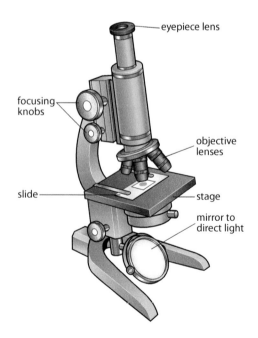

Figure 1.5 Typical compound light microscope.

With a compound microscope, the magnification is the product of both lenses, so if a microscope has a ×10 eyepiece and ×40 objective, the total magnification is ×400.

Printed images of structures seen with a microscope usually show a scale bar or give the magnification, so that the size of an object can be calculated. For example, the magnification of the micrograph in Figure **1.6** is given as ×165.

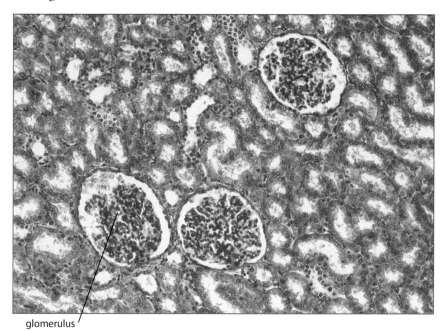

glomerulus

Figure 1.6 Coloured light micrograph of a section through the cortex of a kidney (×165).

In Figure **1.6**, there are three spherical glomeruli present. In the image, each one is approximately 25 mm across. You can check this using a ruler. Thus:

$$\text{actual size of glomerulus} = \frac{\text{size of image}}{\text{magnification}}$$

$$= \frac{25\,\text{mm}}{165}$$

$$= 0.15\,\text{mm}$$

In electron micrographs, most measurements are expressed in micrometres. A micrometre (μm) is 10^{-3} mm, so 1 mm is 1000 μm.

So the diameter of the glomerulus = 0.15 × 1000 = 150 μm.

Worked example

1.1 This image shows a red blood cell. The scale bar shows 2 μm. From this, you can calculate both the size of the cell and the magnification of the image.

Size of the cell
Step 1 Use a ruler to measure the length of the cell (its diameter in this case). This is 30 mm.
Step 2 Use a ruler to measure the length of the scale bar. This is 9 mm.
Step 3 Use the ratio of these two values to work out the actual length of the cell.

$$\frac{2\,\mu m}{9000\,\mu m} = \frac{\text{actual length of cell}}{30\,000\,\mu m}$$

(Remember to convert all the units to μm. 1 mm = 1000 μm.)

Rearranging the equation:

$$\text{actual length of the cell} = 2\,\mu m \times \frac{30\,000\,\mu m}{9000\,\mu m}$$

$$= 6.7\,\mu m$$

Magnification of the image
Use the formula:

$$\text{magnification} = \frac{\text{measured length of the cell}}{\text{actual length of the cell}}$$

So in this case:

$$\text{magnification} = \frac{30\,000\,\mu m}{6.7\,\mu m}$$

$$= \times 4500$$

If you are given a value for the magnification, you can measure the length of the object in the image and then rearrange the equation to work out the actual length of the object.

SI units – International System

1 metre (m) = 1 m

1 millimetre (mm) = 10^{-3} m

1 micrometre (μm) = 10^{-6} m

1 nanometre (nm) = 10^{-9} m

1 centimetre cubed = 1 cm^3

1 decimetre cubed = 1 dm^3

1 second = 1 s

1 minute = 1 min

1 hour = 1 h

concentration is measured in mol dm^{-3}

Becoming multicellular

Surface area to volume ratio

Cells are very small, no matter what the size of the organism that they are part of. Cells do not and cannot grow to be very large and this is important in the way living organisms are built and function. The volume of a cell determines the level of metabolic activity that takes place within it. The surface area of a cell determines the rate of exchange of materials with the outside environment. As the volume of a cell increases, so does its surface area, but not in the same proportion, as Table **1.2** shows for a theoretical cube-shaped cell. So as a cell grows larger, it has proportionately less surface area to obtain the materials it needs and to dispose of waste. The rate of exchange of materials across the outer membrane becomes limiting and cannot keep up with the cell's requirements. Some cells have specialised structures, such as folds and microvilli, to provide a larger surface area relative to their volume but nevertheless there is a limit to the size of a single cell. Beyond this limit, a cell must divide and an organism must become multicellular.

Becoming multicellular has enormous advantages. An organism can grow in size and its cells can **differentiate** – that is, they can take on specific functions, so the organism can grow in complexity as well as size. A multicellular organism may have specialised nerve cells for communication and interaction with the outside, and muscle cells for movement. It may also have special reproductive cells and secretory cells that produce enzymes for digestion. Differentiation allows for new properties to emerge as different cell types interact with each other to allow more complex functions to take place. For example, nerve cells may interact with muscle cells to stimulate movement.

Side of cube/mm	Surface area/mm^2	Volume/mm^3	Ratio of surface area : volume
1	6	1	6 : 1
2	24	8	3 : 1
3	54	27	2 : 1

Table **1.2** Surface area to volume ratios for a cube.

Test yourself

1 Many cells are roughly spherical in shape. The volume of a sphere is $\frac{4}{3}\pi r^3$ and its surface area is $4\pi r^2$. Make a table similar to Table **1.2**, this time for a sphere using a different radii as a starting point. Describe the relationship between surface area and volume in this case.

Take a 2 cm cube of modelling clay. Change its shape so that it becomes a cuboid, a thin cylinder or a sphere. Calculate its surface area each time. Try creating folds in the surface. Which shape produces the greatest surface area?

Emergent properties

One person playing the flute can produce a simple, recognisable tune but if several musicians with other instruments join in and play together as a group, they produce a wide variety of sounds and many different effects. New properties emerge in the cells of multicellular organisms in a similar way. Their cellular components interact so that the organism can carry out a range of more complicated functions. One cell can function on its own, but with other cells in a group, it can produce tissues and organs that carry out a range of roles in the organism. For example, lungs are made of many cells – it is only when all these cells work as a unit that the lungs are able to perform their function. Cells form tissues, tissues form organs, organs form organ systems and organ systems work in synergy so that the whole organism can carry out a complex range of tasks and is greater than the composition of its parts.

The systems approach

A system is defined as an assemblage of parts and the relationships between them that enable them to work together as a functioning whole. The systems approach has long been used in engineering but for many years natural systems were examined from a reductionist point of view. We can see how the two approaches differ if we consider the study of a pond. A reductionist study of the pond would describe the organisms found there in terms of their features and characteristics; for example, whether they are vertebrates or invertebrates, plant or animal. But a reductionist study would not try to consider how the pond worked as a dynamic system.

A systems approach would take a holistic view of the pond and consider interrelationships such as food chains and nutrient cycling that occur between the various components of the pond. In this way a picture of the interdependence of the different parts of the pond – that is, the system's structure – could be built up.

In a study of cells and their components, the systems approach would consider a single cell in terms of the flows of energy and materials between the various structures within in it. On a larger scale, groups of cells, an organ or even a whole organism can be studied using the systems approach so that the parts and the interactions between them can be viewed as a complete functioning entity. Emergent properties in any system can only be studied by means of a systems approach.

Questions to consider
- What are the advantages and disadvantages of the systems approach compared with the reductionist approach to the study of cells?
- In science, the reductionist and the systems approach may use similar methods of study. What is the most important difference between the philosophies of the two approaches?

Differentiation

How do cells in the same organism behave in different ways when they all arose from the same parent cell and so have the same genome (genetic make-up)? In a particular organism, nerve cells and muscle cells all have the same genes but look and behave very differently. The logical answer is that in some cells particular genes are expressed that are not expressed in other cells, and vice versa. For example, a pancreatic cell will express genes for the production of digestive enzymes or insulin, but a skin cell will not. **Differentiation** involves the expression of some genes from the organism's genome in the cell, but not others.

Stem cells

The fertilised egg of any organism contains all the information needed for developing that single cell into a complex organism consisting of many different types of cell. This information is all within the genes, inherited from the maternal and paternal DNA as fine threads called chromosomes. (There is more information on DNA and chromosome structure in Subtopics **2.6** and **3.2**.) A fertilised egg divides rapidly and produces a ball of cells called a blastocyst in which all the cells are alike. Gradually, after this stage, the cells differentiate and become destined to form specialised tissues such as muscle or liver. The process of *differentiation* produces cells for specific purposes – muscle cells for contraction, liver cells for metabolism of toxins, and so on. Once differentiation has happened, it cannot be reversed. Cells in the blastocyst have the potential to turn into a great many different cell types: they are said to be **pluripotent** and are known as **embryonic stem cells**.

Embryonic stem cells are unique in their potential versatility to differentiate into all the body's cell types. However, some adult tissues contain a different form of stem cell – one that can only differentiate into cells associated with that tissue. For example, bone marrow contains stem cells that can form all the different types of blood cell, but not muscle cells or liver cells (Figure **1.7**).

Stem cells differ from most other cells in the following ways.
- They are unspecialised.
- They can divide repeatedly to make large numbers of new cells.
- They can differentiate into several types of cell.
- They have a large nucleus relative to the volume of the cytoplasm.

Ethical issues in stem cell research

Scientists began to investigate and culture stem cells in the 1980s and it soon became apparent that there was enormous potential in using these cells therapeutically. Some of the most recent research aims to grow stem cells to replace damaged or diseased tissue in patients suffering from degenerative conditions such as multiple sclerosis or Alzheimer's disease. Early work concentrated on using embryonic stem cells, but these can only be obtained from discarded embryos from *in vitro* fertilisation (IVF) clinics. There is much debate about the ethics of doing this kind of work, and many people feel that the destruction of an embryo to obtain stem cells is morally unacceptable. Others argue that this type of research will contribute significantly to the treatment of disease and can therefore be fully justified.

The use of adult stem cells is a less controversial area of research. In this case, cells are obtained from bone marrow or other tissue from a donor who has given consent. Cells can be harvested and grown, and used in medical treatments. Bone marrow transplants already help many leukaemia patients to a full recovery.

Although scientists from many countries have cooperated on stem cell research, the laws governing research vary from place to place. In the European Union research using human embryos is permitted in some countries but is illegal in others, such as Germany, Ireland, Italy, Portugal and Austria. In the USA some states fund the research while others ban it; and although Australia permits research, New Zealand restricts it. Laws also differ in their application to embryonic stem cell research and in stems cells taken from adult tissues. Because laws depend on political and religious viewpoints, they may change as new information and research develops.

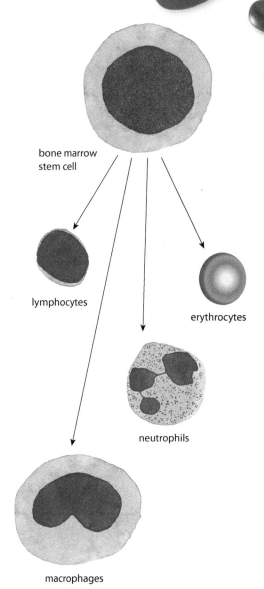

Figure 1.7 Bone marrow cells differentiate into the different types of blood cell.

Plants also contain stem cells. These are found in the meristems just behind the tips of growing stems and roots. These cells can differentiate to become various tissues of the stem and roots.

1 CELL BIOLOGY

Therapeutic use of stem cells

Another source of stem cells, which has been successfully used in medical treatments, is the blood in the umbilical cord of a newborn baby (Figure 1.8). These stem cells can divide and become any type of blood cell. Cord blood can be used to treat certain types of leukaemia, a cancer that causes overproduction of white blood cells in the bone marrow. Cells from the cord blood are collected and their tissue type is determined. After chemotherapy to destroy the patient's own bone marrow cells, stem cells which are the correct match to the patient's tissue are given by transfusion. They become established in the person's bone marrow and start producing blood cells as normal.

This treatment can work well in young children, but there are not enough cells in a single cord to meet the needs of an adult patient. Scientists have been looking for ways to either combine the cells from more than one baby, or to increase the number of cells in the laboratory. Allowing the stem cells to divide in the laboratory produces many blood cells, but not more stem cells. In 2010, scientists at the Fred Hutchinson Cancer Research Center in Seattle, USA, managed to alter a signalling pathway in the stem cells so they could increase in number without losing stem cell properties. As a result of this process, known as therapeutic cloning, umbical cord blood may prove to be an even more valuable source of stem cells in the future.

One of the most recent areas of stem cell research has been in the treatment of Stargardt's disease using retinal pigment epithelium (RPE). Stargardt's disease is an inherited condition, which begins in childhood and leads to macular degeneration (a gradual destruction of cells in the centre of the retina) and eventually causes blindness. Retinal cells can be made from embryonic stem cells. In 2012, as part of a larger trial, the first patients were given transplants of retinal cells developed from human embryonic stem cells, to treat their condition. The cells were injected directly into the retina and researchers found that not only did the stem cells survive but the number of cells increased over a period of 3 months. The cells began to develop important visual pigment and the patients noticed improvement in their vision. Scientists hope that in the future, stem cells will restore some sight not only to people with Stargardt's disease but also many millions of older people suffering from age-related macular degeneration, the most common cause of blindness. Stem cell therapy has also been successfully used in the treatment of type I diabetes, and research is continuing into therapies to treat a range of conditions involving neurological damage, such as multiple sclerosis and Alzheimer's disease.

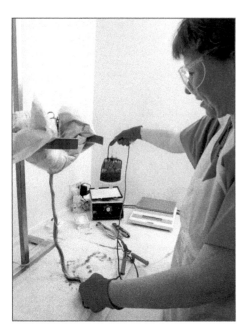

Figure 1.8 This technician is collecting blood from an umbilical cord. This blood is a rich source of stem cells.

Nature of science

Looking for trends and discrepancies – developing theories

In order to explain aspects of the natural world, scientists develop theories. A **theory** is a well-established and widely accepted principle that arises from extensive observation of trends and discrepancies, and incorporates facts, laws, predictions and tested hypotheses. A **hypothesis** is a speculative, specific and testable prediction about what is expected to happen in an investigation. So, a theory predicts events in general terms, while a hypothesis makes a specific prediction about a narrowly defined set of circumstances. If any evidence is collected which contradicts a hypothesis, the hypothesis must be rejected and a new hypothesis formulated.

In developing a theory, it is important to consider not only the main trends of an idea – those observations that are in general agreement – but also to seek out discrepancies which might go against the trend, and perhaps suggest testable hypotheses leading to a change in the theory. So, for example, with regard to cell theory, scientists might ask if observations suggest that the structure of *all* parts of *all* organisms conform to the theory, or if there are any discrepancies.

As we have seen, many millions of cells have been examined and the majority do conform to the principles of cell theory. But fungal hyphae and the fibres in muscles cells could be considered a challenge to it. Mammalian red blood cells have no nucleus (even though similar cells in reptiles and other vertebrates do), which means they cannot carry out all the functions of life and depart from cell theory in that respect. Single-celled protoctista such as *Amoeba* could be thought of as a challenge to the idea that every cell has a specialised function as these organisms carry out all the functions of life. Science must consider observations like these from many different fields of study and be prepared to revise accepted theories if it is necessary.

Questions to consider

- How can evidence be obtained for the principles of the cell theory? Can we prove that cells always arise from pre-existing cells?
- Do the examples of fungal hyphae and muscle cells disprove the cell theory?
- What should happen if evidence is collected that cannot be explained by a theory?
- What happens if evidence is collected that disproves a hypothesis?

Test yourself

2 Calculate how many cells of 100 μm diameter will fit along a 1 mm line.
3 List examples of where the concept of emergent properties can be found in a multicellular animal, such as a bird or a flowering plant.
4 State **one** therapeutic use of stem cells.
5 Explain how cells in multicellular organisms are able to carry out specialised functions.
6 Use the scale bar on Figure **1.2** to calculate the length of the *Paramecium* in the photograph.

Learning objectives

You should understand that:
- Prokaryotes have a simple cell structure with no compartmentalisation.
- Eukaryotes have a compartmentalised cell structure with membrane-bound organelles present in the cytoplasm.
- The resolution of electron microscopes is much higher than that of light microscopes, which allows identification of cell structures and organelles.

1.2 Ultrastructure of cells

Living things are divided into two types – prokaryotes and eukaryotes – according to the structure of their cells. Prokaryotic cells are usually much smaller than eukaryotic cells and have a much simpler structure. Prokaryotes are thought to be the first cells to have evolved. Bacteria are all prokaryotic cells.

Prokaryotic cells

Prokaryotic cells are so called because they have no nucleus ('prokaryote' comes from the Greek, meaning 'before the nucleus'). They also have no organelles (internal structures), so cell functions do not take place in separate compartments within the cytoplasm. From the mid-20th century, when the electron microscope was developed, it became possible to study the internal detail of cells. Figures **1.9** and **1.10** show the main features of a typical prokaryotic cell.

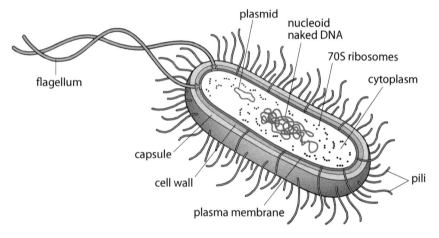

Figure 1.10 The structure of a typical prokaryotic cell.

Figure 1.9 The bacterium *Escherichia coli* is a typical prokaryotic cell. (Coloured TEM × 60 000).

- The **cell wall** surrounds the cell. It protects the cell from bursting and is composed of peptidoglycan, which is a mixture of carbohydrate and amino acids.
- The **plasma membrane** controls the movement of materials into and out of the cell. Some substances are pumped in and out using active transport.
- Cytoplasm inside the membrane contains all the enzymes for the chemical reactions of the cell. It also contains the genetic material.
- The **chromosome** is found in a region of the cytoplasm called the nucleoid. The DNA is not contained in a nuclear envelope and it is also 'naked' – that is, not associated with any proteins. Bacteria also contain additional small circles of DNA called plasmids. Plasmids replicate independently and may be passed from one cell to another.
- **Ribosomes** are found in all prokaryotic cells, where they synthesise proteins. They can be seen in very large numbers in cells that are actively producing protein. Prokaryotes have 70S ribosomes, which are smaller than those found in eukaryotes.

- A **flagellum** is present in some prokaryotic cells. A flagellum, which projects from the cell wall, enables a cell to move.
- Some bacteria have **pili** (singular pilus). These structures, found on the cell wall, can connect to other bacterial cells, drawing them together so that genetic material can be exchanged between them.

Prokaryotic cells are usually much smaller in volume than more complex cells because they have no nucleus. Their means of division is also simple. As they grow, their DNA replicates and separates into two different areas of the cytoplasm, which then divides into two. This is called binary fission. It differs slightly from mitosis in eukaryotic cells (Subtopic **1.6**).

Eukaryotic cells

Eukaryotic organisms have cells that contain a nucleus. Animals, plants, fungi and protoctista all have eukaryotic cells.

The complexity of a eukaryotic cell cannot be fully appreciated using a compound light microscope. But in images made using an electron microscope, which has a much higher resolution, the fine details of many different organelles are visible. Figure **1.11** shows what can be seen of animal and plant cells using a light microscope – compare these images with the electron micrographs and interpretive drawings in Figures **1.12** to **1.15**.

Ribosome sizes

The 'S' unit used to 'measure' ribosomes is a Svedberg unit. It is a measure of the behaviour of particles during sedimentation. 70S and 80S ribosomes are different sizes and so take different times to sediment when they are centrifuged. They are said to have different sedimentation coefficients.

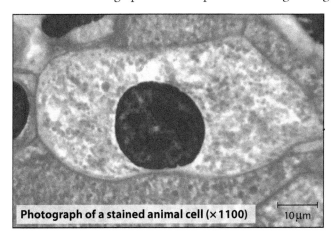
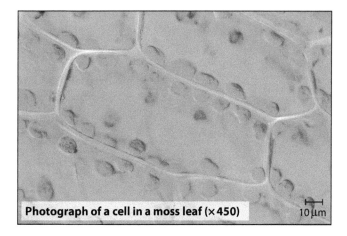
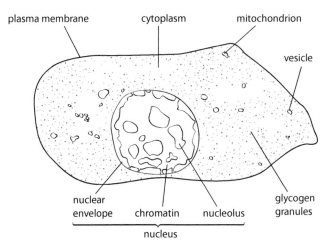
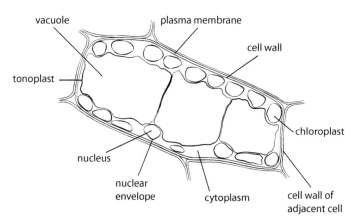

Figure 1.11 Photographs and interpretive drawings to show typical animal and plant cells as they appear using a light microscope.

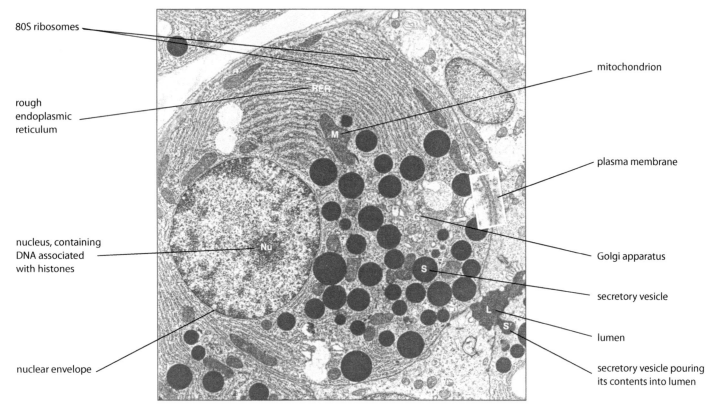

Figure 1.12 Electron micrograph of an exocrine cell from the pancreas (×12 000).

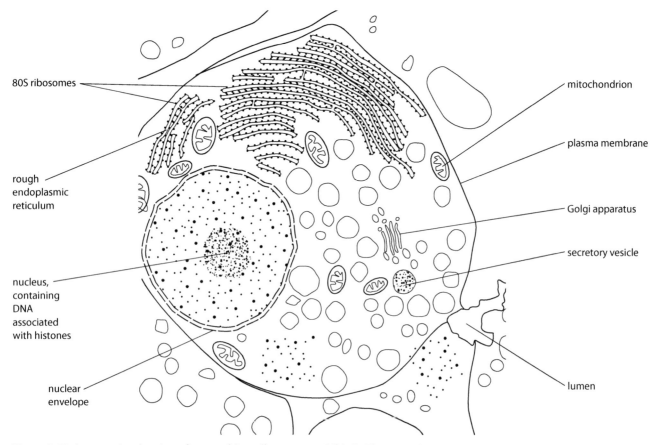

Figure 1.13 Interpretive drawing of some of the cell structures visible in **Figure 1.12**.

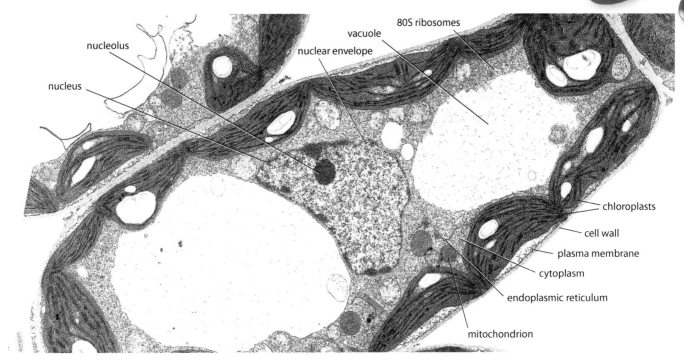

Figure 1.14 Electron micrograph of a palisade mesophyll plant cell (×5600).

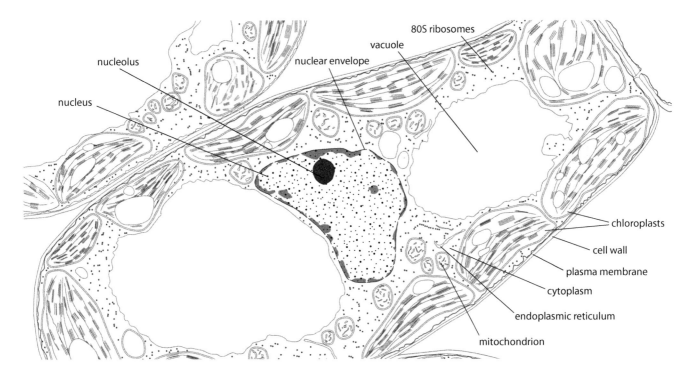

Figure 1.15 Drawing of a palisade mesophyll plant cell made from the electron micrograph in **Figure 1.14**.

Eukaryotic cells contain structures called **organelles**, each of which forms a 'compartment' in which specific functions take place. This compartmentalisation enables a eukaryotic cell to carry out various chemical reactions or processes in separate parts of the cell, which all form part of the same system. Different types of cell have different organelles in different proportions, depending on the role of the cell.

The largest and most obvious structure in a eukaryotic cell is the **nucleus**, which contains the cell's chromosomes. **Chromosomes** are composed of DNA combined with histone protein, forming a material known as chromatin. The nucleus is surrounded by a double-layered membrane, the **nuclear envelope**. Small gaps in the envelope, called nuclear pores, are visible and it is through these that material passes between the nucleus and the rest of the cell. A distinctive feature of the nucleus is the darkly staining **nucleolus**. This is the site of ribosome production.

Associated with the nuclear envelope is a series of membranes known as the **endoplasmic reticulum** (ER). Ribosomes attach to this network to form **rough endoplasmic reticulum** (rER), the site of protein synthesis. As proteins are produced, they collect in the spaces between the membranes, known as the **cisternae**. From here they can be transported in vesicles to other parts of the cell such as the Golgi apparatus. ER that has no ribosomes attached is known as **smooth endoplasmic reticulum** (sER). The membranes of sER have many enzymes on their surfaces. Smooth ER has different roles in different types of cell – in liver cells, it is where toxins are broken down; in the ovaries, it is the site of estrogen production. Smooth ER also produces phospholipids for the construction of membranes and lipids for use in the cell.

The **Golgi apparatus** is similar in appearance to the sER, composed of stacks of flattened, folded membranes. It processes proteins made in the rER, collecting, packaging and modifying them, and then releasing them in vesicles for transport to various parts of the cell or for secretion from the cell. The pancreas contains many secretory cells, which have large areas of Golgi apparatus (Figures **1.12** and **1.13**).

Eukaryotic cells also contain mitochondria (singular **mitochondrion**). These are elongated structures surrounded by a double membrane that are found throughout the cytoplasm. Mitochondria are known as the cell's 'powerhouses' because they are the site of aerobic respiration. The inner membrane is folded to form cristae, which greatly increase the surface area for the production of ATP in the cell. Cells that respire rapidly, such as muscle cells, have numerous mitochondria.

Lysosomes are spherical organelles with little internal structure, which are made by the Golgi apparatus. They contain hydrolytic enzymes for breaking down components of cells. They are important in cell death, in breaking down old organelles and, in white blood cells, digesting bacteria that have been engulfed by phagocytosis. Plant cells do not normally contain lysosomes.

Ribosomes are the site of protein synthesis in cells. They may be free in the cytoplasm or attached to the rER. They are made of RNA and protein but they do not have a membrane around them. Eukaryotic cells contain 80S ribosomes, which are larger than those found in prokaryotes.

As in prokaryotic cells, the **plasma membrane** controls the movement of materials into and out of the cell, and the gel-like **cytoplasm**, which fills much of the volume of the cell, provides a medium for many metabolic reactions.

Plant cells have three additional structures. All plant cells have an outer cellulose cell wall and most have a large central vacuole. Some plant cells, such as palisade mesophyll cells (Figures **1.14** and **1.15**), contain chloroplasts. The **chloroplasts** are found in cells exposed to the light, as they are the sites of photosynthesis. Chloroplasts have a double membrane and are about the same size as bacteria. Both chloroplasts and mitochondria have their own DNA and ribosomes and are able to reproduce independently of the cell.

The large central **vacuole** contains water and salts. The membrane that surrounds it is under pressure from within and exerts a force on the cytoplasm, which in turn exerts a force on the cell wall, making the cell turgid and firm. The outer **cell wall** is composed of cellulose and other carbohydrates such as lignin and pectin, giving plant cells further support and a more rigid structure than animal cells. The cell walls and turgidity of plant cells give strength and support to tissues like leaves, holding them in the optimum position to catch the energy from sunlight for photosynthesis.

Although they are both eukaryotic cells, there are several key differences between animal and plant cells. These are summarised in Table **1.3**.

Animal cells	Plant cells
cell wall absent	cell wall present
small vacuoles sometimes present	large central vacuole present in mature cells
no chloroplasts	chloroplasts often present
cholesterol in plasma membrane	no cholesterol in plasma membrane
centrioles present (see page 36)	centrioles absent
stores glycogen	stores starch

Table 1.3 Differences between animal and plant cells.

Differences between prokaryotic and eukaryotic cells

Comparisons of images of prokaryotic and eukaryotic cells show numerous differences between them. These are summarised in Table **1.4**. Note, for example, the difference in size of ribosomes between prokaryotic and eukaryotic cells.

Structure	Eukaryotic cell	Prokaryotic cell
nucleus	usually present, surrounded by a nuclear envelope and containing chromosomes and a nucleolus	no nucleus, and therefore no nuclear envelope or nucleolus
mitochondria	usually present	never present
chloroplasts	present in some plant cells	never present
endoplasmic reticulum	usually present	never present
ribosomes	relatively large, about 30 nm in diameter, or 80S	relatively small, about 20 nm in diameter, or 70S
chromosomes	DNA arranged in long strands, associated with histone proteins	DNA present, not associated with proteins, circular plasmids may also be present
cell wall	always present in plant cells, made of cellulose, never present in animal cells	always present, made of peptidoglycan
flagella	sometimes present	some have flagella, but these have a different structure from those in eukaryotic cells

Table 1.4 Differences between prokaryotic and eukaryotic cells. The unit 'S' is a Svedberg unit, used to compare sizes of cell organelles.

Nature of science

Scientific advance follows technical innovation – the electron microscope

A typical animal cell is 10–20 µm in diameter, which is about one-fifth the size of the smallest particle visible to the naked eye. Robert Hooke was the first scientist to see and describe cells, although he didn't know what they were. Later, Anton van Leeuwenhoek, who built one of the first microscopes in 1674, was able to see living cells of *Spirogyra* and bacteria.

Can we believe our eyes?

Our own perception is a crucial source of knowledge. The way we see things depends on the interaction between our sense organs and our mind, and what we perceive is a selective interpretation.

When studying material that has been prepared for microscopic examination, we must always bear in mind that staining and cutting cells will alter their appearance. Interpreting images requires care, and what we perceive in a particular image is likely to be influenced by these techniques as well as our own expectations.

Questions to consider

- Consider the shapes of mitochondria in Figure **1.12**. Why do some mitochondria appear cylindrical and others circular?
- Plant cells have a single central vacuole. Examine the plant cell in Figure **1.14**. How many vacuoles can you see? How can you explain this?

It was not until good light microscopes became available in the early part of the 19th century that plant and animal tissues were seen as groups of individual cells and Schleiden and Schwann in 1838 were able to see sufficient structure to propose the cell theory, which incorporated the work of their predecessors (page 1).

Animal cells are tiny and colourless so it was not until the end of the 19th century, when staining techniques were first used, that it was possible to see a little more detail of cell contents. In the early 1940s, far more powerful electron microscopes were used for the first time and organelles and greater complexity of cell structure could be studied. Developments proceeded more rapidly in the 20th century because international communication allowed for more efficient collaboration not only in the designing and building of scientific instrumentation but also in the discussion and understanding of what could be observed.

A light microscope can resolve (view separately) cell details that are about 0.2 μm apart. Resolution is limited by the wavelength of light so that bacteria and mitochondria (500 nm or 0.5 μm) are the smallest objects that can be seen. An electron microscope uses a beam of electrons to probe specimens and in theory it should be able to resolve structures that are 0.002 nm apart (a resolution 10 000 times that of a light microscope). But because of practical problems in preparing specimens the best modern electron microscope resolves about 0.1 nm. For biological material this reduces to about 0.2 nm but, even so, an electron microscope allows a resolution which is 100 times better than a light microscope and its development has led to a greater understanding of cell structures and functions.

Test yourself

7 List **three** differences between prokaryotic and eukaryotic cells.
8 Distinguish between these pairs of terms:
   a 'cell wall' and 'plasma membrane'
   b 'lysosome' and 'ribosome'.
9 State **one** advantage a cell gains from being compartmentalised – that is, from having organelles.
10 Outline the function of the endoplasmic reticulum.

1.3 Membrane structure

The structure of membranes

Membranes not only provide shape for a cell and enclose its contents; there is also considerable activity at membrane surfaces, especially at the plasma membrane in contact with the extracellular space. Our current model of membrane structure, the *fluid mosaic model*, helps to explain how membranes carry out these functions.

Learning objectives

You should understand that:
- Phospholipids form bilayers in water due to the amphipathic properties of the molecules.
- Membranes contain a range of proteins, which differ in their structure, function and position in the phospholipid bilayer.
- Cholesterol is an important component of the membranes of animal cells.

Amphipathic compounds possess both hydrophilic (water-loving, polar) and lipophilic (fat-loving) properties. The amphipathic properties of phospholipids in a membrane explain the way a membrane structure forms. Phospholipids arrange themselves into bilayers, with their polar groups facing the surrounding aqueous (watery) medium, and their hydrophobic chains facing towards the inside of the bilayer. In this way, a non-polar region is formed between two polar ones. Phospholipids are principal constituents of biological membranes, but cholesterol and glycolipids and glycoproteins are also amphipathic and their presence in the bilayer gives membranes different physical and biological properties. You can find out more about the importance of polar groups in Subtopic **2.2**.

High levels of certain types of cholesterol in the blood have been associated with heart disease. You can find out more about this on page 199. The body produces its own cholesterol in the liver, but it is also found in many foods that we eat.

All membranes, wherever they occur in cells, have the same basic structure. Membranes are usually between 7 and 10 nm thick, and are composed of two layers of phospholipid, which form a bilayer. **Phospholipids** are made up of a polar, hydrophilic area containing a phosphate group bonded to glycerol, and a non-polar, lipophilic area containing fatty acids. In the bilayer, the lipophilic or **hydrophobic** (water-hating) parts all point towards each other, and the **hydrophilic** (water-loving) areas point outwards, as Figure **1.16** shows. It is the different properties of each end of the molecule that cause the phospholipids to arrange themselves in this way. The hydrophilic 'heads' of the molecules always appear on the outside of the membrane where water is present, while the hydrophobic 'tails' orientate inside the double layer, away from water. The structure is called a 'mosaic' because, just as a mosaic picture is made up of many small, separate pieces, so the surface of the membrane is composed of the heads of many separate phospholipid molecules. The whole structure is flexible or 'fluid' because the phospholipids can float into a position anywhere in the membrane. Research using radioactively labelled phospholipids shows that these molecules move not only within their own layer, but also between the two layers of the membrane.

Embedded in the bilayer are different molecules that contribute to the functions of membranes. **Cholesterol** is often present in mammal cells and is most commonly found in the plasma membrane. One end of the cholesterol molecule associates with the polar heads of phospholipid molecules while other parts of it are embedded in the membrane next to the non-polar fatty acid chains. This interaction makes the membrane less 'fluid', more rigid, and less permeable to water-soluble molecules.

There are also different types of protein in the bilayer. **Integral proteins** are embedded in the bilayer, whereas **peripheral proteins** are attached to the surface. Many of the proteins on the outer surface are glycoproteins – that is, they have carbohydrate groups attached to them. Some of these serve as hormone binding sites and have special shapes to recognise the specific hormones to which the cell will respond. Others are important in cell-to-cell communication and adhesion. Some integral

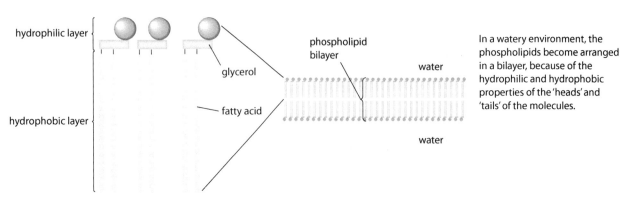

Figure 1.16 A phospholipid molecule includes a phosphate, glycerol and two fatty acids but in diagrams (such as **Figure 1.17**) the molecule is often simplified and shown as a circle with two tails.

proteins are enzymes immobilised within the membrane structure and perfectly placed to carry out sequences of metabolic reactions. Finally, there are proteins that span the bilayer acting as channels for ions and molecules to pass by passive transport, or forming pumps that use active transport to move molecules into or out of the cell.

Models of membrane structure

Our current understanding of the structure of the membrane has arisen from the work of a number of scientists, over many years. Each group refined previous knowledge of membranes, rejecting theories that were not supported by evidence and working to gather new data as microscopy and other techniques improved. The existence of a lipid bilayer was originally proposed and outlined by Gorter and Grendel, in 1925. Their ideas were developed and improved by Hugh Davson and James Danielli, who proposed in 1935 a model of a phospholipid bilayer between two layers of globular protein, the so-called 'fat sandwich' model. The Davson–Danielli model was new and it attempted to explain their observations of the surface tension of lipid bilayers. Since that time, the phenomenon of surface tension in bilayers has been better explained by studying the properties of the phospholipid heads. Nevertheless, the Davson–Danielli model predominated, and was supported by observations using the electron microscope, until 1972 when Singer and Nicolson described the 'fluid mosaic' model. The fluid mosaic model included descriptions of integral proteins that were sited through the membrane, and it rejected the idea of a 'sandwich-like' globular protein layer because it was no longer well supported by experimental evidence (see 'Nature of science', below). Fresh observations obtained using a new technique called freeze-etching were also important.

The model of membrane structure accepted today is based on the Singer and Nicolson fluid mosaic model, illustrated in Figure **1.17**, and has been supported by more recent research, with only minor modifications.

Freeze-etching is a method of preparing membranes to give a three-dimensional view of the surface and detail of the membrane's structures. Cells are rapidly frozen and fractured by breaking them in liquid nitrogen. The fractured surface is shadowed with evaporated heavy metal under vacuum and stabilised. The replicated surface is floated onto fine metal grids so that it can be viewed in the electron microscope to reveal the 3D arrangement of lipids and proteins that are present.

Nature of science

Falsification of theories – developing a model of membrane structure

Scientific theories embody our current understanding of aspects of the real world, and may include models to represent those aspects. However, any scientific theory or model can only exist until it is disproved. The Davson–Danielli model of membrane structure was accepted until new evidence called the model into question. The Davson–Danielli model was very close to what is now accepted, except that it proposed that all membranes are alike. This was disproved, though, when it was found that different organisms transport very different substances across their membranes, using different proteins. For example, mammalian cells transport sodium and potassium ions across their plasma membranes via special protein channels, while methane-producing bacteria move

As you read the evidence that has accumulated and helped our understanding of the structure of membranes, consider the following questions.
- Why is it important for scientists to put forward their ideas in the form of theories?
- How useful are models in developing ideas of biological structures?
- Is it important to learn about theories that have been discredited or superseded?

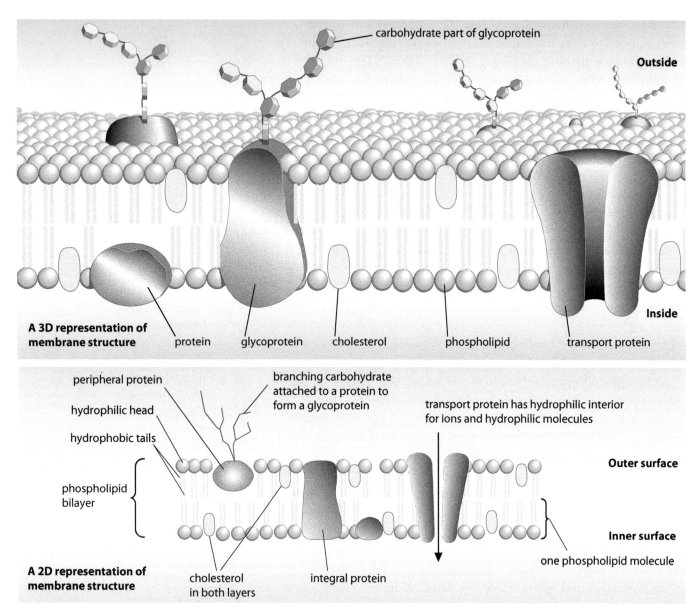

Figure 1.17 Diagrams to show the fluid mosaic model of membrane structure.

methane out of their cells via different protein channels. Evidence such as this showed that all membranes are not alike, and falsified the Davson–Danielli model, which was superseded by our current model.

Test yourself

11 Suggest why the term 'fluid mosaic' is used to describe membrane structure.
12 Suggest why the fatty acid 'tails' of the phospholipid molecules always align themselves in the middle of the membrane.
13 Outline the difference between integral membrane proteins and peripheral membrane proteins.

1.4 Membrane transport

Diffusion, facilitated diffusion and osmosis

Many molecules pass across the plasma membrane. Water, oxygen, carbon dioxide, excretory products, nutrients and ions are continuously exchanged and many cells also secrete products such as hormones and enzymes through the plasma membrane.

The simplest way in which molecules can move into or out of a cell is by **simple diffusion** through the plasma membrane. Diffusion is a passive process, which takes place as molecules move randomly. No energy input is required, and movement occurs by way of a simple concentration gradient. A **concentration gradient** is a difference in concentration of a substance between two regions and diffusion will always occur where such a gradient exists until particles of the substances are evenly distributed and equilibrium is reached. One important example of simple diffusion is its role in the process of cell respiration. Oxygen is needed by cells as it is continuously used up in respiration. As a cell respires, the oxygen concentration inside becomes less than the concentration outside, so oxygen molecules diffuse in. In a similar way, as carbon dioxide is continuously formed during respiration, its concentration builds up inside the cell and it diffuses out through the plasma membrane to an area where the concentration is lower. Simple diffusion occurs where the membrane is fully permeable to the substance or where channel proteins in the membrane are large enough for the substance to pass through.

Large molecules, and charged particles such as chloride ions (Cl^-) and potassium ions (K^+), cannot pass through the membrane by simple diffusion so certain proteins form channels through which they can travel. As in simple diffusion, no energy is used by the cell and the transport relies on the kinetic energy of the particles moving down their concentration gradient. **Channel proteins** have an interior which is hydrophilic (Figure **1.18**) so water-soluble materials can pass though them, and they are specific – that is, they only allow a particular substance to move through. Some of these channels are permanently open, whereas others are **gated** and only open to allow certain ions to pass when they are stimulated to do so. For example, gated channels in the axons of nerve cells open when there is a change in the voltage (potential difference) across the membrane. Gated potassium channels only allow K^+ ions to pass out through the membrane after a nerve impulse has passed along the axon. You can read more about nerve impulses in Subtopic **6.5**.

Other channel proteins allow the movement of substances such as glucose and amino acids, which are polar and cannot diffuse though the lipid layer of the membrane. Substances like these are transported across membranes by **facilitated diffusion**. In this case, a carrier protein first combines with the diffusing molecules on one side of the membrane, carries them through the channel protein and then releases them on the other side (Figure **1.18**). Facilitated diffusion allows a faster diffusion rate

Learning objectives

You should understand that:
- Particles move across membranes by osmosis, active transport, simple diffusion and facilitated diffusion.
- Materials can be taken into cells by endocytosis and leave cells by exocytosis due to the fluid nature of the membrane.
- Within a cell, vesicles move materials around.

Passive transport the movement of substances down a concentration gradient from an area of high concentration to an area of lower concentration without the need for energy to be used

Diffusion one example of passive transport; many molecules pass into and out of cells by diffusion e.g. oxygen, carbon dioxide and glucose

Osmosis another example of passive transport but the term is used only in the context of water molecules; osmosis is the movement of water molecules across a partially permeable membrane from a region of lower solute concentration, where there is a high concentration of water molecules, to a region of higher solute concentration, where the concentration of water molecules is lower

Active transport the movement of substances against the concentration gradient, which always involves the expenditure of energy in the form of ATP

Diffusion through a protein channel

Large or charged substances such as K⁺ and Cl⁻ ions cannot pass easily through membranes. They can pass through special channel proteins if they come in contact with the channel. Only specific ions or molecules can pass and no energy input is required.

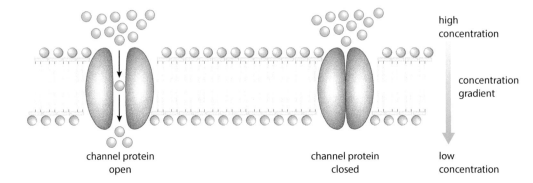

Facilitated diffusion via a carrier protein

Carrier proteins assist some molecules through the membrane, down their concentration gradient, combining with molecules on one side of the membrane and releasing them on the other side. Again, no energy input is required.

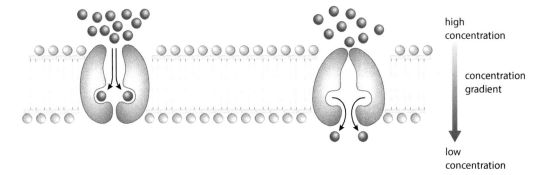

Figure 1.18 Some large or charged ions and molecules pass through the membrane via special channel proteins.

for molecules that particular cells need – for example, the diffusion of glucose into active muscle cells. No energy input is required because the molecules move down their concentration gradient.

A special case of diffusion is **osmosis** (Figure **1.19**). This is the passive movement of water across a partially permeable membrane from an area of lower solute concentration to an area of higher solute concentration.

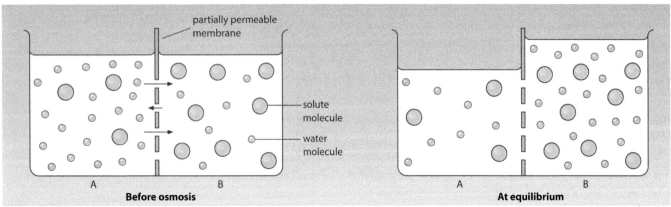

Two solutions are separated by a partially permeable membrane. B has a higher solute concentration than A. The solute molecules are too large to pass through the pores in the membrane but the water molecules are small enough.

As the arrows in the left diagram indicate, more water molecules moved from A to B than from B to A, so the net movement has been from A to B, raising the level to the solution in B and lowering it in A. The water potentials (the tendency of water molecules to move in each direction) in A and B are now the same.

Figure 1.19 Osmosis.

When the solute concentrations inside and outside a cell are the same, the same number of water molecules will pass across the membrane into the cell as those that leave. An animal cell that is placed in pure water will take in water by osmosis until eventually it may burst (Figure **1.20**). Placed in a solution with a very high concentration of solutes, the cell will shrink or 'crenate' as water leaves the cell by osmosis. In either situation, animal cells will not function properly and their metabolism will be affected.

In medical procedures, tissues and organs are bathed in a solution of 'normal saline', which has exactly the same **osmolarity** (a measure of the solute concentration in a solution) as human cell cytoplasm and is said to be isotonic with the cytoplasm. In this case osmosis does not occur and cells are not damaged.

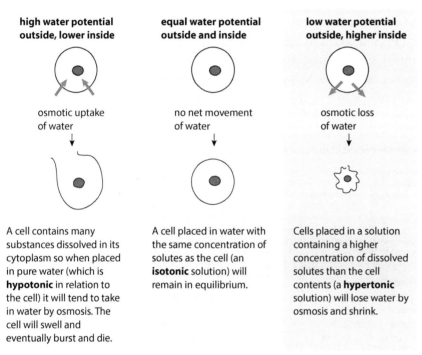

Figure 1.20 Responses of animal cells to solutions of different concentrations (see also Figure **1.21**).

Water potential the tendency of water molecules to move from an area of higher concentration to an area of lower concentration.

Normal saline is a solution of 0.90% w/v (weight by volume) of sodium chloride and is isotonic with human cells. It is used frequently in intravenous drips (IVs) for patients who cannot take fluids orally and are in danger of becoming dehydrated.

Plant cells are also affected by the movement of water into and out of their cells but the presence of a cell wall prevents plant cells being damaged or bursting. If a plant cell is put into water, water will enter by osmosis but the plant cell wall resists the entry of further water once the cell is full. A plant cell that is full becomes firm and rigid – a condition known as turgor (Figure **1.21**).

Active transport

Many of the substances a cell needs occur in low concentrations in the surroundings outside the plasma membrane. For example, plants must take in nitrate ions from very dilute solutions in the soil to build their proteins, and muscle cells actively take in calcium ions to enable them to contract. To move these substances into the cell against a concentration gradient, the cell must use metabolic energy released from the breakdown of ATP to ADP and P_i (Subtopic **2.8**). This is called **active transport** (Figure **1.22**). Specific proteins in the plasma membrane act as transporters or 'carriers' to move substances through. Many of the carrier proteins are specific to particular molecules or ions so that these can be selected for transport into the cell.

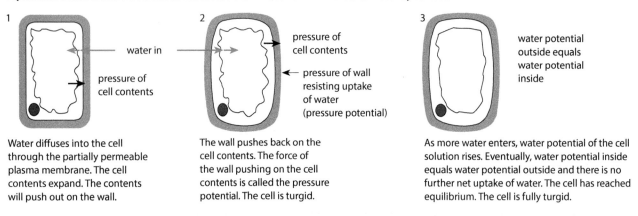

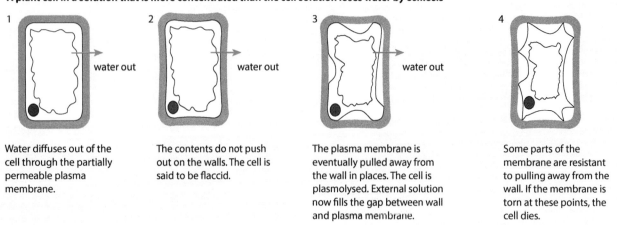

Figure 1.21 Responses of plant cells to solutions of different concentrations. Plant cells are not damaged as water enters by osmosis because their cell wall protects them.

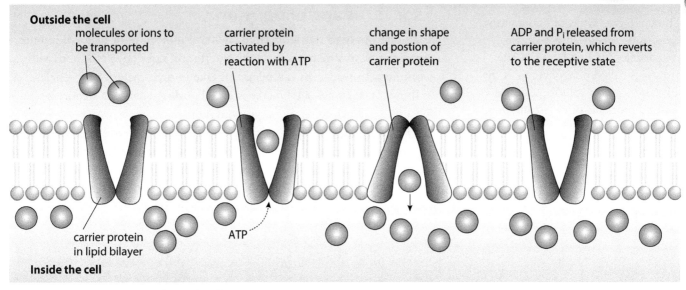

Figure 1.22 Active transport of a single substance.

Figure **1.23** illustrates a very important example of active transport. The sodium–potassium pump maintains the concentration of sodium and potassium ions in the cells and extracellular fluid. Cells are able to exchange sodium ions for potassium ions against concentration gradients using energy provided by ATP. Sodium ions are pumped out of the cell and potassium ions are pumped into the cell.

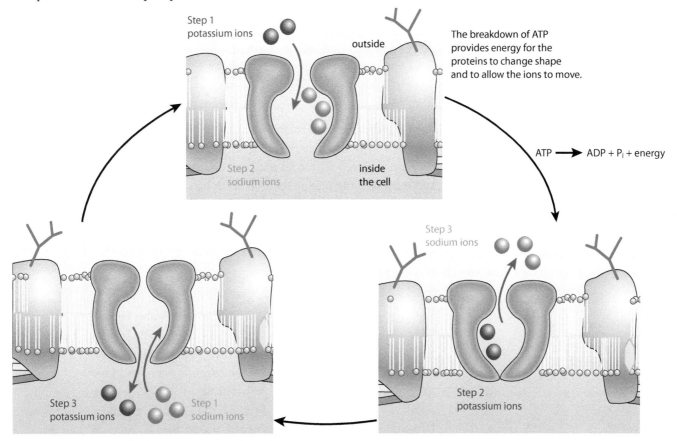

Figure 1.23 An example of active transport – the sodium–potassium pump. Start at step 1 for each ion in turn and work round clockwise.

There are two types of endocytosis. If the substances being taken in are particles, such as bacteria, the process is called phagocytosis. If the substances are in solution, such as the end products of digestion, then it is called pinocytosis.

Exocytosis and endocytosis

Cells often have to transport large chemical molecules or material in bulk across the plasma membrane. Neither diffusion nor active transport will work here. Instead, cells can release or take in such materials in vesicles, as shown in Figure **1.24**. Uptake is called **endocytosis** and export is **exocytosis**. Both require energy from ATP.

During endocytosis, part of the plasma membrane is pulled inward and surrounds the liquid or solid that is to be moved from the extracellular space into the cell. The material becomes enclosed in a vesicle, which pinches off from the plasma membrane and is drawn into the cell. This is how white blood cells take in bacteria (Figure **1.24**).

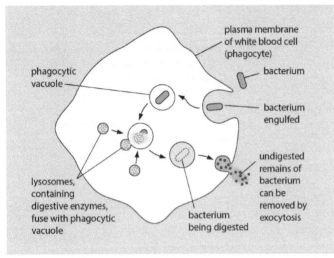

Phagocytosis of a bacterium by a white blood cell – an example of endocytosis.

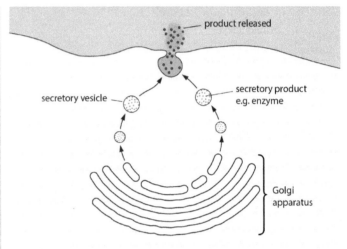

Exocytosis in a secretory cell. If the product is a protein, the Golgi apparatus is often involved in chemically modifying the protein before it is secreted, as in the secretion of digestive enzymes by the pancreas.

Figure 1.24 Examples of endocytosis and exocytosis.

Materials for export, such as digestive enzymes, are made in the rER and then transported to the Golgi apparatus to be processed. From here they are enclosed within a membrane-bound package known as a **vesicle**, and moved to the plasma membrane along microtubules. The arrangement of molecules in the membrane of a vesicle is very similar to that in the plasma membrane. As a vesicle approaches the plasma membrane, it is able to fuse with it and in doing so release its contents to the outside. The flexibility and fluidity of the plasma membrane are essential to enable both endocytosis and exocytosis to happen. Vesicles also help to transfer and organise substances in the cell. They are involved in metabolism, transport and enzyme storage and some chemical reactions also occur inside them.

Nerve impulses are able to pass across synapses (the tiny gaps between one nerve cell and the next) due to exocytosis and endocytosis. Neurotransmitters are secreted at the end of a nerve cell fibre by exocytosis. They stimulate the adjacent nerve and are then reabsorbed by endocytosis to be recycled and reused. You can find out more about the transmission of nerve impulses in Subtopic **6.5**.

Nature of science

Experimental design – accurate quantitative measurement

Whenever experiments are designed, accuracy is important so that the experimenter can be sure that results are valid. Measurements should be obtained using the most suitable equipment with the correct degree of accuracy for the task. In the experiment described below, for example, measurements of the mass of small samples of potato and sucrose crystals are needed. A balance that provides readings accurate to 0.05 g would be most appropriate. Similarly, when measuring a small volume of liquid, a 25 cm³ measuring cylinder would be more appropriate than a 250 cm³ cylinder.

This experiment can be used investigate the process of osmosis and identify the solute concentration (or water potential) of cells. Cubes or chips of potato tissue are weighed accurately and placed in test tubes each containing a sucrose solution of a different solute concentration (molarity) – for example, between 0 mol dm^{-3} and 0.6 mol dm^{-3} – for a suitable period of time. The potato samples are then removed and reweighed and the percentage change in their mass is recorded.

% change = change in mass ÷ original mass × 100

A solution that causes no change in mass of the potato has the same solute concentration as the tissue and the same water potential. It is said to be isotonic. (Solutions of greater solute concentration than the tissue are **hypertonic**, while solutions whose solute concentration is lower than that of the tissue are known as **hypotonic**.)

Figure **1.25** shows the results of such an experiment. From the graph it can be seen that the molarity that causes no change in mass is approximately 0.35 mol dm³. At this concentration, the water potentials inside and outside the cell are equal so there is no net movement of water by osmosis.

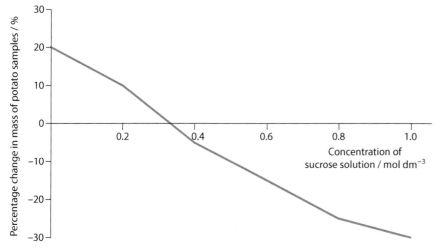

Figure 1.25 Graph to show the results of an experiment with potato samples placed in different sucrose solutions.

> **Test yourself**
>
> 14 Outline the difference between simple diffusion and facilitated diffusion.
> 15 List **three** ways that substances move from one side of a membrane to the other.
> 16 State **one** transport mechanism across a membrane that requires energy from ATP and **one** that does not.
> 17 State **one** difference and one similarity between exocytosis and endocytosis.

1.5 The origin of cells

How are new cells formed?

Cell theory proposes that all organisms are composed of one or more cells, which are the smallest units of life. One of the functions carried out by all living organisms is reproduction. Therefore, the first principle of the cell theory is that cells can only come from pre-existing cells. Louis Pasteur (1822–1895) carried out experiments that provided evidence for this. He showed that bacteria could not grow in a sealed, sterilised container of chicken broth. Only when living bacteria were introduced would more cells appear in the broth. Figure **1.26** summarises Pasteur's experiment.

Learning objectives

You should understand that:
- Cells form by the division of pre-existing cells.
- Non-living material must have given rise to the first cells.
- Endosymbiosis is a theory that explains the origin of eukaryotic cells.

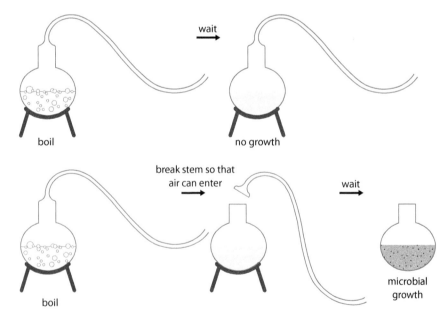

Boiling the flask kills any bacteria present in the broth. The curved neck of the flask prevents the entry of any new organisms from the atmosphere.

If the neck of the flask is broken it is possible for bacteria to enter the broth where they reproduce to produce more cells.

Figure 1.26 Pasteur's experiment demonstrating that living cells cannot 'spontaneously generate', but must originate from pre-existing living cells.

How did the first cells originate?

All prokaryotic and eukaryotic organisms alive today are made up of cells. The very first prokaryotes are thought to have appeared around 3.5 billion years ago and the structures in these first cells must have originated from chemicals present on the Earth at that time. For these structures to have formed, four essential steps must have occurred.

- Living things are made of organic molecules, so simple organic molecules such as amino acids, sugars, fatty acids, glycerol and bases must have formed.
- Organic molecules in living organisms (such as triglycerides, phospholipids, polypeptides and nucleic acids) are large, so single molecules must have been assembled to make these more complex molecules.
- All living things reproduce, so molecules must have formed that could replicate themselves and control other chemical reactions. This is the basis of inheritance.
- Finally, cells have membranes, so the mixtures of these molecules must have been enclosed within membrane-bound vesicles.

The endosymbiotic theory

The endosymbiotic theory explains how eukaryotic cells could have developed from a simple cell or prokaryote. The theory suggests that some organelles found inside eukaryotes were once free-living prokaryotes. There is evidence to suggest that some prokaryotes were engulfed by larger cells, and were retained inside their membranes where they provided some advantages to the larger cell (Figure **1.27**).

Evidence for this theory includes the fact that two important organelles, mitochondria and chloroplasts, share many characteristics with prokaryotic cells. Both chloroplasts and mitochondria:

- contain ribosomes that are smaller than those found in other parts of eukaryotic cells but are identical in size to those found in bacteria
- contain small circular pieces of DNA resembling bacterial plasmids in their basic structure

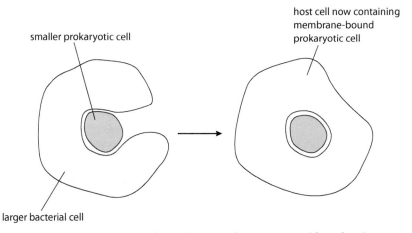

Figure 1.27 Organelles such as chloroplasts may have originated from free-living prokaryotes that were engulfed by larger cells.

- have their own envelope surrounding them, on the inner membrane of which are proteins synthesised inside the organelle, suggesting that they may have used this ability long ago when they were independent organisms
- can replicate themselves by binary fission.

This evidence supports the theory that these organelles are modified bacteria that were taken in by phagocytosis, early in the evolution of eukaryotic cells. Here they became useful inclusions. The double outer envelope of chloroplasts and mitochondria may have originated from the bacterial plasma membrane together with the membrane of an engulfing phagocytic vesicle. Perhaps some of the enclosed bacteria carried pigment molecules on their membranes and used light energy to make organic molecules and release oxygen – these may have become chloroplasts. It may be that others became efficient at using the oxygen molecules for aerobic energy production, and these became mitochondria.

Critics of the endosymbiotic theory might argue that, even if prokaryotes were engulfed by larger cells, there is no certainty that they could be passed on to both daughter cells when the larger cell divided, because there is no special mechanism to ensure this. However, when a cell divides by binary fission each daughter cell contains some cytoplasm from the parent and so at least one of the daughter cells would contain the engulfed prokaryotes. Both mitochondria and chloroplasts have retained the ability to self-replicate and so their numbers can increase in the cytoplasm prior to cell division, which increases the chance of both daughter cells containing some. Critics also note that mitochondria and chloroplasts are not able to survive on their own if they are isolated from a cell, which they might be expected to do if they originated from free-living cells. But perhaps over time they have lost the ability to synthesise one or more essential molecules and have come to depend on the 'host' cell to provide them.

Nature of science

General principles underlying the natural world – evidence of the unbroken chain of life from the first cells

Further evidence for a common origin for all life on Earth comes from the genetic material, in the form of chromosomes, which is inherited by every cell. Chromosomes are made from nucleic acids, such as DNA (deoxyribonucleic acid), and built into the chromosomes is a code, which is used by the cell to assemble all the molecules it needs to live.

A sequence of four key molecules, known as bases, along a DNA molecule forms this code. The bases are adenine (A), cytosine (C), guanine (G) and thymine (T). The code determines the sequence of the amino acid components that are bonded together as protein molecules are synthesised (Subtopic **2.7**). The code is 'read' in sets of three bases known as 'codons' and 64 different codons can be made from combinations of the four bases A, C, G and T. A single codon represents the code for one amino acid unit in a protein.

Endosymbiosis

Symbiosis means 'life together'. Endo means 'inside' and so **endosymbiosis** describes a relationship taking place inside a cell.

gene a length of DNA at a specific location on a chromosome that controls a specific heritable characteristic

The genetic code is said to be *universal* because all living organisms share the same genetic code, with only a few minor variations, which are due to mutations that have occurred over the millions of years of evolution. All life – from plants and animals to bacteria and fungi – relies on the same 64 codons to carry the biological instructions for their bodies. What varies from one organism to another is not the code's structure or the way in which it is translated, but the individual **genes** which are formed from lengths of DNA (Subtopic **3.1**). The Human Genome Project and similar projects have deciphered gene sequences and compared sequences and codes in different species. Scientists have found whole 'sentences' of identical DNA 'text'.

The study of the genetic code, together with studies of molecular processes such as respiration, photosynthesis and protein production, show that these vital processes are similar in all living cells. This provides strong evidence that all life on Earth had a common origin billions of years ago.

Test yourself

18 Explain how Pasteur's experiment supports the idea that life does not arise by spontaneous generation of cells.
19 Define the term 'endosymbiosis'.

1.6 Cell division

New cells are needed to replace cells that have died or to allow an organism to grow. The nucleus and cytoplasm of a cell divide in processes known as **mitosis** and **cytokinesis**, which are phases in a series of events known as the **cell cycle**.

The cell cycle

The cycle of a cell's life can be divided into three stages, as shown in Figure **1.28**:
1 interphase
2 mitosis (division of the nucleus)
3 cytokinesis (division of the cytoplasm).

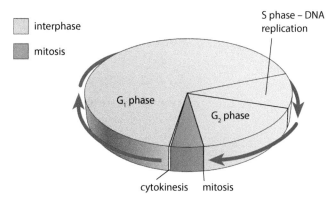

Figure 1.28 The cell cycle.

Learning objectives

You should understand that:
- Different stages of the cell cycle can be identified. One of these is interphase in which many process occur in the cytoplasm and nucleus.
- Mitosis describes the division of the nucleus in a cell into two genetically identical daughter nuclei.
- During mitosis chromosomes condense by supercoiling.
- Cytokinesis is a phase that occurs after mitosis. It is different in plant and animal cells.
- Cyclins are involved in controlling the cell cycle.
- The development of primary and secondary tumours can involve mutagens, oncogenes and metastasis.

Summary of the cell cycle

G_1, S and G_2 are the three stages of the part of the cell cycle known as interphase.

G_1 phase
- cell grows
- DNA is transcribed
- protein is synthesised

S phase
- DNA is replicated

G_2 phase
- cell prepares for division

mitosis
- cell nucleus divides

cytokinesis
- cytoplasm divides

Root tip squash preparations

Squash preparations of onion cell root tips where cells are actively dividing can be prepared by softening the tissue in $1 \, mol \, dm^{-3}$ hydrochloric acid. One drop of toluedine blue stain is added and the cells are squashed by placing them between a cover slip and microscope slide, which is gently tapped with a pencil. The root tip will spread out as a mass on the slide and cells will separate from one another so that stages of mitosis can be seen under high power using a light microscope.

Interphase

During most of the life of a cell, it performs the task for which it has been pre-programmed during differentiation. This period is called **interphase**. Part of interphase is spent in preparation for cell division (the G_2 phase) and part of it is the period immediately after division (the G_1 phase). The two stages of cell division are the separation and division of the chromosomes (mitosis), and the division of the cell into two daughter cells (cytokinesis).

If a cell is examined during interphase using a light microscope, little appears to be happening, but this is a very active phase of the cell cycle when the cell carries out its normal activities and also prepares itself for mitosis. In the nucleus, the DNA in the chromosomes is replicated (**S phase**) so that after cell division there will be exactly the same number of chromosomes in the two daughter cells. During interphase many proteins necessary for the division need to be synthesised at the ribosomes in the cytoplasm. The number of mitochondria increases so that the respiratory rate can be rapid enough to provide energy for cell division. In the case of plant cells with chloroplasts, the number of chloroplasts increases so there are sufficient for each daughter cell.

Mitosis

The two new cells that will be formed after mitosis and cytokinesis are genetically identical. These processes allow an organism to grow more cells, or to repair injured tissue by replacing damaged cells, or to make new cells to replace old ones. Mitosis is also the way in which an embryo grows from a fertilised egg during development. Many organisms reproduce themselves using mitosis; examples include the unicellular organisms such as *Amoeba*, *Paramecium* and yeast. Reproducing in this way is known as **asexual reproduction** as no gametes (Subtopic **3.3**) are involved and the offspring are genetically identical to the parent.

There are four distinct stages in mitosis, though the process is continuous, with each stage running into the next. There are no intervals in between stages. Figures **1.29** and **1.30** show in detail the stages of mitosis.

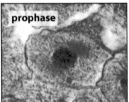 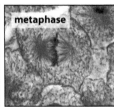 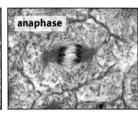 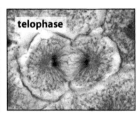 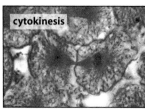

Figure 1.29 Stages of mitosis in stained onion cells, as seen in a root squash preparation (×900 at 10 centimetres wide each).

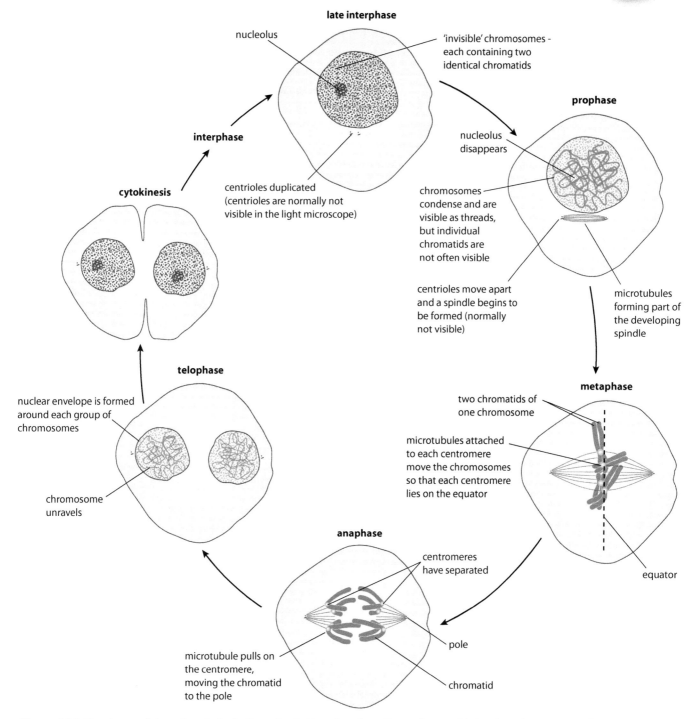

Figure 1.30 The stages of the cell cycle, including mitosis. Note that the cells are shown with just four chromosomes here, to make it easier to understand the process.

Prophase

During **prophase**, the chromosomes become visible using a microscope. During interphase they have been drawn out into long threads, allowing the cellular machinery access to the genes but now the chromosomes coil round themselves several times to produce a **supercoil** (Figure **1.31**). Supercoiling not only makes the chromosomes shorter and thicker, it also reduces the space that they take up and enables them to take part in the processes that follow. We can follow these processes because supercoiled chromosomes can be seen using a microscope. Because the DNA was replicated during interphase, at this stage each chromosome consists of two identical copies. These two copies are called the **sister chromatids** and are attached to each other at a place called the **centromere**. Also visible at this time are structures known as **centrioles**, which move to opposite sides of the cell as microtubules form between them. This microtubule structure is called the **spindle**. As prophase draws to a close, the nuclear envelope breaks down.

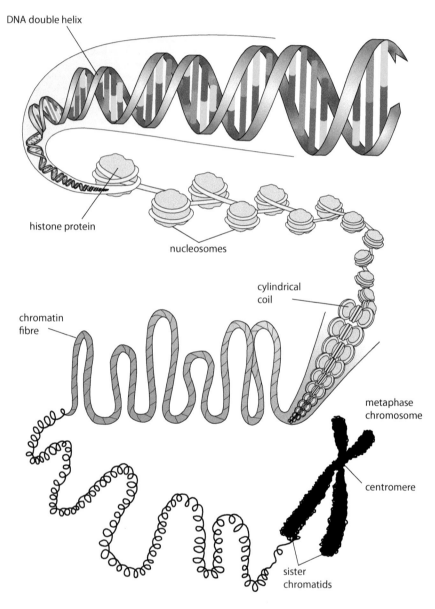

Figure 1.31 Supercoiling produces condensed, compact chromosomes in preparation for the next stages of mitosis.

Metaphase

Metaphase begins when the nuclear envelope has broken down. As it disappears, more space is created so that the chromosomes can move into position during their division. The sister chromatids align themselves on the microtubules in the middle, or equator, of the spindle and are attached by their centromeres.

Anaphase

During anaphase, the centromeres split and the sister chromatids are pulled apart and move towards the centrioles at opposite sides, or poles, of the cell as the spindle fibres shorten. Each sister chromatid is now called a chromosome again.

Telophase

Once the two sets of chromosomes reach their opposite poles, the spindle fibres break down and a nuclear envelope forms around each set of chromosomes. At the same time, the chromosomes uncoil and become invisible through a light microscope.

Following telophase, in animal cells, the plasma membrane pinches in and the two new nuclei become separated. Eventually, during cytokinesis, the two sides of the plasma membrane meet and two completely new cells are formed. Each has a complete set of chromosomes, cytoplasm, organelles and a centriole.

In plant cells, the cytoplasm divides in a slightly different way. Firstly, a cell plate forms along the centre of the cell, separating the cytoplasm into two regions. Vesicles accumulate at the edges of the cell plate and release cellulose and pectins, which are needed to form a new cell wall. Gradually a cell wall builds up along the cell plate separating the two nuclei and dividing the cytoplasm to form two new cells.

Cyclins

Cyclins are compounds that are involved in the control of the cell cycle. Cyclins interact with other proteins called CDKs (cyclin-dependent kinases) to form enzymes that direct cells through the cell cycle and control specific events such as microtubule formation and chromatid alignment. Cyclins were discovered by Timothy Hunt in 1982 when he was studying the cell cycle of sea urchins. In 2001, Hunt, together with Lee Hartwell and Paul Nurse who also contributed to the discovery, were awarded the Nobel Prize in Physiology or Medicine for their work. Understanding factors that control the cell cycle is important in the study of cancer and the way cell division can be disrupted.

Cyclins are divided into four types based on their behaviour in vertebrate and yeast cells (Figure **1.32**) but some cyclins have different functions in different types of cell.

Mitotic index

When studying cells under the microscope, the ratio of the number of cells undergoing mitosis to the total number of cells in view is called the mitotic index. The mitotic index in an important prognostic tool used in predicting the response of cancer cells to chemotherapy – a low mitotic index indicates a longer survival time, and suggests that the treatment is working. It can be less accurate when used with elderly patients, whose cells divide more slowly. In such patients, a low mitotic index may not indicate that a treatment is working.

In the laboratory, it is possible to work out the mitotic index of growing and dividing cells from an electron micrograph.

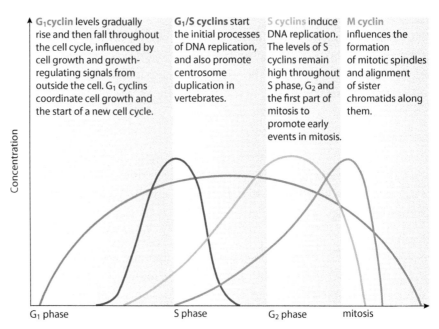

Figure 1.32 Cyclins can be divided into four types, which are important at different stages of the cell cycle: G1/S cyclins (green), S cyclins (blue), M cyclins (orange) and G1 cyclins (red).

Primary and secondary tumours

In most cases, mitosis continues until a tissue has grown sufficiently or repairs have been made to damaged areas. Most normal cells also undergo a programmed form of death known as apoptosis as tissues develop. But sometimes mitosis does not proceed normally. Cell division may continue unchecked and produce an excess of cells, which clump together. This growth is called a **tumour**. Tumours can be either benign, which means they are restricted to that tissue or organ, or malignant (cancerous), where some of the abnormal cells migrate to other tissues or organs and continue to grow further tumours there. If they are allowed to grow without treatment, tumours can cause obstructions in organs or tissues and interfere with their functions.

Cancer occurs when cells from a **primary tumour** (Figure 1.33) migrate to other tissues and form new **secondary tumours** in a process known as **metastasis**. Cancer is caused by damage to genetic material, producing cells that undergo uncontrolled, abnormal mitosis, but it cannot be thought of as a single disease. Cancer can take different forms in different tissues and the DNA damage that leads to cancer can be caused by a range of factors. DNA may be modified by physical, chemical and biological agents known as mutagens. Mutagens include ionising radiation – such as X-rays, gamma rays and ultraviolet light – and also chemical compounds, such as those found in tobacco smoke and aflatoxins produced by certain fungi. The DNA changes caused by mutagens are called mutations, and not all are harmful. However, because some of them cause cancer, some mutagens are said to be **carcinogens** (cancer causing) (Subtopic **3.4**). The development of a primary tumour can also

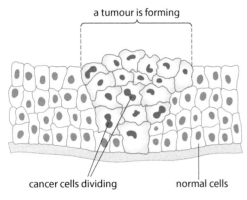

Figure 1.33 Formation of a primary tumour. If cells from a primary tumour become detached and form a new tumour in another part of the body, then the cells are said to be cancerous.

be caused by mistakes in copying DNA, or a genetic predisposition as a result of inheritance.

Many tumours are caused by activated **oncogenes**, which are special genes with the potential to cause cancer. Oncogenes may either be normal genes that have become altered, or they may be genes that are expressed at abnormally high levels. Activated oncogenes can cause cells that should die during apoptosis to survive and divide instead. Most oncogenes become active as a result of some additional process such as mutation in another gene (often those which regulate cell growth or differentiation), direct exposure to a mutagen or another environmental factor such as a viral infection.

 Because of their importance in human cancers, oncogenes are specifically targeted in many new cancer treatments that are being developed in laboratories all over the world.

Smoking and cancer

Smoking is a major cause of several types of cancer. There is strong evidence to show that it increases the risk of cancer of the bladder, cervix, kidney, larynx and stomach, and smokers are seven times more likely to die of these cancers than non-smokers. In the UK, approximately 85% of lung cancers in both men and women are related to smoking.

The risk of contracting lung cancer increases with the number of cigarettes that a person smokes and the number of years that they continue to smoke. If a person gives up smoking, their risk of developing cancer decreases (Figure **1.34**).

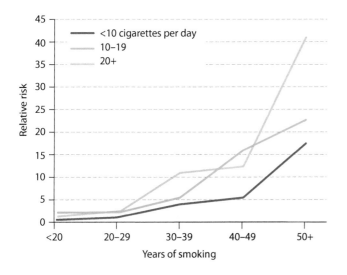
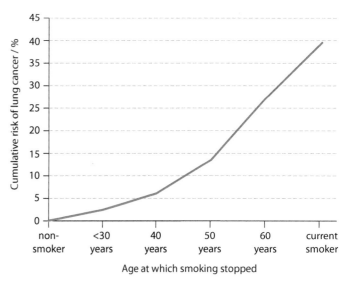

Figure 1.34 Graphs to show the relationship between smoking and lung cancer, and the cumulative risk of lung cancer among men in the UK at age 75 according to the age at which they stopped smoking (data from Cancer Research UK).

Lung cancer develops slowly and it takes years before the effects of smoking become obvious. The number of men who suffered lung cancer in the UK was at its highest levels in the early 1970s. This was as a result of a peak in smoking 20–30 years earlier. Cancer in women increased through the 1970s and 1980s and numbers are now stable. Statisticians predict that cancer in women will increase to reach the same levels as those in men over the next decade but that if people give up smoking as a result of new laws and health campaigns the number of deaths in both groups should decrease.

Test yourself

20 State **three** substances that can act as carcinogens.
21 State the result of uncontrolled cell divisions.
22 List in order the **four** stages of mitosis.
23 State what is meant by 'apoptosis'.

Nature of science

Serendipity in science – the discovery of cyclins

Serendipity is a term derived from an old name for Sri Lanka. It is said to come from a Persian fairy tale about 'The Three Princes of Serendip' who made discoveries by accident. It has come to describe the role of chance in science and indicate how unexpected discoveries are sometimes made. Working scientifically, researchers often benefit from serendipity or 'happy accidents' as new discoveries are made by chance or from apparently unrelated findings.

The discovery of cyclins is one example of a serendipitous discovery. Hunt, Hartwell and Nurse were all working on separate areas of the cell cycle and with different organisms but by chance the three strands of their work coincided. Hartwell worked with baker's yeast in the 1970s and discovered 'checkpoint' genes, which seemed to start the cell cycle. In the early 1980s Nurse, working with a different species of yeast, found a gene that if mutated stopped the cell cycle or initiated early cell division, and he identified CDK. In 1982 Hunt, who worked with sea urchin eggs, discovered the other key factor that drives the cell cycle, the protein cyclin. Cyclin regulates the function of the CDK molecule and increases and decreases as cell division occurs. You can read more about their discoveries on the Nobel Prizes website by visiting www.nobelprize.org and searching for 'cyclins'.

Serendipity could also be called luck, chance or even fluke. To what extent might serendipitous discoveries be the result of intuition rather than luck?

Exam-style questions

1 Prokaryotic cells differ from eukaryotic cells because prokaryotic cells:

    A have larger ribosomes
    B have smaller ribosomes
    C contain mitochondria
    D have more than one nucleus [1]

2 The correct order of the stages in the cell cycle is:

    A cytokinesis → mitosis → interphase
    B interphase → cytokinesis → prophase
    C mitosis → prophase → cytokinesis
    D cytokinesis → interphase → mitosis [1]

3 Explain how the properties of phospholipids help to maintain the structure of the plasma membrane. [2]

4 Outline the evidence for the theory of endosymbiosis. [3]

5 **a** Some ions can move across the membrane by passive or active transport. Distinguish between active transport and facilitated diffusion of ions. [2]

   b Digestive enzymes leave the cell by exocytosis. Describe the process of exocytosis. [2]

6 **a** The mitotic index is defined as the total number of cells in mitosis in an observed sample divided by the total number of cells in the sample.

Calculate the mitotic index for a sample of onion cells, using the data in the table below.

Show your working.

Stage	Number of cells
interphase	460
prophase	21
metaphase	24
anaphase	7
telophase	17

[2]

   b Outline the role of cyclins in the control of cell division. [3]

7 a Name the cell organelle shown in the micrograph below. [1]

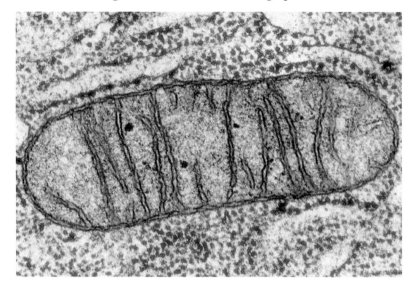

b Explain what is meant by 'compartmentalisation' in a cell's structure. [3]

c Electron microscopes have a higher resolution than light microscopes. Outline what is meant by the term 'resolution' in relation to microscope images. [2]

Molecular biology 2

Introduction

Molecular biology considers the chemical substances that are important to life and explains living processes in terms of these chemicals. Living things are built up of many chemical elements, the majority of which are bonded together in organic, carbon-containing compounds. Most organic compounds in living things are carbohydrates, proteins, nucleic acids or lipids. Other inorganic, non-carbon-containing substances are also important but are present in much smaller quantities.

Learning objectives

You should understand that:
- Molecular biology explains the roles of the chemical substances involved in life processes.
- Carbon atoms are important because they form four covalent bonds and allow the formation of many different, stable compounds.
- Carbon compounds including carbohydrates, lipids, proteins and nucleic acids form the basis for life.
- Metabolism is defined as the series of reactions, catalysed by enzymes, which occur in an organism.
- Macromolecules are built up from monomers by condensation during anabolic reactions.
- Macromolecules are broken down to monomers by hydrolysis during catabolic reactions.

2.1 Molecules to metabolism

Elements in living things

Carbon, hydrogen, oxygen and nitrogen are the four most common elements found in living organisms.

Carbon, hydrogen and oxygen are found in all the key *organic* molecules – proteins, carbohydrates, nucleic acids and lipids. Proteins and nucleic acids also contain nitrogen.

Any compound that does not contain carbon is said to be *inorganic*. A variety of inorganic substances are found in living things and are vital to both the structure and functioning of different organisms. Some important roles of inorganic elements are shown in Table 2.1. Molecular biology explains the life processes that we observe in terms of all the chemical substances that are involved and the reactions that occur between them.

Carbon atoms

Carbon is found in all organic molecules and forms a wide range of diverse compounds. Figure 2.1 shows how other elements can be added to a carbon atom in one of four different directions so that complex 3D molecules can be built up.

Element	Example of role in prokaryotes	Example of role in plants	Example of role in animals
sulfur (S)	a component of two amino acids	a component of two amino acids	a component of two amino acids, needed to make some antibodies
calcium (Ca)	co-factor in some enzyme reactions	co-factor in some enzyme reactions	important constituent of bones, needed for muscle contraction
phosphorus (P)	a component of ATP and DNA	a component of ATP and DNA	a component of ATP and DNA
iron (Fe)	a component of cytochrome pigments	a component of cytochrome pigments	a component of hemoglobin and cytochrome pigments
sodium (Na)	important in membranes, changes solute concentration and affects osmosis	important in membranes, changes solute concentration and affects osmosis	important in membranes, changes solute concentration and affects osmosis; also important in transmission of nerve impulses

Table 2.1 Roles of inorganic elements in living things.

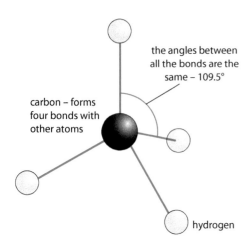

Figure 2.1 Carbon atoms can form four covalent bonds. Here carbon is bonded to four hydrogen atoms producing methane but many other atoms bond with carbon to produce a wide range of diverse compounds.

Organic compounds are a vast group of compounds that includes gases, liquids and solid substances. Every organic compound contains two or more atoms of carbon. As carbon atoms can easily bond with each other, organic compounds can be formed from carbon chains that differ in shape and length. Carbon atoms can also form double and triple bonds with other atoms, so increasing the variety in the molecular structure of organic compounds. Carbohydrates, proteins, lipids and nucleic acids are the main types of carbon-containing molecules on which life is based.

Carbon compounds – the building blocks of life

Carbon compounds form the basic molecules for life – carbohydrates, lipids, proteins and nucleic acids. Carbohydrates are compounds that contain only the elements carbon, hydrogen and oxygen and are the most abundant group of biological molecules. Lipids contain the same three elements but with much less oxygen than a carbohydrate of the same size. Lipids may also contain small amounts of other elements such as phosphorus. Proteins, unlike carbohydrates and lipids, always contain nitrogen. Sulfur, phosphorus and other elements are also often present.

Many organic molecules are very large and complex but they are built up of small subunits, which can be relatively simple. Figure **2.2** shows some of these building blocks. Small subunits called **monomers** are built into larger complex molecules called **polymers** in a process known as polymerisation.

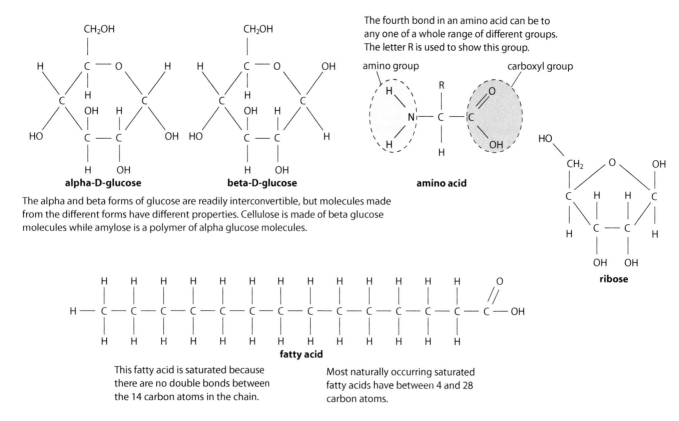

Figure 2.2 The basic structures of glucose, amino acids, fatty acids and ribose – the building blocks of organic molecules.

Carbohydrates

Carbohydrates are the most abundant category of molecule in living things. In both plants and animals they have an important role as a source of energy, and in plants they also have a structural function. Carbohydrates occur in different forms. **Monosaccharides**, with the general formula $(CH_2O)_n$ where n = the number of carbon atoms in the molecule, are monomers – single sugars made up of just one subunit. **Disaccharides** are sugars that have two subunits joined together by a condensation reaction (see page 46); and **polysaccharides** are long molecules consisting of a chains of monosaccharides linked together.

Lipids

Lipids are insoluble in water but do dissolve in organic solvents. Lipids are used as energy storage molecules in plants and animals. One group known as triglyceride lipids includes fats and oils. Those that are solid are generally referred to as fats, while those that are liquid are known as oils. Animals store energy as fat whereas plants store oils – familiar examples include linseed oil and olive oil. Lipid contains about twice as much energy per gram as carbohydrate (Table **2.2**) but each type of storage molecule has its own advantages (Subtopic **2.3**). The second group of lipids includes steroids (Figure **2.3**), which consist of four interlinked rings of carbon atoms. Vitamin D and cholesterol are the two best-known examples of steroids.

Molecule	Approximate energy content per gram / kJ
carbohydrate	17
lipid	39
protein	18

Table 2.2 Energy content of carbon compounds.

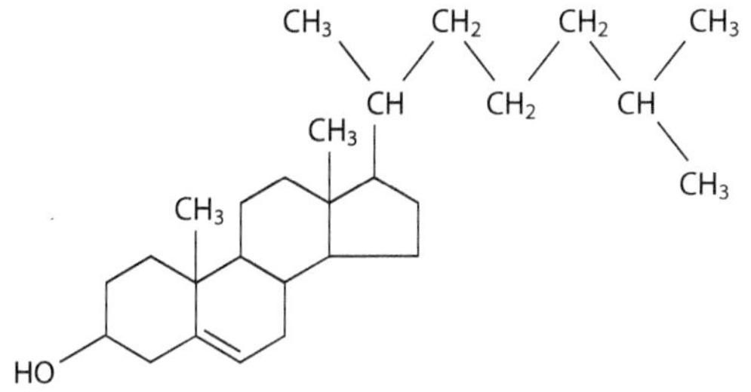

Figure 2.3 Like all steroids, cholesterol has four rings of carbon atoms. Other steroids differ in the side groups attached to them.

Protein

Proteins are built up of building blocks called **amino acids**. The atoms occurring at the fourth bond (shown as the R group in Figure **2.2**) differ in different amino acids and give each one its own properties. The simplest amino acid is glycine, in which R is a hydrogen atom, while the R group in the amino acid alanine is CH_3. There are more than 100 naturally occurring amino acids but only 20 are used in building the bodies of living things. Proteins are built up of amino acids in condensation reactions (see page 46).

Metabolism the networks of chemical reactions that occur in an organism. Metabolic reactions are catalysed by enzymes. Respiration and photosynthesis are two metabolic reactions that are vital to life and which consist of many interrelated chemical reactions. Anabolic reactions such as condensation and catabolic reactions such as hydrolysis (digestion) are simpler metabolic processes but they too are catalysed by enzymes.

Nucleic acids

Nucleic acids are found in all living cells and in viruses. Two types of nucleic acid found in cells are deoxyribonucleic acid (DNA) and ribonucleic acid (RNA). DNA is found in the nucleus, mitochondria and chloroplasts of eukaryotes while RNA may occur in the nucleus but is usually found mainly in the cytoplasm. Nucleic acids are vital to inheritance and development, which are discussed in Topics **3** and **7**. Nucleic acids are long molecules consisting of chains of units called nucleotides. Each nucleotide consists of a **pentose** sugar – ribose in RNA and deoxyribose in DNA – linked to phosphoric acid and an organic base. Nucleic acid chains are longer and more complex than those found in proteins.

Condensation

In a **condensation reaction**, two molecules can be joined to form a larger molecule, held together by strong **covalent bonds**. Condensation is an example of an **anabolic** reaction, which is a type of reaction that builds up monomers to form macromolecules. Each condensation reaction requires an enzyme to catalyse the process and it produces one molecule of water. The condensation of two monosaccharide monomers produces a disaccharide. For example:

glucose + galactose → lactose + water
(monosaccharide) (monosaccharide) (disaccharide)

If further monosaccharides are added to a disaccharide, a polysaccharide is formed, as you can see in Figure **2.4**.

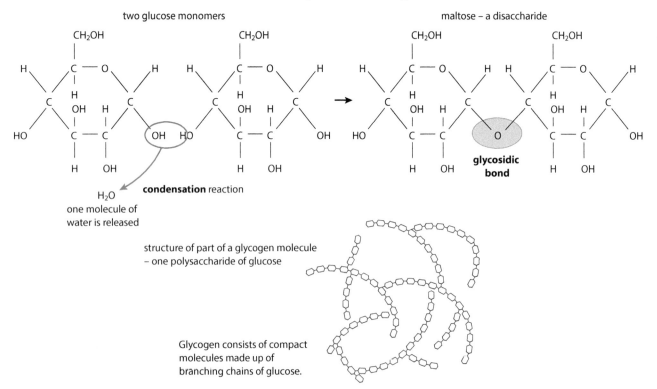

Figure 2.4 Monosaccharide subunits (glucose in this case) are joined in a condensation reaction, forming a disaccharide (maltose) and water. Glycogen is a polysaccharide, formed from long chains of glucose subunits.

In a similar way, two amino acids can be linked to form a **dipeptide** (Figure **2.5**):

amino acid + amino acid → dipeptide + water

When more than two amino acids are joined in this way, a **polypeptide** is formed. Polypeptide chains form protein molecules.

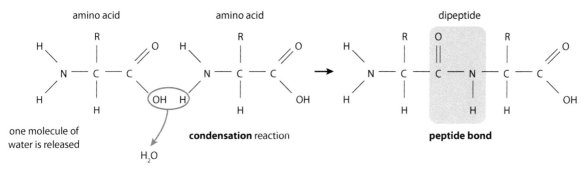

Figure 2.5 Two amino acids combine to form a dipeptide.

In another condensation reaction, glycerol links to fatty acids to produce triglyceride **lipid** molecules (Figure **2.6**):

glycerol + 3 fatty acids → triglyceride lipid + water

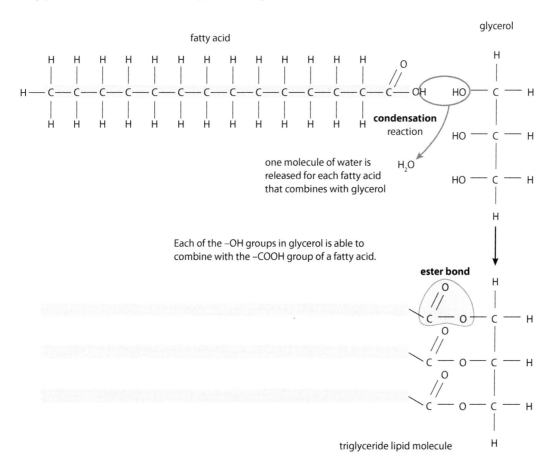

Figure 2.6 How a triglyceride lipid is formed from glycerol and three fatty acids in a condensation reaction.

Hydrolysis

Hydrolysis reactions occur every time food is digested. These reactions involve breaking down polysaccharides, polypeptides and triglycerides into the smaller units of which they are made. Hydrolysis is an example of a **catabolic** reaction, in which macromolecules are broken down into monomers. Water molecules are used in hydrolysis reactions which are the reverse of condensation reactions. Once again, enzymes are required to catalyse the reactions.

- Hydrolysis of starch (a polysaccharide) uses water and produces many molecules of glucose.
- Hydrolysis of protein (made of polypeptide chains) uses water and produces many amino acids.
- Hydrolysis of a triglyceride (a lipid) uses water and produces fatty acids and glycerol molecules.

Nature of science

Falsification of theories – how the synthesis of urea helped to falsify the theory of vitalism

Vitalism is a belief that living things have a distinctive 'spirit' contained in their bodies, which gives them life. The theory proposes that living organisms are fundamentally different from non-living things because they contain this 'life force', and that the organic substances upon which life is built cannot be synthesised artificially from inorganic components. Vitalism dates back to Aristotle and Galen in the 3rd century BCE and also forms part of traditional healing that views illness as an imbalance in 'vital forces'. Hippocrates associated vital forces with four 'temperaments' and in Eastern philosophy the imbalance was said to block the body's 'qi'.

Vitalism became less accepted from the 19th century as microscopy revealed the structure of cells and their components, and showed that they obeyed the laws of physics and chemistry. In addition, in 1828 Friedrich Wöhler successfully synthesised the organic molecule urea (Figure **2.7**), thus disproving the vitalist theory, which held that organic substances could not be synthesised from inorganic components.

No modern discovery has disproved the observation that parts of an organism behave in a co-ordinated way and that this is different from the way the parts behave in isolation. For example, a single cell behaves differently when it is within an organism from the way it behaves when alone. In an organism it is in direct contact with other parts of the body, which in turn interact with other parts. Furthermore, no evidence has yet contradicted the laws of physical and chemical interaction between cellular components. So far, no observations suggest that we need another 'force' to account for biological phenomena. In the 21st century, vitalism is no longer a generally accepted idea.

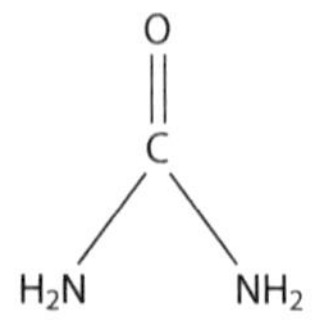

Figure 2.7 Urea is produced from the breakdown of amino acids in the mammalian liver. It is excreted in urine.

Test yourself

1. State the number of covalent bonds formed by a carbon atom.
2. Define anabolism.
3. State what is meant by the term 'monomer'.

2.2 Water

Structure and properties of water

Water is the main component of living things. Most human cells are approximately 80% water. Water provides the environment in which the biochemical reactions of life can occur. It also takes part in and is produced by many reactions. Its many important properties described below are due to its molecular structure, which consists of two hydrogen atoms each bonded to an oxygen atom by a covalent bond (Figure **2.8**).

Hydrogen bonds

The water molecule is unusual because it has a small positive charge on the two hydrogen atoms and a small negative charge on the oxygen atom. Because of this arrangement, water is said to be a **polar** molecule. Polar molecules are those that have an unevenly distributed electrical charge so that there is a positive region and a negative region. Sugars and amino acids are also polar molecules.

A weak bond can form between the negative charge of one water molecule and the positive charge of another, as shown in Figure **2.9**. This type of bond, known as a hydrogen bond, is responsible for many of the properties of water.

Cohesion and adhesion

Hydrogen bonds between water molecules hold them together in a network, resulting in a phenomenon known as cohesion. Cohesive forces give water many of its biologically important properties. For example, they enable water to be drawn up inside the xylem of a plant stem in a continuous column. Strong pulling forces, produced as water evaporates from the leaves at the top of tall trees, draw water and dissolved minerals up great distances to the tips of branches high above the ground. Cohesion is also responsible for surface tension, which enables some small organisms to 'walk on water', and contributes to the thermal properties of water too.

Water also tends to be attracted and adhere to the walls of its container. There are forces of attraction, known as adhesive forces, which occur between water molecules and different molecules in vessels that contain the water. Adhesion attracts water molecules to the sides of the xylem and is important as water is drawn up the stem of a plant. Because adhesive

Learning objectives

You should understand that:
- Hydrogen bonds form between water molecules, because they are polar.
- The cohesive, adhesive, thermal and solvent properties of water can be explained by hydrogen bonding.
- A substance may be classified as hydrophilic or hydrophobic depending on its solubility in water.

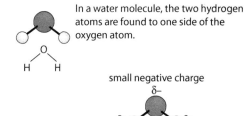

In a water molecule, the two hydrogen atoms are found to one side of the oxygen atom.

The oxygen atom pulls the bonding electrons towards it, which makes the oxygen slightly negatively charged. The hydrogen atoms have small positive charges.

Figure 2.8 The structure of a water molecule.

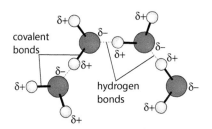

Figure 2.9 Hydrogen bonding in water.

Methane – a compound without hydrogen bonds

Methane is the smallest and simplest hydrocarbon, consisting of one carbon atom bonded to four hydrogen atoms (Figure 2.1). Unlike water, a methane molecule does not have hydrogen bonding between its H atoms and those of nearby molecules – as a result, very little energy is needed to separate its molecules, which move freely apart. Methane therefore exists as a gas at room temperature and standard pressure – its boiling point is −161 °C. Water, on the other hand, is liquid at room temperature and standard pressure, even though its molecules are a similar size to those of methane. Its boiling point is much higher, at 100 °C, because of the large input of energy needed to break the many hydrogen bonds between its molecules, and convert it from a liquid to a vapour (gas).

forces are greater in a narrow tube where relatively more water molecules are in contact with the sides, adhesive forces are able to 'hold up' and support a substantial mass of water in the fine xylem vessels. The water column is held together by cohesive forces.

Thermal properties

Water also has unusual **thermal properties**. It is unusual among small molecules because it is a liquid at a normal range of temperatures. A large amount of energy is needed to break the many weak hydrogen bonds between the water molecules. This gives water a high specific heat capacity – it can absorb or give out a great deal of heat energy without its temperature changing very much. A stable temperature is important to living things because the range of temperatures in which biological reactions can occur is quite narrow. The thermal properties of water allow it to keep an organism's temperature fairly constant. Within the body, water can act as a temperature regulator – for example, water, which is a major component of blood, carries heat from warmer parts of the body such as the liver, to cooler parts such as the feet.

In order for liquid water to evaporate and become vapour, many hydrogen bonds between the molecules must be broken, so evaporation requires a lot of energy. As a result, water is a liquid at most temperatures found on Earth, and it has a high boiling point. When it evaporates, it carries a great deal of heat with it – so, for example, when sweat evaporates from the skin surface of a mammal it acts as a coolant for the body. The properties of water are summarised in Table 2.3.

Property	Reason	Consequence/Benefits to living organisms
cohesion	Hydrogen bonds hold water molecules together.	Water can travel in continuous columns – for example, in the stems of plants – and act as a transport medium.
adhesion	Water molecules are attracted to other different molecules.	A column of water can be held up in the narrow xylem of a plant.
solvent	The polar molecules of water can interact with other polar molecules.	Ions dissolve easily. Large molecules with polar side groups, such as carbohydrates and proteins, can also dissolve. So water acts as an excellent transport medium and as a medium for metabolic reactions.
thermal	Water has a high heat capacity. Large amounts of energy are needed to break hydrogen bonds and change its temperature.	The temperature of organisms tends to change slowly. Fluids such as blood can transport heat round their bodies.
	Water has a high boiling point compared with other solvents because hydrogen bonds need large amounts of energy to break them.	Water is liquid at most temperatures at which life exists, so is a useful medium for metabolic reactions.
	Water evaporates as hydrogen bonds are broken and heat from water is used.	Sweating and transpiration enable animals and plants to lose heat. Water acts as a coolant.

Table 2.3 Summary of the properties of water.

Solvent properties

Water is sometimes known as a universal **solvent**. Its polarity makes it an excellent solvent for other polar molecules. Most inorganic ions, such as sodium, potassium and chloride ions, dissolve well as their positive or negative charges are attracted to the charges of water molecules (Figure 2.10). Polar organic molecules, such as amino acids and sugars, are also soluble in water. Water is the medium in which most biochemical reactions take place since almost all the substances involved dissolve well in it. Protein synthesis and most of the reactions of photosynthesis and respiration take place in an aqueous (water) solution.

The solvent properties of water also make it an excellent medium for transporting substances around the bodies of all organisms. In plants, the xylem carries dissolved minerals from the roots to the leaves, while the phloem transports soluble sugars up and down the plant. Many animals have blood as their transport medium. Blood is predominantly water, and the blood plasma carries many dissolved solutes including amino acids, sodium and chloride ions, sugars such as glucose, and carbon dioxide.

Substances are classified into two groups according to their solubility in water. Hydrophilic substances such as sugars and salts dissolve well, as do amino acids with polar side groups. Hydrophobic or 'water-hating' substances do not dissolve in water. They are usually uncharged, and examples include fats and oils, and large proteins that do not carry any polar groups.

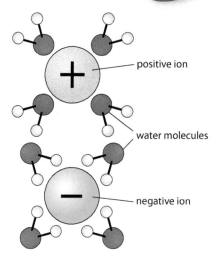

Figure 2.10 The positive and negative charges of water molecules attract ions with negative or positive charges, which means that the ions dissolve.

Uncharged and non-polar substances are not very soluble in water because water molecules would rather remain hydrogen-bonded to each other than allow such molecules to come between them. Various gases, such as oxygen and carbon dioxide, are not very soluble in water because they are essentially non-polar. Oxygen can dissolve in water (and does so to sustain aquatic life) but its solubility is very low. Almost all oxygen carried in the blood is bound to hemoglobin, and not in solution. Most of the carbon dioxide in the blood, on the other hand, is carried in the form of bicarbonate ions (HCO_3^-) dissolved in the plasma (Option **D**).

Many large molecules such as lipids and proteins are also largely non-polar but are sufficiently soluble to be carried through the aqueous environment of the blood because they have some polar groups exposed. On the outside of soluble proteins, for example, are polar groups that are able to interact with the polar water molecules and make the entire protein soluble.

Cholesterol is only slightly soluble in water and dissolves in the blood in very small amounts. For this reason, cholesterol is transported in the circulatory system within **lipoproteins**, which have an outer surface made up of amphipathic proteins and lipids. The outward-facing surfaces of these molecules are water-soluble (hydrophilic) and their inward-facing surfaces are lipid-soluble (lipophilic). Triglycerides (fats) are carried inside such molecules, while phospholipids and cholesterol, being amphipathic themselves, are transported in the surface layer of lipoprotein particles. There are different types of lipoproteins in blood – low-density lipoprotein (LDL) and high-density lipoprotein (HDL). The more lipid and less protein a lipoprotein has, the lower its density.

Memory of water

The 'memory of water' is a phrase that is usually associated with homeopathy. It was coined by Jacques Benveniste (1935–2004) who claimed that water retains a 'memory' of substances that have once been dissolved in it. Homeopathic remedies are prepared by diluting solutions to such an extent that, in some cases, no molecules of the original solute are found in them. Nevertheless, healing effects are claimed for these remedies, based on the idea that the water retains properties from the original substance. There is no scientific evidence to support the claim that water has such a 'memory' and the subject has drawn a lot of controversy, with many scientists rejecting it completely.

Question to consider
- What criteria can be used to distinguish scientific claims from pseudoscientific claims?

Nature of science

Using theories to explain natural phenomena – hydrogen bonds and the properties of water

The nature of liquid water and how the molecules in it interact are questions that have been studied by scientists for many years. Techniques including infrared absorption, neutron scattering and nuclear magnetic resonance imaging (NMRI) have been used to study the structure of water, and the results – along with data from theoretical calculation – have led to the development of a number of models, which try to describe the structure of water and explain its properties.

Water is a small, simple molecule in which each hydrogen atom is covalently bonded to the central oxygen atom by a pair of electrons that are shared between them (Figure **2.11**). Only two of the six outer-shell electrons of each oxygen atom are used to form these covalent bonds, leaving four electrons in two non-bonding pairs. These non-bonding pairs remain closer to the oxygen atom and exert a strong repulsion against the two covalently bonded pairs so that the two hydrogen atoms are pushed closer together. Overall, water molecules are electrically neutral, but this model of the water molecule results in small positive and negative charges unevenly distributed over the molecule. When the H_2O molecules are crowded together in liquid water, the forces between the atoms produce the effects that give water the properties we observe.

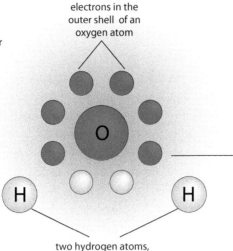

Each oxygen atom has 8 electrons in total, which we can imagine orbiting the nucleus in concentric 'shells'. The 6 electrons in the 'outer shell' are the ones involved in bonding with other atoms.

The most stable state for an atom is to have 8 electrons in its outer shell. In oxygen, the outer shell contains 6 electrons. In a water molecule, two of these electrons pair up with the single electron of a hydrogen atom, so that there are effectively 8 electrons in the outer shell.

Where a pair of electrons is shared between two atoms, the atoms are held together in a strong **covalent bond**.

Figure 2.11 The two hydrogen atoms in a water molecule are pushed together on one side because of the repulsive effect of the two pairs of non-bonding electrons in the outer shell of the oxygen atom.

Scientists continue to test theories and refine their models, deepening our understanding of the world around us. About 50 years ago it was assumed that water consisted of hydrogen-bonded clusters aggregated together, but more recent evidence from modelling does not support this view. Present-day thinking based on computer-generated molecular modelling is that, for very short time periods of time (less than a picosecond), water has a gel-like structure that consists of a single, huge hydrogen-bonded cluster.

Test yourself

4 Explain why water makes a good coolant for animals.
5 Define the term 'hydrophilic'.
6 Explain how hydrogen bonding affects the force of cohesion.

Learning objectives

You should understand that:
- Monosaccharides are linked together by condensation reactions to form polymers such as disaccharides and polysaccharides.
- A fatty acid molecule may be saturated, monounsaturated or polyunsaturated.
- Unsaturated fatty acids occur in two forms, or isomers: either *cis* or *trans*.
- Triglycerides are formed by the condensation of three fatty acid molecules and one glycerol molecule.

2.3 Carbohydrates and lipids

Carbohydrates

Carbohydrates are the most abundant category of molecule in living things. Different types of carbohydrate are produced by linking together monosaccharide monomers to build up polymers. The condensation reactions involved in this process result in the production of either disaccharides, which consist of two monomers, or polysaccharides, formed from long chains of monosaccharide monomers (Figure 2.4). The covalent bond between two monomers in a carbohydrate is known as a glycosidic link and a water molecule is released in the condensation reaction.

Glucose is the most common monosaccharide and it has the chemical formula $C_6H_{12}O_6$. The structures of alpha-D-glucose and beta-D-glucose are shown in Figure 2.2. These two forms of the molecule (known as isomers) have slightly different arrangements of the side groups, giving them slightly different properties.

When a bond is formed between two glucose monomers, a disaccharide called maltose is produced. Maltose is found in seeds such as barley. Other monosaccharides include fructose, found in fruits, and galactose, which is present in milk. Different combinations of these monomers produce a range of disaccharides, which are shown in Table 2.4.

Form of carbohydrate	Examples	Example of use in plants	Example of use in animals
monosaccharide	glucose, galactose, fructose	fructose is a component of fruits, making them taste sweet and attracting animals to eat them, thereby dispersing the seeds inside	glucose is the source of energy for cell respiration – it is obtained from the digestion of carbohydrate foods
disaccharide	maltose, lactose, sucrose	sucrose is transported from leaves to storage tissues and other parts of the plant to provide an energy source	lactose is found in milk and provides energy for young mammals
polysaccharide	starch, glycogen, cellulose	cellulose is a structural component of plant cell walls starch is used as a food store	glycogen is the storage carbohydrate of animals, found in the liver and muscles

Table 2.4 Examples and roles of carbohydrates.

Polysaccharides may contain from 40 to over 1000 monomers. Starch, glycogen and cellulose are all polymers of glucose monomers. Glycogen and starch are storage carbohydrates – glycogen in animals and starch in plants. They both have a compact shape and are insoluble. Glycogen is made of branching chains of glucose monomers (Figure 2.4) while starch has long chains of glucose that coil into a helical shape (Figure 2.12). Cellulose is used to build the cell walls of plants and is made up of long, straight chains of glucose molecules. The arrangement of the molecules means that hydrogen bonds can form cross-links between adjacent, parallel chains, which gives the polysaccharide its structural properties (Figure 2.12).

Starch Starch is made up of alpha-glucose units, linked by 1–4 glycosidic bonds, which causes the molecule to form a helical shape.

Cellulose Cellulose is made up of straight chains of beta-glucose units, with OH groups forming hydrogen bonds between chains.

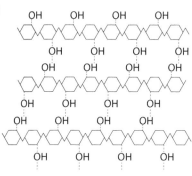

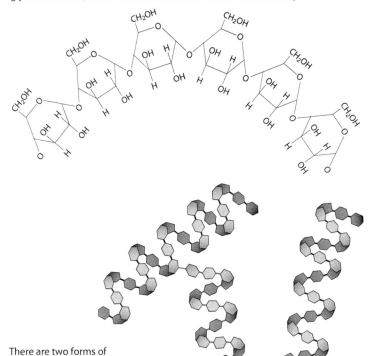

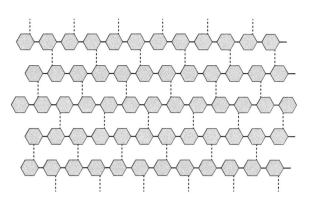

There are two forms of starch. Amylose contains only 1–4 glycosidic bonds and forms linear helices. Amylopectin also contains some 1–6 glycosidic bonds, which causes branching.

amylopectin   amylose

The hydrogen bonding between chains in cellulose causes the formation of strong, straight fibres.

Figure 2.12 Starch and cellulose are two polysaccharides found in plants.

Lipids

Fats and oils are one of the two main groups of lipid and are compounds of glycerol and fatty acids (Figure **2.6**). Glycerol has just one structural form but fatty acids have a wide variety of structures, which give the lipids that contain them their different physical and chemical properties. These molecules have important roles in the bodies of plants and animals and these are summarised in Table **2.5**), with the properties of lipids that make them suitable for these functions.

Key properties of lipids	
energy content	Lipids contain more energy per gramme than carbohydrates, so lipid stores are lighter than carbohydrates storing an equivalent amount of energy.
density	Lipids are less dense than water, so fat stores help large aquatic animals to float.
solubility	Lipids are non-polar, insoluble molecules so they do not affect the movement of water in and out of cells by osmosis.
insulation	Lipids are also important in providing heat insulation. Fat stored under the skin reduces heat loss and is vital for animals such as seals, polar bears and whales, which live in cold conditions.

Table 2.5 The important properties of lipids that suit them to particular roles in living organisms.

Fatty acids

Fatty acids consist of a long chain of carbon atoms that are joined to hydrogen atoms (Figure **2.6**). Chains like this are called **hydrocarbon chains**. The number of carbon atoms in a fatty acid is usually an even number most commonly between 14 and 22 although shorter and longer chain fatty acids are found. At one end of the chain is a carboxyl group (COOH) and at the other a methyl (CH_3) group. If the carbon chain is linked to the maximum number of H atoms with no double bonds it is said to be **saturated** because no more H atoms can be added. If the chain contains a double bond between two of the carbon atoms it is said to be **unsaturated** (Figure **2.13**). A chain with just one double bond is **monounsaturated**, while one with two or more double bonds is said to be **polyunsaturated**. Polyunsaturated fatty acids tend to be liquids at 20° C and are mainly derived from plant sources. Examples are sunflower oil, corn oil and olive oil.

Unsaturated fatty acids may be either a *cis* or a *trans* **configuration**. If the 'spaces' where additional hydrogen atoms could bond are both on the same side of the hydrocarbon chain, the fatty acid is known as a *cis* fatty acid and the carbon chain is slightly bent. If the spaces are on opposite sides, it is a *trans* fatty acid, which has a straight chain (Figure **2.14**).

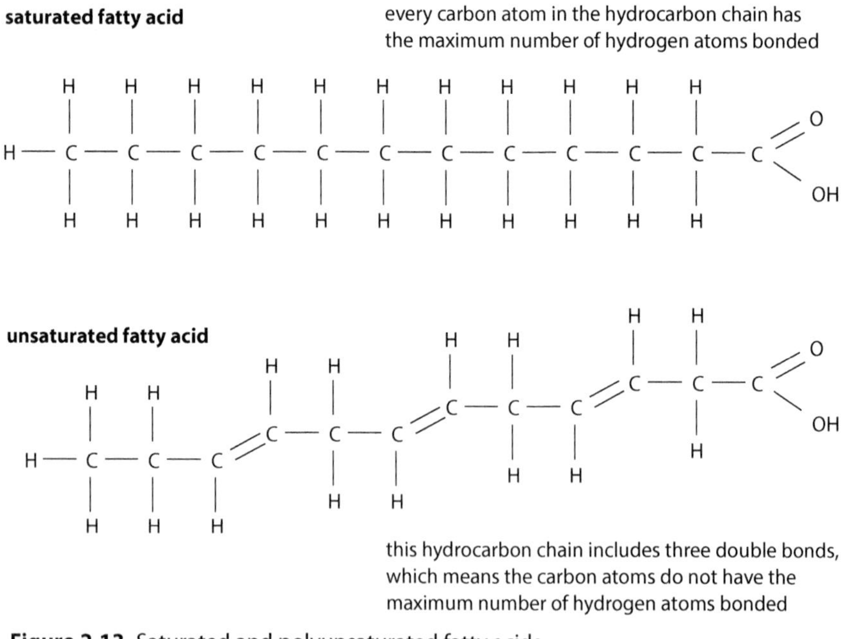

Figure 2.13 Saturated and polyunsaturated fatty acids.

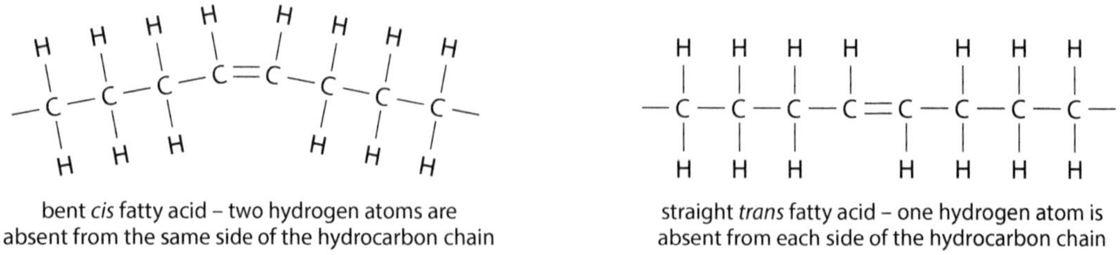

Figure 2.14 *Cis* and *trans* fatty acids.

One type of *cis* fatty acid is the omega-3 group. These have a double bond at the third bond from the CH_3 end of the molecule. Omega-3 fatty acids in our diet come from eating fish such as salmon and pilchards, and from walnuts and flax seeds. Another group, the omega-6 fatty acids, have a double bond at the sixth position and come from vegetable oils.

The relative amounts of different types of fatty acid in a person's diet can, in some cases, be correlated with health issues. Eating a diet that is high in saturated fatty acids, such as is prevalent in some western countries, has been shown to have a positive correlation with an increased risk of coronary heart disease (CHD). Saturated fatty acids can be deposited inside the arteries, and if the deposits combine with cholesterol they may lead to atherosclerosis, a condition that reduces the diameter of the arteries and leads to high blood pressure (Subtopic **6.2**). Reliable evidence suggests that in countries where the diet is high in saturated fatty acids and many high-fat foods, animal products and processed foods are eaten there is likely to be a high incidence of CHD. Since all fatty acids are high in energy, an excess of these foods in the diet can also lead to obesity, which places a further strain on the heart.

On the other hand, people who eat a Mediterranean-style diet, rich in unsaturated fatty acids from olive oil and fresh vegetables, tend to have a low incidence of CHD. These fats do not combine with cholesterol and so arteries tend to remain unblocked and healthy.

Some polyunsaturated fats are modified or 'hydrogenated' so they can be used in processed foods. These hydrogenated fats become *trans* fatty acids. There is a positive link between a high intake of these *trans* fatty acids and CHD, but many other factors must also be considered.

In animals, omega-3 fatty acids are used to synthesise long-chain fatty acids found in the nervous system. It has been suggested that a lack of omega-3 fatty acids could affect brain and nerve development but no conclusive evidence has yet been found.

Health issues – trans fatty acids and saturated fat

Artificial **trans fats** are formed when vegetable oil is hydrogenated, a process that solidifies the oil. The substance produced is known as hydrogenated fat and can be used for frying or in the manufacture of processed foods. Trans fats are included in biscuits and cakes to extend their shelf life but in recent years many food manufacturers have removed trans fats from their products.

Eating a diet containing high levels of trans fats has been shown to lead to high cholesterol levels, which in turn can lead to CHD and strokes. But most people do not eat large amounts of trans fats. In the UK, for example, it has been estimated that most people eat only about half the maximum recommended level of these fats. Most health professionals advise that saturated fats are a greater risk to health because of their contribution to atherosclerosis.

Question to consider

- How do we decide between different views about the relative harms and benefits of foods in our diets?

Triglycerides

Triglycerides are formed by condensation reactions between three fatty acids and one glycerol molecule (Figure **2.6**). Three molecules of water are released and the bonds between the fatty acids and the glycerol are known as ester bonds. Triglycerides play a major role in the structure of membranes (Subtopic **1.3**) when they combine with a phosphate group in the form of phosphoric acid to form phospholipids. The phosphoric acid combines with one of the three OH (hydroxyl) groups of glycerol and two fatty acid chains attach to the other two (Figure **1.16**).

Nature of science

Evaluating claims – cause and correlation in health statistics

Looking for **correlation** is one of the most common and useful statistical analysis techniques. Correlation describes the degree of relationship between two variables. For example, in the last 30 years, the number of people taking a holiday each year has increased. In the last 30 years, there has also been an increase in the number of hotels at holiday resorts. This data shows a **positive correlation**.

We could also consider annual deaths from influenza and the number of influenza vaccines given. In this case, there is a **negative correlation**. With these examples, we might feel safe to say that one set of data is linked to the other and that there is a **causal relationship** – because there are more tourists, more hotels have been built; greater use of the influenza vaccine has resulted in fewer deaths from influenza.

However, just because the data shows a **trend**, it does not necessarily mean that there is a causal relationship. If we consider the number of people using mobile phones in the last 10 years against the area of Amazon rainforest cut down there would be a positive correlation. But this does not mean that the use of mobile phones has *caused* rainforest to be cut down – nor does it mean that a reduction in rainforest area results in more mobile phone use.

Observations without experiments can show a correlation but usually experiments must be used to provide evidence to show the cause of the correlation. It is not ethically possible to conduct experiments to find evidence for correlation between diet and human health. We cannot restrict different groups of subjects to diets containing different amounts of saturated fats to assess the effects on their health and so observational or **epidemiological** data is all we have to go on. We must think about how the data is gathered and what other variables, such as lifestyle and genetics and family history, are important. There are difficulties in collecting objective data that accounts for all possible variables – indeed, it may never be possible to say that one type of diet or fatty acid is 'good' and another is 'bad' because individual subjects vary in so many different ways.

Correlation and cause

When studying the occurrence of medical conditions that may be related to diet, it is important to distinguish between **correlation** and **cause**. A correlation between two variables, such as a high incidence of CHD and a high intake of saturated fatty acids, does not mean that the CHD is caused by the fat intake.

? Test yourself

7 State **two** examples of disaccharides.
8 Outline the difference between a saturated and an unsaturated fatty acid.
9 State the type of reaction that leads to the formation of triglycerides.

2.4 Proteins

Polypeptides

Polypeptides are built up from amino acid monomers during condensation reactions (Figure **2.5**). Two amino acids are joined with a reaction between the amino (NH_2) group of one amino acid and the carboxyl (COOH) group of the other forming a peptide bond and producing a dipeptide. If further condensation reactions occur, a series of amino acids can become joined to form a polypeptide. The covalent bonds linking the amino acids produce what is known as the primary structure of any protein that is formed from the polypeptide.

In living cells, polypeptides are synthesised by ribosomes in the cytoplasm. Twenty different amino acids are used to construct polypeptides. Other amino acids do exist but these are not used in the biosynthesis of protein. Polypeptides can consist of up to 400 amino acids and, because these can be linked together in any sequence, there is a huge range of possible polypeptides. Some amino acids may also be modified once the polypeptide has been incorporated into a protein molecule so that even more different structures can be formed.

The sequence of amino acids in a polypeptide is coded for by an organism's genes. Genes consist of a series of codons, each of which carries the specific code for one amino acid (Subtopic **2.7**). The sequence of codons in a gene is used as a template to direct the sequence in which the amino acids will be assembled.

Building a protein

A protein may either consist of one polypeptide or several linked together. The basic chain of amino acids in a polypeptide folds and becomes a 3D shape once it is complete. The shape, known as secondary structure, results from the formation of hydrogen bonds between different parts of the chain. The most common shape is an alpha helix, held together by weak hydrogen bonds between the amino acids that form the turns in the structure. Keratin, a structural protein found in hair, is an alpha helix. In other proteins, such as silk, polypeptides in parallel chains are linked to form flat, folded shapes known as beta pleated sheets.

Further folding of polypeptides can occur due to the interactions between the R groups of the amino acids present. A complex three-dimensional shape known as tertiary structure results and is held together by **ionic bonds** and disulfide bridges. Tertiary structure is important in enzymes because the shape of the molecule determines where substrate molecules can bind to them.

Some proteins are composed of two or more polypeptides linked together and are said to have quaternary structure. The pigment hemoglobin found in red blood cells has four subunits and the positioning of these subunits is very important for the role of hemoglobin in carrying oxygen (Figure **7.20**, Subtopic **7.3**).

Learning objectives

You should understand that:
- Amino acids are linked in condensation reactions to form polypeptide chains.
- Twenty different amino acids found in polypeptides, which are synthesized by ribosomes.
- Because amino acids can be linked in any sequence it is possible to make a huge range of different polypeptides.
- The sequence of amino acids in a polypeptide is determined by the genetic code.
- Proteins consist of one or more polypeptides linked together.
- The sequence of amino acids in a polypeptide determines the three-dimensional shape of a protein.
- Living things synthesise many different proteins with many different functions.
- Every individual organism has a unique proteome.

Denaturation

Denaturation destroys the complex structure of a protein. A protein's structure is determined by the different types of bonds it contains. Heat or the presence of strong acids or alkalis can all disturb the bonds between the different parts of a protein molecule and disrupt its structure. The primary structure of the protein will remain but secondary, tertiary and quaternary structures are usually lost.

A denatured protein has different properties from the origin molecule. For example, enzymes are easily denatured by extremes of pH or temperature (Subtopic **2.5**) and lose the ability to function as catalysts. Some proteins lose their solubility or aggregate to form clumps as they denature and this can be observed during cooking. The heat used to cook meat denatures the proteins found in it so that its texture is changed, and eggs become hardened as they cook and their proteins are denatured.

Functions of proteins

The function of a protein is determined by the shape of its molecule. Proteins are divided into two main types: globular and fibrous proteins.

Fibrous proteins are long, insoluble molecules made up of parallel polypeptide chains. The chains are cross-linked along their lengths. Keratin found in hair, nails and hooves and collagen found in bones, muscles and tendons are two abundant fibrous proteins.

Globular proteins have polypeptide chains that are folded into compact, almost spherical shapes. The hormone insulin is a globular protein and so are enzymes. Each enzyme has its own 3D shape, which enables it to work as a catalyst for a particular reaction.

Examples of some important proteins and their functions are summarised in Table **2.6**.

As Table **2.6** shows, a wide range of proteins is found in different organisms and each protein has its own structural or biochemical function. Every individual organism has its own unique proteins, which are determined by its unique genome. The proteins found in an organism are known as its **proteome**, a term derived from a combination of the words 'protein' and 'genome'.

Protein	Function
Rubisco	an enzyme involved in carbon fixation in photosynthesis
insulin	a hormone produced by the pancreas, which stimulates the liver to take up glucose from the blood and store it as glycogen
immunoglobulin	a large Y-shaped protein (antibody) produced by the immune system to fight infection
rhodopsin	a protein linked to a pigment found in the photoreceptor cells in the retina of the eye
collagen	a structural protein which builds muscle, tendons, ligaments and the skin of vertebrates
spider silk	a protein fibre spun by spiders, which is tough and elastic and used for constructing a spider's web

Table 2.6 The function of some different proteins.

Nature of science

Looking for trends and discrepancies – do all organisms use only 20 amino acids?

Humans can produce 10 of the 20 amino acids we need to build proteins but we do not have the enzymes needed for the biosynthesis of the others. Plants and other autotrophs, on the other hand, must be able to make all the amino acids they require.

Researchers have investigated the trends in amino acid compositions of proteins found in species of the important kingdoms of Archaea, Bacteria and Eukaryotes. The international databases 'Proteomes' and 'Swiss-Prot' (which contain information about the structure and composition of proteins) can be used to compare amino acid frequencies for 195 known proteomes and all recorded sequences of proteins. Such comparisons have shown that the amino acid compositions of proteins do differ substantially for different kingdoms.

In addition to variations in amino acid sequence in proteins, some microorganisms and plants are able to make so called 'non-standard' amino acids by modifying standard amino acids. Some species are also able to synthesise many uncommon amino acids. For example, some microbes synthesise lanthionine, which is a modified version of the amino acid alanine. Many other proteins are modified after they have been produced. This 'post-translational modification' involves the addition of extra side groups to the amino acids in a protein.

Considering all the evidence, it seems that, although we can observe many similar proteins in different species, we cannot always say that the same amino acids are used in their construction. The range of amino acids in proteins can vary considerably from species to species.

Test yourself

10 Distinguish between hydrolysis and condensation reactions.
11 State how the amino acid sequence in polypeptide is determined.
12 Define the term 'proteome'.

Learning objectives

You should understand that:
- Enzymes have an area on their molecule, known as the active site, to which specific substrates bind.
- During enzyme-catalysed reactions, molecules move about, and substrate molecules collide with the active sites on enzyme molecules.
- The rate of enzyme activity is influenced by temperature, pH and substrate concentration.
- Enzyme molecules can be denatured.
- Immobilised enzymes are used in many industrial processes.

Enzyme a globular protein that functions as a biological catalyst of chemical reactions

Denaturation irreversible changes to the structure of an enzyme or other protein so that it can no longer function

Active site region on the surface of an enzyme molecule where a substrate molecule binds and which catalyses a reaction involving the substrate

2.5 Enzymes

Enzymes and active sites

An **enzyme** is a biological **catalyst**. Like all catalysts, enzymes speed up biochemical reactions, such as digestion and respiration, but they remain unchanged at the end of the process. All enzymes are proteins with long polypeptide chains that are folded into 3D shapes. The arrangement of these shapes is very precise and gives each enzyme the ability to catalyse one specific reaction. If the 3D shape of an enzyme is destroyed or damaged, it can no longer carry out its job and is said to be **denatured**. Extremes of temperature, heavy metals and, in some cases, pH changes can cause permanent changes in an enzyme.

The three-dimensional shape of an enzyme is crucial to the way it works. In the structure of every enzyme is a specially shaped region known as an **active site** (Figure **2.15**). It is here that the substrate and enzyme bind together. The substrates are the chemicals involved in the reaction catalysed by the enzyme. The shapes of the enzyme and substrate are complementary, so that they fit together perfectly like a key fits into a lock. The 'lock-and-key hypothesis' is a way of explaining how each enzyme can be so specific. To unlock a door requires just one special key. To catalyse a reaction requires one special enzyme. Just as only one key fits perfectly into the lock, only one substrate fits perfectly into the active site of an enzyme.

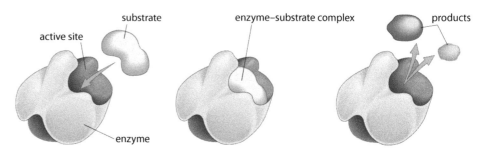

a An enzyme has a cleft in its surface called the active site. The substrate molecule has a complementary shape.

b Random movement of enzyme and substrate brings the substrate into the active site. An enzyme–substrate complex is temporarily formed. The R groups of the amino acids in the active site interact with the substrate.

c The interaction of the substrate with the active site breaks the substrate apart. The two product molecules leave the active site, leaving the enzyme molecule unchanged and ready to bind with another substrate molecule.

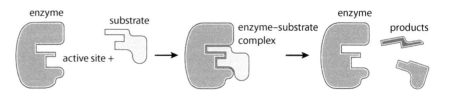

Figure 2.15 How an enzyme catalyses the breakdown of a substrate molecule into two product molecules.

Enzyme and substrate molecules move freely in solution and in most cases eventually collide with one another. When a substrate molecule collides with the active site of an enzyme it binds with it to form an **enzyme–substrate complex**. Once in place in an active site, substrates may be bonded together to form a new substance or they may be broken apart in processes such as digestion and respiration. For example, one type of enzyme bonds amino acids together to form a polypeptide, while very different enzymes are involved in digesting them.

Factors affecting enzyme action

Enzymes work in many different places in living organisms and they require special conditions to work at their greatest, or optimum, efficiency. Temperature, pH and the concentration of the substrates involved all affect the rate at which enzymes operate and produce their products.

Temperature

Enzymes and their substrates usually meet as a result of random collisions between their molecules, which move freely in body fluids or cytoplasm. In the human body, most reactions proceed at their greatest rate at a temperature of about 37 °C and deviations from this **optimum temperature** affect the reaction rate, as the graph in Figure **2.16** shows.

Below 37 °C, molecules in solution move more slowly, so the likelihood of collision between them is reduced. This slows down the production of products. At very low temperatures, enzymes hardly work at all and the rate of reaction is very low. As the temperature rises, molecular collisions are more frequent and energetic, and therefore the rate of the enzyme-controlled reaction increases.

As the temperature rises above the optimum, the enzyme and substrate molecules move faster – but atoms within the enzyme molecule itself also move more energetically, straining the bonds holding it together. Eventually, these bonds may be stressed or broken to such an extent that the enzyme loses its 3D shape and the active site can no longer receive substrate molecules. At these high temperatures, the structure is permanently destroyed – the enzyme is denatured and can no longer catalyse the reaction.

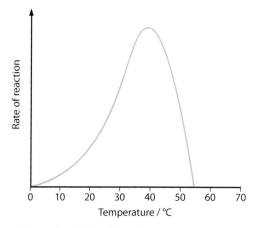

Figure 2.16 The effect of temperature on the rate of an enzyme-controlled reaction. An enzyme works most efficiently at its optimum temperature.

pH

pH is a measure of the relative numbers of H^+ and OH^- ions in a solution. A solution with a low pH value has many free H^+ ions and is acidic, whereas a high pH value indicates more OH^- ions and a basic solution. Pure water is neutral and has a pH value of 7 indicating that the number of OH^- and H^+ ions is equal.

Enzyme action is influenced by pH because the amino acids that make up an enzyme molecule contain many positive and negative regions, some of which are around the active site. An excess of H^+ ions in an acidic solution can lead to bonding between the H^+ ions and negative charges in the active site or other parts of the enzyme. These interactions can inhibit the matching process between the enzyme and its substrate, and slow down

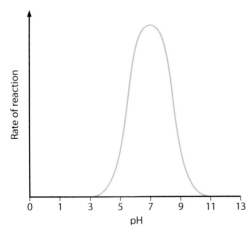

Figure 2.17 The effect of pH on the rate of an enzyme-controlled reaction.

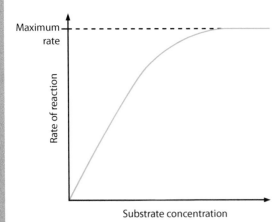

Figure 2.18 The effect of substrate concentration on the rate of an enzyme-catalysed reaction.

or even prevent enzyme activity. A similar effect occurs if a solution becomes too basic – the excess of negative ions upsets the enzyme in the same way. At extremes of pH, the enzyme may even lose its shape and be denatured.

Not all enzymes have the same **optimum pH**. Proteases (protein-digesting enzymes) in the stomach have an optimum pH of 2 and work well in the acidic conditions there, but proteases in the small intestine have an optimum of pH 8. Most enzymes that work in the cytoplasm of body cells have an optimum pH of about 7. The graph in Figure 2.17 shows how reaction rate varies with pH for this type of enzyme.

Concentration of substrate

If there is a set concentration of enzyme present in a reaction mixture, and the concentration of substrate increases, the rate of production of the products will increase because of the greater chance of collisions between substrate and enzyme molecules. More collisions mean that the enzyme is able to process or 'turn over' more substrate molecules. But there is a limit to this increase in reaction rate. If the concentration of substrate increases too much, it will exceed the maximum rate at which the enzyme can work. When this happens, at any one moment all the active sites are occupied by substrate or product molecules, and so adding further substrate has no effect. The rate reaches its limit – you can see this as the plateau in the graph in Figure 2.18.

Enzymes in industry

Enzymes work as catalysts in biological reactions but they are not used up so they can work over and over again. People have used enzymes from microorganisms for thousands of years in baking, cheese production and brewing. Today, many microbes are used as a source of enzymes for a wide range of industrial processes. Enzymes from fungi are used to produce biological detergents and in the textile industry to smooth fabrics. Bacterial enzymes are used in the leather industry to soften hides, in brewing and in the production of medicines.

Industrial enzymes are mainly isolated from microbes that are grown on a large scale in fermenters. The enzymes are separated from the microbial culture and purified before they are used. Once enzymes have been produced they can be used in processes such as those shown in Table 2.7. Using purified enzymes improves the efficiency of reactions and makes it easier to provide optimum conditions for enzyme activity.

Industry	Enzymes	Uses
biological detergents	protease and amylase	removal of protein and starch stains
baking	amylase	allows continuous dough production, converts starch to sugar
	protease	breaks down gluten to produce gluten-free products
biosensors	glucose oxidase	testing for glucose in blood samples
dairy	rennin	forms curd in cheese production
	lactase	production of lactose-free foods
confectionery	invertase	smoothing agent in confectionery production
medicine	streptokinase	treating bruises and blood clots
fruit juice production	cellulase and pectinase	speed the extraction of juice from fruit and prevent cloudiness

Table 2.7 Some industrial applications of enzymes.

Immobilised enzymes

For the greatest efficiency in producing the required products, enzymes are 'immobilised' by attaching them to a substance that remains stationary in a column or other container. This method allows the enzyme to be retained so that it can be re-used many times with fresh batches of substrate.

Enzymes are immobilised by adsorption on to solid resins, charcoal or similar substances, or by bonding them to collagen or synthetic polymers. Another method that can be used, and one which is easy to replicate on a laboratory scale, is to *enclose* the enzyme in inert, insoluble calcium alginate beads. In this method, the enzyme is mixed with a suspension of sodium alginate and calcium chloride. As the mixture turns from a liquid to a gel, it can be formed into small spheres, which contain the enzyme. The alginate beads are usually enclosed in a vertical column so that substrate can be poured in at the top and products collected at the base (Figure 2.19). The size of the beads and the rate of flow of the substrate through them are crucial factors that determine the maximum enzyme efficiency and production of products. The substrate must be able to diffuse freely through the beads and in and out of the lattice that they form. The rate of flow of the substrate is very important to optimise the rate of reaction. Enzyme and substrate must remain in contact for sufficient time for reactions to occur but not too long so that the production process is too slow to be economically viable.

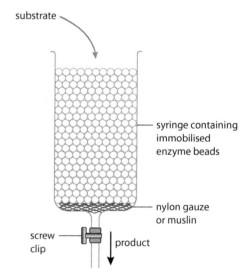

Figure 2.19 The alginate beads are enclosed in a 20 cm³ syringe in a laboratory demonstration, such as that shown here. An industrial process uses the same principles but on a much larger scale.

Advantages of using immobilised enzymes include:
- ease of harvesting the product – after the reaction the output usually contains only solvent and products
- ease of recovering the enzyme – the immobilised enzyme can easily be collected and re-used
- continuous production is possible – immobilised enzymes are more stable and less sensitive to variation in temperature than enzyme solutions
- immobilisation extends production time especially when using proteases, which might digest each other in reactions in solution.

Production of lactose-free milk

Lactose-free milk is produced using immobilised enzymes. Milk contains the sugar lactose, which is digested in the intestine by an enzyme called lactase. This produces two simple sugars (glucose and galactose) that can be absorbed into the body. Lactase can be obtained commercially from yeast that grows in milk and this enzyme is used to produce lactose-free milk.

Lactose-free milk is useful because some people are lactose intolerant and cannot digest lactose properly. If they take more than a small amount of milk, they suffer symptoms such as cramps or diarrhoea. Milk that contains glucose and galactose rather than lactose also tastes sweeter and so manufacturers need to add less sweetener to milk products such as yogurt made with lactose-free milk. Glucose and galactose are also more soluble than lactose, so they produce smoother textures in dairy products such as ice cream.

Nature of science

Experimental design – accuracy and reliability

The **accuracy** of each measurement in any experiment depends on the quality of the measuring apparatus and the skill of the scientist taking the measurement. Faulty or inappropriate apparatus, or mistakes in using it, can produce inaccurate measurements. The level of accuracy of a measuring instrument determines the closeness of its measurement to the quantity's true value. For example, a micrometer measures length to a greater level of accuracy than a ruler.

For the data to be **reliable**, the variation within the values of the results must be small. There is always some variation in any set of measurements, whatever is being measured. These may be due to small variations in the substances used in an experiment or the way the measuring apparatus is used.

When designing experiments with immobilised enzymes, such as the one shown in Figure **2.19**, accurate measurement of volumes of enzyme solutions, alginate and calcium chloride are important in producing the alginate beads. For example, it would not be appropriate to use a $20\,cm^3$ syringe to measure $2\,cm^3$ of solution because the degree of error would be too great. The smallest possible syringe should be used so that the error is $\pm 0.05\,cm^3$, and not $\pm 0.5\,cm^3$.

Also important in all experimental design is reliability. Reliable results are those that can be accepted as being trustworthy, with no readings that are erroneous. Unreliable data may be due to misreading a stopwatch or a probe or recording a value incorrectly. To minimise the risk of errors like these it is important to take a number of readings (replicates) of the results for each value of the controlled (independent) variable. Three similar results can usually be accepted as reliable and an average value calculated from them. But if the results differ widely, further replicates must be carried out. Repeating the whole experiment with a different batch of the same chemicals is a good way of testing reliability further – if the repeat experiment produces very similar results, they can be said to be reliable.

Development of techniques that benefit some people more than others

The distribution of lactose intolerance around the globe shows considerable variation. Only 4% of the Scandinavian population is affected, while countries around the Mediterranean have incidences of the order of 50–75% and in Africa the figure reaches 80%. Asia is affected even more with about 90% of the population suffering from lactose intolerance.

The control of lactase production by the human digestive system was disputed by scientists for many years. Some researchers in the 1960s argued that lactase production was stimulated in the presence of its substrate, lactose from milk. They proposed that populations that did not use milk as adults lost the ability to produce lactase, whereas groups that did consume milk continued to make the enzyme. More recent studies have cast doubt on this theory and shown that lactase production is controlled by a gene that is located on chromosome 2.

Development of lactose-free products, which required significant financial investment, occurred in western, developed countries despite the fact that the benefits were more widely applicable in other parts of the world.

Questions to consider

- Should scientific knowledge always be shared, even if it involved significant financial investment, and there are likely to be substantial rewards in keeping control of it?
- Should techniques developed in one part of the world be freely shared when they are more applicable to another?
- How could international cooperation benefit the efforts to combat diseases such as malaria?

Test yourself

13 Define the 'active site'.
14 Outline the effect of increasing temperature on an enzyme-catalysed reaction.
15 State **one** use of immobilised enzymes in industry.

Learning objectives

You should understand that:
- DNA and RNA are nucleic acids, which are polymers of nucleotides.
- There are three essential differences between DNA and RNA: the number of strands, the composition of bases and the type of pentose sugar present in the molecule (Table **2.8**).
- The DNA molecule is a double helix with nucleotides arranged in two antiparallel strands that are linked by hydrogen bonds between complementary pairs of bases.

2.6 Structure of DNA and RNA

DNA (deoxyribonucleic acid) molecules make up the genetic material of living organisms. DNA is an extremely long molecule but, like proteins and carbohydrates, it is built up of many monomer subunits. The subunits of DNA are called nucleotides. RNA (ribonucleic acid) is also built up of many nucleotides but these differ from DNA nucleotides in the type of pentose sugar they contain and the bases that are attached to them.

Each nucleotide consists of three parts – a pentose (five-carbon) sugar (deoxyribose or ribose), a phosphate group and a nitrogenous base (Figure **2.20**). DNA contains four different bases: adenine, guanine, cytosine and thymine. These are usually known by their initial letters: A, G, C and T (Figure **2.21**). RNA also contains four bases but in an RNA molecule thymine is not present and is replaced by uracil (U).

To form a DNA molecule, nucleotide monomers are linked together. The phosphate group of one nucleotide links to the deoxyribose of the next molecule to form a chain of nucleotides, as shown in Figure **2.21**. The sugar and phosphate groups are identical all the way along the chain and form the 'backbone' of the DNA molecule. The sequence of bases in the chain will vary and it is this sequence that forms the genetic code determining the characteristics of an organism.

Two strands of nucleotides are linked by hydrogen bonds that form between the bases and this two-stranded structure makes up the double helix of a complete DNA molecule. Adenine always pairs with thymine and is bonded with two hydrogen bonds, while cytosine is paired with guanine by three hydrogen bonds. The arrangement is known as complementary base pairing. Notice that the two DNA chains run in opposite directions and are said to be **antiparallel**.

You can imagine the molecule rather like a rope ladder with the sugar–phosphate backbone being the sides of the ladder and the rungs being formed by the hydrogen-bonded base pairs. To form the characteristic double helix of a DNA molecule, the ladder must be twisted to resemble a spiral staircase.

To form a molecule of RNA, nucleotide monomers are linked in a similar way to those of DNA. In the case of RNA the molecule remains **single stranded** and the bases it contains do not bond with bases in other RNA molecules.

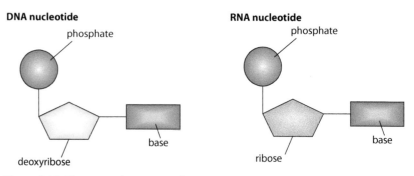

Figure 2.20 The general structure of DNA and RNA nucleotides.

Part of an RNA molecule, which is a single strand

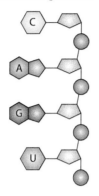

An RNA single helix, with a sugar–phosphate backbone and bases projecting inwards

How nucleotides join together in both DNA and RNA. Each nucleotide is linked to the next by covalent bonds between phosphates and sugars.

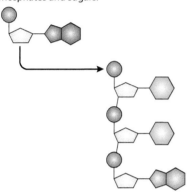

Part of a DNA molecule. Two DNA strands, running in opposite directions, are held together by hydrogen bonds between the bases.

C links to G by three hydrogen bonds

one DNA strand

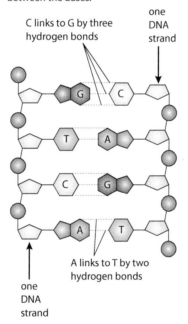

A links to T by two hydrogen bonds

one DNA strand

The DNA double helix. It has a 'backbone' of alternating sugar–phosphate units, with the bases projecting into the centre creating base pairs.

sugar–phosphate 'backbone'

complementary base pair

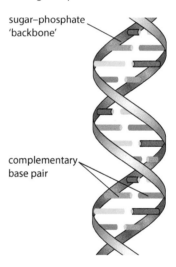

Figure 2.21 The structure of the bases in DNA and RNA.

DNA	RNA
contains the five-carbon sugar deoxyribose	contains the five-carbon sugar ribose
contains the bases adenine, guanine, cytosine and thymine (A, G, C, T)	contains the bases adenine, guanine, cytosine and uracil (instead of thymine) (A, G, C, U)
a double-stranded molecule	a single-stranded molecule

Table 2.8 A comparison of the structure of DNA and RNA.

2 MOLECULAR BIOLOGY

Collaboration versus competition

The story of the discovery of DNA structure illustrates how important collaboration can be in scientific discovery. Cooperation and competition can both occur between research groups.

Questions to consider:
- To what extent is keeping research discoveries secret 'anti-scientific'?
- How are shared and personal knowledge related in scientific research?

Nature of science

Careful observation – the discovery of DNA

James Watson and Frances Crick, together with Maurice Wilkins, were awarded the Nobel Prize in 1962 for their discovery of the structure of DNA (Figure **2.22**). Watson and Crick put forward their theory for DNA structure in 1953, basing their ideas on the work of an American chemist, Erwin Chargaff, who calculated the proportions of the bases in DNA. Watson and Crick suggested that DNA was composed of two parallel strands held together by pairs of bases, A pairing with T, and C with G. At the same time, in different laboratories, other researchers were trying to work out DNA's three-dimensional structure using X-ray diffraction. Rosalind Franklin (Figure **2.23**) and Maurice Wilkins spent many hours trying to interpret photographs of diffraction patterns produced by DNA. (You can read more about their work in Subtopic **7.1**.) From careful observation and calculation of the positions of certain markers on the X-ray photographs, Watson and Crick finally worked out that the two

Figure 2.22 Watson and Crick built a 3D model to help formulate their proposal for the structure of DNA.

Figure 2.23 Rosalind Franklin was an expert in the field of X-ray crystallography. Her skill and careful observations enabled her to work out that the phosphate groups of DNA are found on the outside of the molecule. She died at the age of 37 before the Nobel Prize was awarded to Watson, Crick and Wilkins. Nobel prizes cannot be awarded posthumously.

chains of DNA were wrapped into a double helix, linked at regular intervals by the bases of the nucleotides. Furthermore they were able to see from a model of DNA that there are 10 nucleotides per turn of the helix. It is important to remember that their achievement would not have been possible without the data which was provided by the other scientists working on DNA structure at the time.

Test yourself
16 State **two** differences between DNA and RNA.
17 Outline the structure of a nucleotide.

2.7 DNA replication, transcription and translation

Learning objectives

You should understand that:
- DNA replication is semi-conservative and depends on complementary base pairing.
- The first stage of the replication process involves the enzyme helicase, which unwinds the DNA molecule, separating the strands by breaking hydrogen bonds.
- DNA polymerase joins nucleotides to form new strands using each original strand as a template.
- The synthesis of mRNA (messenger RNA) copies of DNA sequences, by the enzyme RNA polymerase, is called transcription.
- The synthesis of polypeptides by ribosomes is called translation.
- The sequence of amino acids in a polypeptide is determined by the sequence of bases in the mRNA, according to the genetic code.
- Three bases (a codon) on a mRNA molecule correspond to one amino acid in a polypeptide.
- The process of translation depends on complementary base pairing between mRNA codons and anticodons on tRNA (transfer RNA) molecules.

DNA replication

An essential feature of DNA is that it must be able to replicate itself accurately, so that when a cell divides, the genetic code it carries can be passed on to the daughter cells. **DNA replication** copies DNA precisely so that new molecules are produced with exactly the same sequence of bases as the original strands. DNA replication takes place in the nucleus during interphase of the cell cycle when DNA is not tightly coiled (Subtopic **1.6**).

Taq DNA polymerase is used in the polymerase chain reaction (PCR) to produce multiple copies of DNA for forensic examination of small DNA samples (Subtopic **3.5**). Taq polymerase is used because it is stable at high temperatures. It is named after the thermophilic (heat tolerant) bacterium *Thermus aquaticus* from which it was originally isolated.

As Figure **2.24** shows, this process does not occur in a haphazard manner. An enzyme called helicase unzips one region of the DNA molecule and nucleotides are added in a step-by-step process that links them to one another and to their complementary bases in an area known as the replication fork.

1. The first step in the process is the 'unzipping' of the two strands. **DNA helicase** moves along the double helix, unwinding the two strands, which separate from one another as the relatively weak hydrogen bonds between the bases are broken.
2. The unpaired nucleotides are exposed and each single strand now acts as a template for the formation of a new complementary strand. Free nucleotides move into place: C pairs with G and A pairs with T.
3. The free nucleotide bases form complementary pairs with the bases on the single DNA strands. **DNA polymerase** is the enzyme involved in linking the new nucleotides into place. Finally, the two new DNA molecules are rewound, each one forming a new double helix.

The two new DNA strands that are produced are absolutely identical to the original strands. Complementary base pairing between the template strand and the new strand ensures that an accurate copy of the original DNA is made every time replication occurs. DNA replication is said to be **semi-conservative** because no DNA molecule is ever completely new. Every double helix contains one 'original' and one 'new' strand.

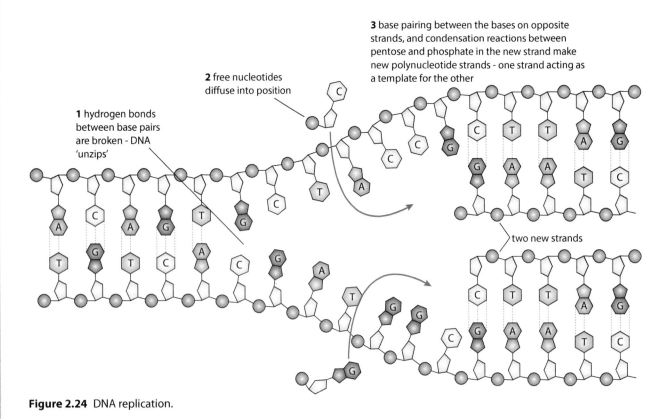

Figure 2.24 DNA replication.

Nature of science

Obtaining evidence – Meselson and Stahl's experiment

The research of Meselson and Stahl demonstrates the importance of making and testing a hypothesis in science. They investigated the two hypotheses about DNA replication that were current in the 1950s. The first hypothesis proposed that when DNA is replicated the original helix is conserved unchanged and the newly produced helix contains all new material. This conservative hypothesis was in contrast to the semi-conservative hypothesis, which proposed that one of the original DNA strands from a helix would always be found as one half of the new double helix produced after replication. Meselson and Stahl designed their experiments using *Escherichia. coli*. The bacteria were grown on a medium containing nitrogen ^{15}N, which is a heavy isotope of the normal ^{14}N. After many generations the bacteria incorporated ^{15}N into their cells so that their DNA became 'labelled' with the heavy isotope and could be identified easily. The bacteria were then transferred to a new medium containing the lighter isotope ^{14}N, and allowed to grow for a period of time that corresponded to the length of a generation. Figure 2.25 shows how the labelled DNA would be distributed among the daughter molecules after one and two replications, according to the semi-conservative theory and the conservative theory. Meselson and Stahl's careful measurements of the amounts of ^{15}N in the daughter molecules after one replication showed that all the helices contained one strand of labelled DNA and one strand of normal DNA. Their results therefore supported the theory of semi-conservative replication.

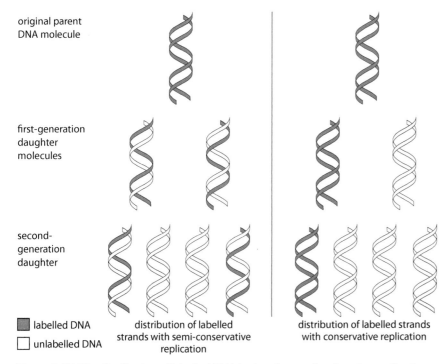

Figure 2.25 The distribution of labelled DNA in daughter molecules after replication, according to the semi-conservative theory and the conservative theory of replication.

Transcription and translation

During DNA replication, genetic material is accurately copied but the main role of DNA is to direct the activities of the cell. It does this by controlling the proteins that the cell produces. Enzymes, hormones and many other important biochemical molecules are proteins, which control what the cell becomes, what it synthesises and how it functions. Protein synthesis can be divided into two sets of reactions: the first is **transcription** and the second, **translation**. In eukaryotes, transcription occurs in the nucleus and translation in the cytoplasm.

The sections of DNA that code for particular proteins are known as **genes**. Genes contain specific sequences of bases in sets of three, called **triplets**. Some triplets control where transcription begins and ends.

Transcription

The first stage in the synthesis of a protein is the production of an intermediate molecule that carries the coded message of DNA from the nucleus into the cytoplasm where the protein can be produced. This intermediate molecule is called **messenger RNA** or **mRNA**. RNA (ribonucleic acid) has similarities and differences with DNA and these are shown in Table **2.8**.

As we have seen, the building blocks for RNA are the RNA nucleotides that are found in the nucleus. Complementary base pairing of RNA to DNA occurs in exactly the same way as in the replication process but this time uracil (U) pairs with adenine since the base thymine (T) is not found in RNA. Transcription results in the copying of one section of the DNA molecule, not its entire length. Figure **2.26** describes the process.

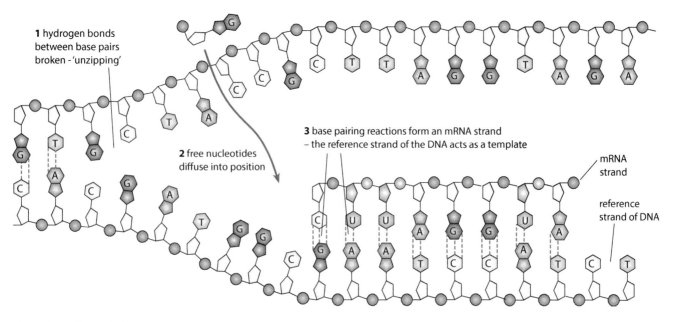

Figure 2.26 Transcription.

1 DNA is unzipped by the enzyme RNA polymerase and the two strands uncoil and separate.
2 Free nucleotides move into place along one of the two strands.
3 The same enzyme, RNA polymerase, assembles the free nucleotides in the correct places using complementary base pairing. As the RNA nucleotides are linked together, a single strand of mRNA is formed. This molecule is much shorter than the DNA molecule because it is a copy of just one section – a gene. The mRNA separates from the DNA and the DNA double helix is zipped up again by RNA polymerase.

Once an mRNA molecule has been transcribed, it moves via the pores in the nuclear envelope to the cytoplasm where the process of translation can take place.

Translation

Due to complementary base pairing, the sequence of bases along the mRNA molecule corresponds to the sequence on the original DNA molecule. Each sequence of three bases, called a triplet, corresponds to a specific amino acid, so the order of these triplets determines how amino acids will be assembled into polypeptide chains in the cytoplasm.

The mRNA codons that code for each amino acid are shown in Table **3.1**. From this table you should be able to deduce which amino acid corresponds to any codon.

Translation is the process by which the coded information in mRNA strands is used to construct polypeptide chains, which in turn make functioning proteins. Each triplet of mRNA bases is called a **codon** and codes for one amino acid. Translation is carried out in the cytoplasm by structures called **ribosomes** (Figure **2.27**) and molecules of another type of RNA known as **transfer RNA** or **tRNA** (Figure **2.28**).

Ribosomes have binding sites for both the mRNA molecule and tRNA molecules. The ribosome binds to the mRNA and then draws in specific tRNA molecules with **anticodons** that match the mRNA codons. Only two tRNA molecules bind to the ribosome at once. Each one carries with it the amino acid specified by its anticodon. The anticodon of the tRNA binds to the complementary codon of the mRNA molecule with hydrogen bonds.

When two tRNA molecules are in place on the ribosome, a **peptide bond** forms between the two amino acids they carry to form a dipeptide. Once a dipeptide has been formed, the first tRNA molecule detaches from both the amino acid and the ribosome. The ribosome moves along the mRNA one triplet to the next codon.

These processes, shown in Figure **2.27**, are repeated over and over again until the complete polypeptide is formed. The final codon that is reached is a 'stop' codon, which does not code for an amino acid but tells the ribosome to detach from the mRNA. As it does so, the polypeptide floats free in the cytoplasm.

The genetic code in part of an mRNA molecule
the mRNA molecule is read in this direction ⟶

the genetic code in the mRNA molecule ⟶

A   U   G   G   A   U   U   C   C   U   G   C   U   A   A

This codon represents the amino acid methionine; 'start' codon.

This codon represents the amino acid aspartate.

This codon represents the amino acid serine.

This codon represents the amino acid cysteine.

This codon represents 'end'; 'stop' codon.

Translation on a ribosome

1 Complementary base pairing between codon and anticodon.

2 Another amino acid is brought in attached to its tRNA.

3 A condensation reaction forms a peptide bond.

4 The ribosome moves along the mRNA by one triplet and a tRNA is released.

5 Another amino acid is brought in.

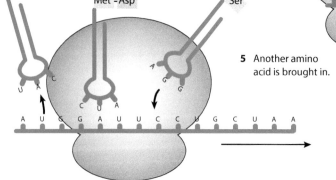

Figure 2.27 Translation.

The genetic code that is responsible for the production of polypeptides in all organisms is said to be 'universal'. This means that the mRNA codons (shown in Table **3.1**, page 94) used by bacteria are exactly the same as those used in humans or any other species. A gene from one species can be transferred into another to create a **transgenic** organism, which will transcribe and translate the inserted gene. Transgenic microorganisms are used as 'biofactories' to produce human proteins. For example, transgenic bacteria are used to produce a range of medicinal proteins, including insulin and growth hormones (Subtopic **3.5**).

'One gene, one polypeptide' hypothesis

In the 1940s, scientists proposed that each gene was responsible for the production of one protein. Later, the hypothesis was modified to state that one gene produces one polypeptide, when it was discovered that some proteins are composed of more than one polypeptide subunit and that each subunit is coded for by its own specific gene. An example of this is hemoglobin, which is composed of two pairs of subunits and is coded for by two genes.

Today, it is generally agreed that each gene does code for a single polypeptide, but that there are some exceptions to the rule. For example, some DNA sequences act as regulators for the expression of other genes and are not transcribed or translated themselves. Others code for mRNA or tRNA but not for proteins. Most recently, researchers have found that some genes code for single mRNA strands which are then modified in the cytoplasm. Variations in the modifications can lead to the production of different polypeptides when the mRNA is translated. For example, when antibodies are produced, lymphocytes splice together sections of RNA in different ways to make a range of antibody proteins. In a few cells, there is differential expression of certain genes, influenced by the type of tissue in which the cells are found. One example of this is the expression of genes that produce the insulin-like growth factors (IGF-1 and IGF-2) in the liver.

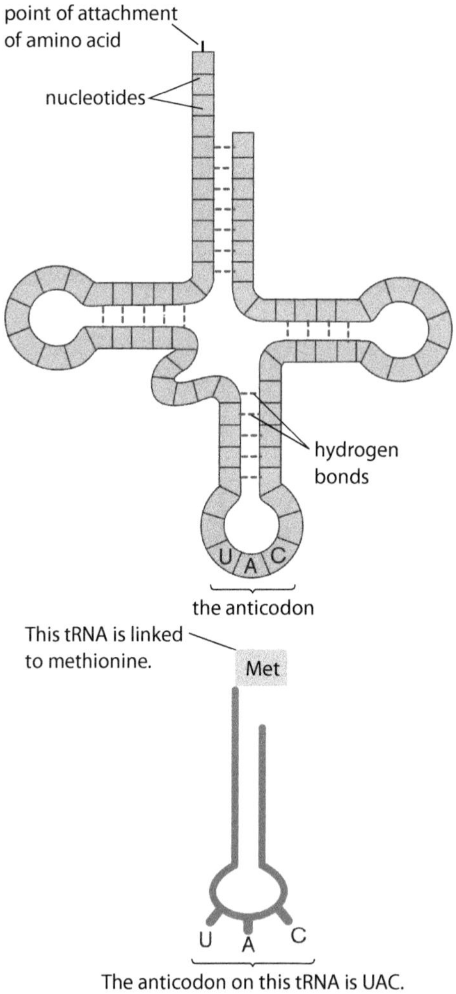

Transfer RNA (tRNA) is a single strand of RNA that is folded into a 'clover leaf' shape (Figure **2.28**). Within the molecule, sections are bonded together by complementary base pairing but one particular area is exposed to reveal a triplet of bases called an **anticodon**. This triplet corresponds to one of the codons found in mRNA. At the opposite end of the tRNA molecule is a binding site for one amino acid, which corresponds to the codon on mRNA that matches the anticodon of the tRNA.

Figure 2.28 The structure of a tRNA molecule.

? Test yourself

18 State what is meant by 'semi-conservative' replication of DNA.
19 Outline the function of DNA polymerase.
20 Define the term 'transcription' of DNA.
21 Outline what is meant by 'complementary base pairing'.
22 Where in a cell does translation take place?

Learning objectives

You should understand that:
- Cell respiration is the controlled release of energy from organic substances to produce ATP (adenosine triphosphate).
- ATP from respiration is an immediately available source of energy in the cell.
- During anaerobic cell respiration, a small amount of ATP is produced from the breakdown of glucose.
- During aerobic respiration, a process which requires oxygen, there is a large yield of ATP from the breakdown of glucose.

Cell respiration the controlled release of energy in the form of ATP from organic compounds in a cell

ATP (adenosine triphosphate) is the immediately available energy currency of a cell. It is needed for every activity that requires energy. Cells make their own ATP in mitochondria. When energy is used, ATP is broken down to ADP (adenosine diphosphate) and inorganic phosphate. This conversion releases energy for use and a cyclic process will reform the ATP during respiration.

$$\text{ADP} + \text{P}_i \xrightleftharpoons[\text{during metabolic activity}]{\text{during respiration}} \text{ATP} + \text{H}_2\text{O}$$

2.8 Cell respiration

Cell respiration and ATP

All living cells need energy to stay alive. The energy is used to power all the activities of life including digestion, protein synthesis and active transport. A cell's energy sources are the sugars and other substances derived from nutrients, which can be broken down to release the energy that holds their molecules together.

Cell respiration is the gradual breakdown of nutrient molecules such as glucose and fatty acids in a series of reactions that ultimately release energy in the form of ATP.

Glucose is the most commonly used source of energy. Each glucose molecule is broken down by enzymes in a number of stages, which release energy in small amounts as each covalent bond is broken. If there is insufficient glucose available, fatty acids or amino acids can be used instead.

Glycolysis

The first stage in cell respiration is **glycolysis**. Glucose that is present in the cytoplasm of a cell is broken down by a series of enzymes, to produce two molecules of a simpler compound called pyruvate. As this occurs, there is a net production of two molecules of ATP (Figure **2.29**).

$$\underset{\text{six-carbon sugar}}{\text{glucose}} \rightarrow \underset{2 \times \text{three-carbon sugar}}{2 \text{ pyruvate}} + 2 \text{ ATP}$$

Aerobic and anaerobic respiration

The next stage of cell respiration depends on whether or not oxygen is available. In the presence of oxygen, aerobic respiration can take place; without it, respiration must be anaerobic.

Aerobic respiration is the most efficient way of producing ATP. Aerobic respiration is carried out by cells that have mitochondria and it produces a great deal of ATP. Pyruvate molecules produced by glycolysis enter the mitochondria and are broken down, or oxidised, in a series of reactions that release carbon dioxide and water and produce ATP.

The two pyruvate molecules from glycolysis first lose carbon dioxide and become two molecules of acetyl CoA in the **link reaction**, as shown in Figure **2.29**. Acetyl CoA then enters a stage called the **Krebs cycle** and is modified still further, releasing more carbon dioxide. Finally, products of the cycle react directly with oxygen and the result is the release of large amounts of ATP. The original glucose molecule is completely broken down to carbon dioxide and water so the equation for aerobic respiration is often shown as:

$$\text{glucose} + \text{oxygen} \rightarrow \text{carbon dioxide} + \text{water} + 38 \text{ ATP}$$
$$C_6H_{12}O_6 + 6O_2 \rightarrow 6CO_2 + 6H_2O + 38 \text{ ATP}$$

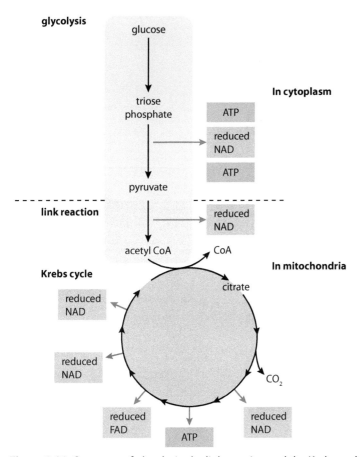

Figure 2.29 Summary of glycolysis, the link reaction and the Krebs cycle.

Glycolysis actually uses two molecules of ATP to get the process underway, but produces four molecules of ATP in total, per molecule of glucose. Thus, we say there is a net production of two ATPs.

Anaerobic respiration occurs in the cytoplasm of cells. In animal cells, the pyruvate produced by glycolysis is converted to lactate (Figure 2.30), which is a waste product and is taken out of the cells. In humans, anaerobic respiration occurs if a person is doing vigorous exercise and their cardiovascular system is unable to supply sufficient oxygen for aerobic respiration to provide ATP at the necessary rate. Although anaerobic respiration releases far less energy per molecule of glucose than aerobic respiration, the extra ATP enables the person to continue exercising for a short period, at a time of great exertion, so as to maximise power output. One consequence of the build-up of lactate in the muscles that occurs during anaerobic respiration is the sensation of cramp, so this type of respiration cannot be sustained for very long.

pyruvate → lactate

In other organisms, such as yeast, anaerobic respiration is also known as fermentation, and produces a different outcome. The pyruvate molecules from glycolysis are converted to ethanol (alcohol) and carbon dioxide (Figure 2.31).

pyruvate → ethanol + carbon dioxide

Lactate is taken by the blood to the liver, where it is converted back to pyruvate. This may either be used as a fuel, producing carbon dioxide and water, or be converted back to glucose using energy.

2 MOLECULAR BIOLOGY

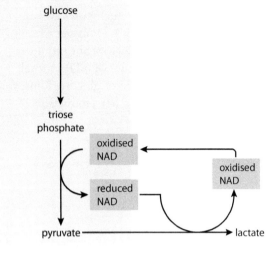

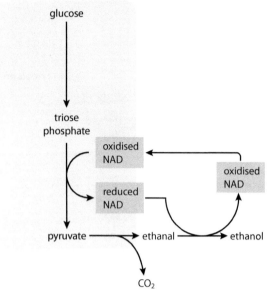

Figure 2.30 Anaerobic respiration in animal cells.

Figure 2.31 Anaerobic respiration in yeast cells.

Anaerobic respiration can only be sustained for short periods of time. A 100 m sprinter can run an entire race anaerobically, but a long-distance runner will use only aerobic respiration for maximum efficiency.

No further ATP is produced by the anaerobic respiration of pyruvate, so this type of respiration gives only a small yield of ATP from glucose. Respiration is described in greater detail in Subtopic **8.2**.

Anaerobic respiration in food production

Anaerobic respiration of yeast has been used by people in baking and brewing for thousands of years. Today, many different types of yeast are used in the production of bread, wine and beer. The strains of yeast used for baking and brewing are different and each has been selected for its different characteristics. Baking yeasts feed on sugar and flour in bread dough and grow more quickly than brewing yeasts, which are slow growing but able to tolerate higher alcohol concentrations. In bread making, the yeast initially respires aerobically, releasing carbon dioxide gas and water into the dough in a very short period of time. Carbon dioxide in the dough causes it to rise as the gas becomes trapped in pockets between gluten fibres in the flour. When oxygen in the dough has been depleted the yeast continues to respire anaerobically, producing ethanol which evaporates during baking, the yeast is also killed by the high temperature of the oven.

Nature of science

Assessing ethics in science – using invertebrates in a respirometer

A simple respirometer, such as the one shown in Figure **2.32**, can be used to monitor respiration in small organisms such as woodlice or in germinating seeds. The apparatus can demonstrate that oxygen is used and carbon dioxide produced during respiration. Test organisms are placed in two large boiling tubes as shown, so that one contains living

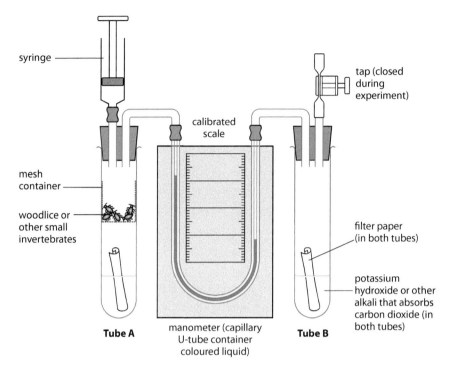

Figure 2.32 A simple respirometer.

organisms (tube A) and the other, which acts as a control, contains either dead organisms or is left empty (tube B). Soda lime or another alkali such as potassium hydroxide absorbs carbon dioxide. As oxygen is used by the living things in tube A, the level of liquid rises in the arm of the manometer attached to tube A. If required, measurements of time can be made so that the rate of respiration can be estimated. The temperature in the apparatus is kept constant by immersing the tubes in a water bath. This minimises any change in volume due to temperature change.

Questions to consider

- How can we ensure that the invertebrates used in experiments like this are treated ethically?
- What measures would you use to minimise the distress and disturbance to the organisms and also to the habitat from which they are taken?
- How can we know whether the organisms are experiencing distress?

Test yourself

23 State the **two** products of anaerobic respiration in muscles.
24 State the site of aerobic respiration in a eukaryotic cell.
25 State the site of glycolysis in a cell.
26 Outline the role of anaerobic respiration in baking.

Learning objectives

You should understand that:
- Photosynthesis uses light energy to make carbon compounds within cells.
- Light from the Sun is composed of a range of wavelengths (colours): violet is the shortest and red the longest.
- Chlorophyll is the main photosynthetic pigment and it absorbs most red and blue light, but reflects green light more than the other colours.
- Oxygen is released during photosynthesis as a result of the splitting (photolysis) of water molecules.
- Energy is needed to produce carbohydrates from the fixation of carbon dioxide.
- Temperature, light intensity and carbon dioxide concentration affect the rate of photosynthesis and can be limiting factors.

Photosynthesis means 'making things with light'. Glucose is the molecule most commonly made.

2.9 Photosynthesis

Photosynthesis and light

The Sun is the source of energy for almost all life on Earth. Light energy from the Sun is captured by plants and other photosynthetic organisms, and converted into stored chemical energy. Photosynthesis uses light energy to combine simple inorganic compounds – water and carbon dioxide – to produce carbon-containing molecules such as glucose, which contain stored chemical energy. Oxygen is released as a waste product. The series of reactions that occurs during photosynthesis is summarised as:

$$\text{carbon dioxide} + \text{water} \rightarrow \text{glucose} + \text{oxygen}$$
$$6CO_2 + 6H_2O \rightarrow C_6H_{12}O_6 + 6O_2$$

The energy stored in molecules such as glucose provides a source of food for organisms that cannot use light energy directly.

Visible light is composed of a spectrum of colours, which can be separated using a prism (Figure **2.33**). A prism bends rays of light and separates the colours because each one has a slightly different wavelength and is refracted (bent) to a slightly different degree. Visible light has a range of wavelengths that are between 400 and 700 nm. Violet light has the shortest wavelength and red the longest, but the most important regions of the spectrum for photosynthesis are red and blue.

The colour of any object is determined by the wavelength of the light that it reflects back into our eyes. A blue shirt appears blue because it reflects blue light, which our eyes can perceive, but the shirt absorbs other wavelengths that fall on it and we do not see those colours. A black object absorbs all wavelengths of light, while something white reflects them all.

Most plants have green leaves. This tells us that they do not absorb the green part of the spectrum well – green light is reflected and makes the leaf appear green. Looking closely at the structure of plant cells such as those shown in Figure **1.14** we can see that the green colour is due to the chloroplasts, which contain a green pigment called **chlorophyll**. Chlorophyll is unable to absorb green light, which it reflects, but it does absorb other wavelengths well. Red and blue light are absorbed particularly well and provide the energy needed for photosynthesis. The top graph in Figure **2.34** shows that the red and blue ends of the visible spectrum are the wavelengths that the photosynthetic pigments in plants absorb most efficiently. The bottom graph shows that the rate of photosynthesis is highest when plants absorb these wavelengths.

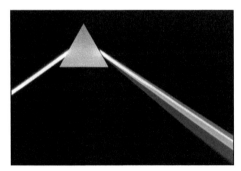

Figure 2.33 'White light', such as sunlight, is composed of a range of wavelengths, which become separated as they pass through a glass prism.

Photosynthetic pigments

Chloroplasts contain a number of different pigments that are associated with light absorption. Figure **2.34** shows **absorption spectra** for two types of chlorophyll and carotenoid pigments found in green plants. Chromatography is a simple technique used to separate different substances in a mixture and it can be used to separate the pigments in extracts from plant leaves.

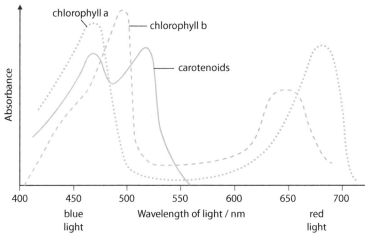

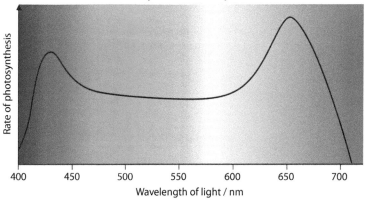

Figure 2.34 These graphs show the wavelengths (colours) of light absorbed by plants and the rate of photosynthesis that occurs at each wavelength.

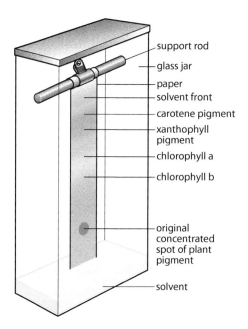

Figure 2.35 A chromatogram can be used to identify different photosynthetic pigments. Different pigments are found in different plants and the pigments may vary with the seasons.

Two techniques are commonly used: paper chromatography, which uses a special high-grade paper with carefully controlled spaces between the cellulose fibres, and thin-layer chromatography (TLC), which is carried out on a thin plate of glass or plastic coated with a layer of adsorbent material such as silica gel or cellulose (known as the stationary phase).

During chromatography a solvent moves up the paper or plate by capillary action and carries pigments with it by mass flow. Smaller molecules are able to move more easily and so can travel further than larger molecules. After a period of time, photosynthetic pigments from chloroplast extracts become separated (Figure **2.35**) and can be compared and measured.

Nature of science

Careful observation – Engelmann's experiment

Theodor Wilhelm Engelmann (1843–1909) was a German botanist who used the filamentous alga *Spirogyra* to demonstrate not only that oxygen is evolved during photosynthesis but also that different wavelengths of light affect the rate of photosynthesis. *Spirogyra* is an alga that has cylindrical cells containing spiral-shaped chloroplasts. Engelmann mounted a sample of *Spirogyra* under a microscope and, after a period of darkness, illuminated it with different colours of light. He carefully watched the movement of motile bacteria (*Pseudomonas*) that he had added to the water and noticed that, after a period of oxygen deprivation, they moved towards areas where there was a higher concentration of oxygen around the alga's chloroplasts (Figure **2.36**).

Test yourself

27 Looking at Figure **2.36a**, explain why the bacteria moved towards certain areas of the *Spirogyra* and not towards others when a spot of red light was used.

28 Explain why there are no bacteria between the areas of the *Spirogyra* illuminated with light of 500–600 nm (Figure **2.36b**).

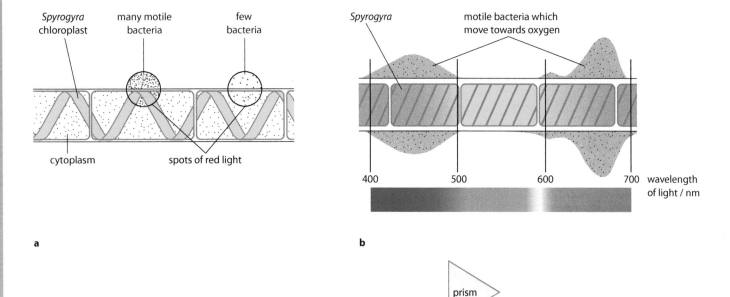

Figure 2.36 Engelmann used bacteria in two experiments **a** and **b** to measure rates of photosynthesis and determine which wavelengths are most effective for photosynthesis.

The chemistry of photosynthesis

Photosynthesis is a complex series of reactions catalysed by a number of different enzymes. To aid understanding, we can consider photosynthesis in two stages: the light-dependent reactions and the light-independent reactions.

Light-dependent reactions

The first stage is known as the 'light-dependent reactions' because light is essential for them to occur.

Chlorophyll absorbs light energy and this energy is used to produce ATP. The energy is also used to split water molecules into hydrogen and oxygen in a process called photolysis. Hydrogen ions and electrons (from the hydrogen part of water) and oxygen are released. Oxygen is a waste product of photosynthesis but is vital to sustain the lives of aerobic organisms once it has been released into the atmosphere. The ATP, hydrogen ions and electrons are used in the light-independent reactions.

Light-independent reactions

ATP, hydrogen ions and electrons are used in the second stage of photosynthesis, the 'light-independent reactions'.

During the 'light-independent reactions', carbon dioxide, taken in from the environment, is combined with hydrogen and ATP to form a range of organic molecules for the plant. The conversion of inorganic carbon dioxide to organic molecules such as glucose is known as carbon fixation. ATP provides the energy for the process.

The series of reactions that occurs during photosynthesis is summarised as:

$$\text{carbon dioxide} + \text{water} \rightarrow \text{glucose} + \text{oxygen}$$
$$6CO_2 + 6H_2O \rightarrow C_6H_{12}O_6 + 6O_2$$

Measuring the rate of photosynthesis

The equation above shows that when photosynthesis occurs, carbon dioxide is used and oxygen is released. The mass of the plant (its biomass) will also increase as glucose is used to produce other plant materials. Any of these three factors can be used to measure how quickly the reactions of photosynthesis are occurring.

Aquatic plants release bubbles of oxygen as they photosynthesise and if the volume of these bubbles is measured for a period of time, the rate of photosynthesis can be determined directly (Figure 2.37).

Aquatic plants also remove carbon dioxide from their environment, causing the pH of the water to rise. Carbon dioxide dissolves in water to form a weak acid so as it is removed, the pH will go up. Therefore, another way of determining the rate of photosynthesis experimentally is to monitor the change in pH of the water surrounding an aquatic plant over a period of time.

Terrestrial plants also remove carbon dioxide from their surroundings but this is difficult to measure. It can be done experimentally by supplying a confined plant with radioactive carbon dioxide, which can be measured as it is taken up and released from the plant.

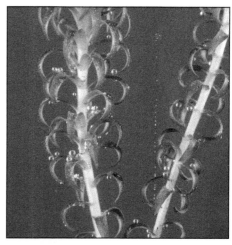

Figure 2.37 The rate of oxygen production can be used as a direct measure of the rate of photosynthesis.

A third method of measuring the rate of photosynthesis in plants is to determine their biomass at different times. This is an indirect method. Samples of the plants can be collected and measured at different times and the rate of increase in their biomass calculated to determine their rate of photosynthesis.

Limits to photosynthesis

The rate at which a plant can photosynthesise depends on factors in the environment that surrounds it. On a warm, sunny afternoon, photosynthesis will be more rapid than on a cool, shady morning. More oxygen will be produced and more carbon dioxide used. Temperature, light intensity and carbon dioxide concentration are all possible limiting factors on the rate of photosynthesis. But photosynthesis cannot increase beyond certain limits. The effect of light, temperature and carbon dioxide in the environment can be measured experimentally, varying one factor while keeping the others the same, and graphs such as those in Figure **2.38** can be drawn.

An increase in light intensity, when all other variables are unchanging, will produce an increase in the rate of photosynthesis that is directly proportional to the increase in light intensity. However, at a certain light intensity, enzymes will be working at their maximum rate, limited by

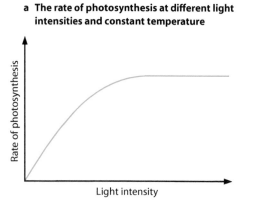

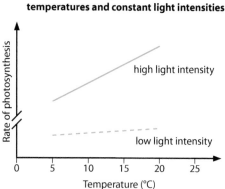

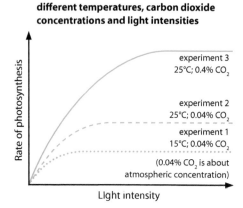

Figure 2.38 These graphs show the effects on photosynthesis of varying light intensity, carbon dioxide concentration and temperature.

temperature and the availability of carbon dioxide. At very high light intensities, light absorption (and therefore the rate of photosynthesis) reaches its maximum and cannot increase further. At this point, the graph reaches a plateau (Figure **2.38a**).

Increasing temperature also increases the rate of photosynthesis as the frequency and energy of molecular collision increases (Figure **2.38b**). Photosynthesis has an optimum temperature above which the rate will decrease sharply as enzymes are denatured, or the plant wilts and is unable to take in carbon dioxide.

An increase in the concentration of carbon dioxide causes the rate of photosynthesis to increase, as carbon dioxide is a vital raw material for the process. At very high concentrations, the rate will plateau as other factors such as light and temperature limit the rate of reaction (Figure **2.38c**).

The effects of temperature, light and carbon dioxide concentration are well known to horticulturalists who grow crops in glasshouses. Commercial producers of cucumbers and tomatoes keep their glasshouses warm and well lit. They may also introduce carbon dioxide to boost photosynthesis to its maximum rate, thereby increasing crop production and profits.

Photosynthesis is described in more detail in Subtopic **8.3**.

Influence of photosynthesis on the Earth's atmosphere, oceans and rocks

Living organisms started to have a major impact on the Earth's environment after the evolution of photosynthetic organisms. Today, most scientists agree that the first oxygen-producing photosynthesis evolved about 2.4 billion years ago.

These early photosynthetic life forms reduced the carbon dioxide content of the atmosphere and they also started to produce oxygen. At first, oxygen was taken up by rocks in the Earth's mantle, where oxygen combined with reduced iron compounds to form iron oxides. It was probably only about 1 billion years ago that these rocks became saturated with oxygen and free oxygen remained in the atmosphere. The Earth's rocks still contain huge reservoirs of oxygen in the form of oxides.

As the level of free oxygen built up and became sufficient to support life, a period known as the Biological Era in the evolution of the atmosphere began, which had three important consequences.

- Once the level of oxygen reached about 1% of its current level, the metabolism of eukaryotes became possible. According to the fossil record, this probably began to happen about 2 billion years ago.
- With greater availability of oxygen, metabolic pathways became more efficient, and aerobic organisms proliferated in the oceans. Carbon dioxide removed from the atmosphere ultimately ended up in the shells and decomposed remains of sea creatures, and built up marine sediments, which eventually became incorporated in sedimentary and, later, metamorphic rocks.
- As oxygen accumulated in the upper atmosphere, sunlight would have been able to act on it and form ozone. The ozone layer helped

to maintain a more stable temperature on the Earth so that organisms were able to colonise the land. Evidence suggests that this occurred about 400 million years ago.

Sometime just before the Cambrian era (about 550 million years ago), atmospheric oxygen reached levels close enough to today's to allow for the rapid evolution of the higher life forms. For the rest of geological time, the oxygen in the atmosphere has been maintained by the photosynthesis of the green plants of the world, including the green algae in the surface waters of the ocean. It is important to remember that oxygen became a component of the atmosphere, rocks and oceans entirely as a result of life processes.

Nature of science

Experimental design – controlling variables

Any investigations involving living organisms must be designed so that all the possible variables are controlled. Consider this diagram of apparatus (Figure **2.39**) that has been set up in a laboratory to estimate the rate of photosynthesis of a pond plant.

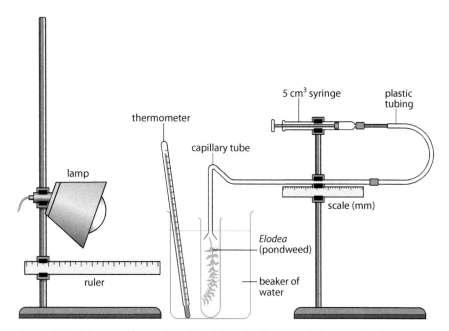

Figure 2.39 Diagram of experiment to determine the rate of photosynthesis.

Questions to consider

- Which variable is being controlled by the presence of the beaker of water?
- Why must this variable be controlled?
  Photosynthesis is a metabolic reaction controlled by enzymes.
- List the factors that affect enzyme action.
- How could each one of these be controlled in this experiment?

Test yourself

29 If you wanted to make plants grow as well as possible, state which colour of light you should shine on them.
30 Describe what would happen to a plant's growth if it were kept in green light.
31 State the colours of the visible spectrum that are used in photosynthesis.
32 Describe **two** ways in which the rate of photosynthesis can be measured.

Exam-style questions

1 Which of the following statements is correct?

    A The most frequently occurring chemical elements in living organisms are carbon, hydrogen, oxygen and calcium.
    B A water molecule can form hydrogen bonds with other water molecules due to its polarity.
    C Sweating cools an animal because body heat is required to break the covalent bonds in the water in the sweat, allowing it to evaporate.
    D Increasing substrate concentration causes enzymes to denature. [1]

2 Which of the following statements is correct?

    A The components of a nucleotide are a sugar molecule attached to two phosphate groups and a base.
    B In a molecule of DNA, the bases thymine and uracil are held together by hydrogen bonds.
    C During the process of transcription, tRNA molecules bond to mRNA using complementary base pairing.
    D During DNA replication, new nucleotides are added using the enzyme DNA polymerase. [1]

3 Outline how monosaccharides are converted into polysaccharides. [2]

4 State why each step in a biochemical pathway often requires a separate enzyme. [2]

5 Study the graph below which shows the rate of enzyme activity for pepsin, a protease enzyme found in the stomach. Enzyme activity is measured by the rate of breakdown of protein to amino acids.

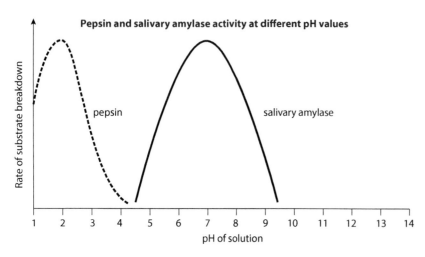

   a What is the optimum pH of the enzyme pepsin? [1]
   b Within what range does pepsin act at over 50% of its maximum rate? [2]
   c What would you expect the pH of the stomach contents to be? [1]

6 a Draw a nucleotide of DNA to show its structure [2]
   b State **three** differences in the structure of a DNA and RNA molecule [3]
   c Outline what is meant by 'complementary base pairs' in a DNA molecule. [2]

The table below shows the mRNA codons for some amino acids.

mRNA codon	Amino acid
AGU	serine
GCU	alanine
UGC	cysteine
ACG	threonine
GUC	valine
CUA	leucine

   d State the DNA sequence that corresponds to the mRNA coden for the amino acid cysteine. [1]
   e Name the amino acid coded for by the DNA sequence TCA. [1]

7 The pigments found in a leaf can be extracted with a solvent and separated using thin layer chromatography to produce a chromatogram shown in the diagram below.

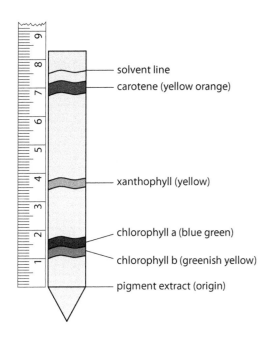

The distance moved by each pigment divided by the distance moved by the solvent is unique for each pigment and is known as the Rf value.

Rf = distance moved by pigment/distance moved by solvent

a Calculate the Rf value for
   i   chlorophyll a [1]
   ii carotene. [1]

b The graph below shows the absorption spectrum for the total plant extract.
   i   State which wavelengths and colours are absorbed best. [2]
   ii Outline the relationship between the absorption and action spectra. [2]

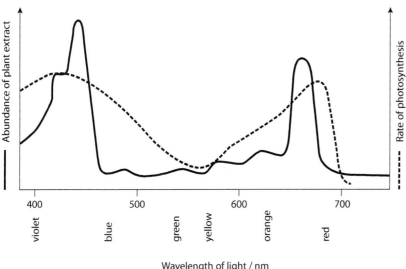

c The graph below shows the absorption spectrum for pure solutions of chlorophyll a and carotene.

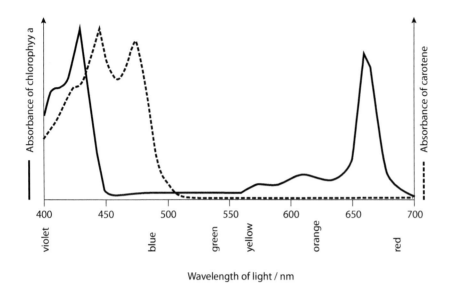

Suggest a role of carotene in capturing light energy for photosynthesis. [2]

8 Body mass index (BMI) is a measure used to indicate whether a person has an appropriate body mass. The table below indicates the range of BMI measurements for adults over the age of 20 years.

BMI = mass in kg ÷ (height in m)^2

BMI	Status
Below 18.5	underweight
18.5 – 24.9	normal weight
25.0 – 29.9	overweight
30.0 and above	obese

i Calculate the BMI of a male Olympic weight lifter who has a body mass of 99 kg and is 1.75 m tall [1]
ii Calculate the BMI of a female office worker who has a body mass of 70 kg and is 1.5 m tall. [1]

BMI is used to indicate whether a person is overweight and likely to be storing excess fat in their body, but fitness training builds up large muscles which weigh more than fat.

iii Use this information to comment on the status of the weight lifter and the office worker. [2]

Genetics 3

Introduction

Chimpanzees are set apart from all other organisms because their parents were chimpanzees and their offspring will also be chimpanzees. Every living organism inherits its own blueprint for life in the chromosomes and genes that are passed to it from its parents. The study of genetics attempts to explain this process of heredity and it also plays a very significant role in the modern world, from plant and animal breeding to human health and disease.

3.1 Genes

Chromosomes, genes and mutations

A DNA molecule comprises a pair of strands, each strand consisting of a linear sequence of nucleotides, with weak hydrogen bonds between the bases holding the two strands together. This linear sequence of bases contains the genetic code in the form of triplets of bases. A **gene** is a particular section of a DNA strand that, when transcribed and translated, forms a specific polypeptide. Some of the polypeptides will form structural proteins such as collagen while others become enzymes or pigments such as hemoglobin (Subtopic **2.4**) and it is the translation of the genes which gives each individual organism its own specific characteristics. (Transcription and translation are described in Topic **2** and in more detail in Topic **7**.) Each gene is found at a specific position on a chromosome and so, for example, it is possible to say that the gene for human insulin is always found on chromosome number 11.

Organisms that reproduce sexually almost always have pairs of chromosomes, one of each pair coming from each parent. The members of the pair carry equivalent genes, so that – for example – in humans, both versions of chromosome number 11 carry the insulin gene. But there may be slight differences in the version of the gene on each chromosome. These slightly different forms of the gene are known as **alleles**. Alleles differ from one another by one or only a few bases and it is these differences in alleles that give rise to the variation we observe in living organisms.

When the genetic code is used to build proteins, a triplet of bases in the DNA molecule is transcribed into a triplet of bases in the mRNA molecule (also called a **codon** – Subtopic **2.7**), which is then translated into a specific amino acid, as shown in Figure **3.1**.

The process of DNA replication is complex and mistakes sometimes occur – a nucleotide may be left out, an extra one may be added, or the wrong one inserted. These mistakes are known as **gene mutations**. Mutations may occur spontaneously, as a result of errors in copying

Learning objectives

You should understand that:
- Genes are heritable factors that each consist of a section of DNA and influence a specific characteristic.
- Each gene occupies a specific position on one chromosome.
- Alleles are the various specific forms of a gene.
- Alleles differ from one another by only a few bases.
- Mutation causes the formation of new alleles.
- The whole of the genetic information of an organism is known as its genome.
- The Human Genome Project worked out the entire base sequence of human genes.

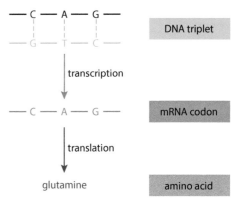

Figure 3.1 The base sequence in DNA is decoded via transcription and translation.

Gene a heritable factor that controls a specific characteristic, or a section of DNA that codes for the formation of a polypeptide

Allele a specific form of a gene occupying the same gene locus or position as other alleles of that gene, but differing from other alleles by small variations in its base sequence

Genome the whole of the genetic information of an organism

Gene mutation a change in the sequence of bases in a gene

There are several thousand human disorders that are caused by mutations in single genes and about 100 'syndromes' associated with chromosome abnormalities.

DNA, or they can be caused by factors in the environment known as mutagens (Subtopic **1.6**). The insertion of an incorrect nucleotide is called a **base substitution mutation**. When the DNA containing an incorrect nucleotide is transcribed and translated, errors may occur in the polypeptide produced.

Table **3.1** shows the amino acids that are specified by different mRNA codons. Most amino acids are coded for by more than one codon and so many substitution mutations have no effect on the final polypeptide that is produced. These are said to be neutral mutations. For example, a mutation in the DNA triplet CCA into CCG would change the codon in the mRNA from GGU to GGC but it would still result in the amino acid glycine being placed in a polypeptide. Some substitution mutations, however, do have serious effects – for example, one important human condition that results from a single base substitution is sickle-cell anemia.

Sickle-cell anemia – the result of a base substitution mutation

Sickle-cell anemia is a blood disorder in which red blood cells become sickle shaped and cannot carry oxygen properly (Figure **3.2**). It occurs most frequently in people with African ancestry – about 1% suffer from the condition and between 10% and 40% are carriers of it. Sickle-cell anemia is due to a single base substitution mutation in one of the genes that make **hemoglobin**, the oxygen-carrying pigment in red blood cells.

Hemoglobin is made up of four subunits, as shown in Figure **3.3** – two α-chains and two β-chains. The β-chains are affected by the sickle-cell

First base		Second base								Third base
		U		C		A		G		
	U	UUU	phenylalanine	UCU	serine	UAU	tyrosine	UGU	cysteine	U
		UUC		UCC		UAC		UGC		C
		UUA	leucine	UCA		UAA	'stop'	UGA	'stop'	A
		UUG		UCG		UAG		UGG	tryptophan	G
	C	CUU	leucine	CCU	proline	CAU	histidine	CGU	arginine	U
		CUC		CCC		CAC		CGC		C
		CUA		CCA		CAA	glutamine	CGA		A
		CUG		CCG		CAG		CGG		G
	A	AUU	isoleucine	ACU	threonine	AAU	asparagine	AGU	serine	U
		AUC		ACC		AAC		AGC		C
		AUA		ACA		AAA	lysine	AGA	arginine	A
		AUG	methionine or 'start'	ACG		AAG		AGG		G
	G	GUU	valine	GCU	alanine	GAU	aspartic acid	GGU	glycine	U
		GUC		GCC		GAC		GGC		C
		GUA		GCA		GAA	glutamic acid	GGA		A
		GUG		GCG		GAG		GGG		G

Table 3.1 Amino acids and their associated mRNA codons.

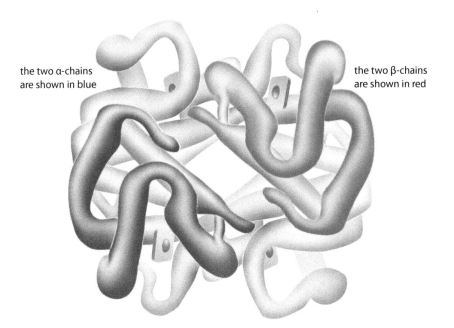
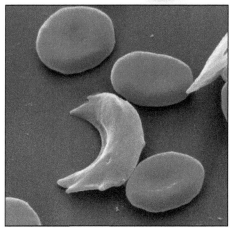

Figure 3.2 Coloured scanning electron micrograph showing a sickle cell and normal red blood cells (×7400).

Figure 3.3 The structure of a hemoglobin molecule showing the 3D arrangement of the subunits that makes it up.

mutation. To form a normal β-chain, the particular triplet base pairing in the DNA is:

```
— G — A — G —
    |   |   |
— C — T — C —
```

The C–T–C on the coding strand of the DNA (in blue here) is transcribed into the mRNA triplet G–A–G, which in turn is translated to give glutamic acid in the polypeptide chain of the β-chain.

If the sickle-cell mutation occurs, the adenine base A is substituted for thymine T on the coding strand, so the triplet base pairing becomes:

```
— G — T — G —
    |   |   |
— C — A — C —
```

C–A–C on the coding strand of the DNA is now transcribed into the mRNA triplet G–U–G, which in turn is translated to give the amino acid valine. Valine replaces glutamic acid in the β-chain.

Valine has different properties from glutamic acid and so this single change in the amino acid sequence has very serious effects. The resulting hemoglobin molecule is a different shape, it is less soluble and when in low oxygen concentrations, it deforms the red blood cells to give them a sickle shape. Sickle cells carry less oxygen, which results in anemia. They are also rapidly removed from the circulation, leading to a lack of red blood cells and other symptoms such as jaundice, kidney problems and enlargement of the spleen.

Only one strand of DNA is transcribed into mRNA for any gene. The transcribed strand is called the coding strand.

The genome and Human Genome Project

The **genome** of an organism is defined as the whole of its genetic information and every cell has a complete copy of the organism's genome. Genome analysis is an important field of modern biological research.

Comparative genomics is an area that analyses genomes from different species. Genomes are compared to gain a better understanding of how species have evolved and to work out the functions of genes and also of the non-coding regions of the genome. Researchers look at many different features such as sequence similarity, gene location, the length and number of coding regions within genes and the amount of DNA that does not code for proteins. They use computer programs to line up genomes from different organisms and look for regions of similarity. There are many databases that store this information on DNA base sequences (Table **3.3**). Sequences for commonly occurring proteins such as cytochrome c, which is involved in cell respiration in many organisms, can be compared and have been useful for establishing the evolutionary relationships between different species.

The Human Genome Project, which has sequenced all the bases in the entire complement of a human's genetic information, was the first step in attempting to understand the human genome. The project revealed that all humans, from all over the world, share the vast majority of their base sequences, regardless of their race. At the same time, every human being (apart from identical twins) is genetically unique due to differences in single nucleotides, which together account for the enormous human diversity we can see. Although the project is now complete, there remain many unanswered questions. The functions of most human genes and the roles of non-coding regions and repeated sequences are still unknown.

Humans and most mammals have roughly the same number of nucleotides (about 3 billion base pairs) in their genomes. Although the number of genes is not necessarily proportional to the total amount of DNA present in a genome, the work of many scientists has provided evidence to support the theory that most mammal genomes do in fact contain more or less the same number of genes. In addition to the human genome, the genomes of about 800 organisms have been sequenced in recent years. These include the mouse *Mus musculus*, the fruit fly *Drosophila melanogaster*, the worm *Caenorhabditis elegans*, the bacterium *Escherichia coli*, the yeast *Saccharomyces cerevisiae*, the plant *Arabidopsis thaliana*, and many microbes. Some information about the genomes of these organisms is shown in Table **3.2**.

Organism	Estimated number of genes	Number of Chromosomes
Mus musculus (mouse)	25 000	40
Arabidopsis thaliana (plant)	23 000–25 000	10
Homo sapiens (human)	22 000–25 000	46
Caenorhabditis elegans (roundworm)	19 000	12
Drosophila melanogaster (fruit fly)	13 000–14 000	8
Saccharomyces cerevisiae (yeast)	6000	32
Escherichia coli (bacterium)	1700–3000	1

Table 3.2 Comparison of the estimated numbers of genes in different organisms. The number of genes shown in this table is only approximate as data is being refined all the time. Note that the number of genes does not correlate with evolutionary status, nor is the number of genes proportional to the total amount of DNA present in an organism.

Sickle-cell anemia – cause and correlation

The global distribution of sickle-cell anemia is correlated with the incidence of malaria. In parts of Africa, up to 1 in 25 children are born with sickle-cell anemia and its geographical distribution follows the pattern of prevalence of malaria.

We might expect that the allele responsible for sickle-cell anaemia would die out since people who suffer from the condition tend to die younger and have few, if any, children. But the evidence shows that individuals who are heterozygous for sickle-cell anemia (that is, they have one sickle-cell allele and one normal allele) have an advantage over both those without the sickle-cell allele and those with two copies of it, who are sufferers. Heterozygotes are less susceptible to the malaria parasite, which spends part of its life cycle living in red blood cells.

Questions to consider

- Can we establish whether there is a causal link between cases of sickle-cell anemia and prevalence of malaria?
- Is the relationship simple a correlation?
- What do you understand by the term 'heterozygous advantage'?
- How has the world distribution of sickle-cell anemia changed with international travel as populations have become more mobile?

Nature of science

Scientific advance follows technical innovation – gene sequencers and databases

Different methods of DNA sequencing are used to work out either the sequence of individual genes, groups of genes or even entire chromosomes and genomes. Since the advances in technology made during the Human Genome Project many new and automated methods of sequencing have been developed.

Some processes involve cutting DNA into short segments using restriction enzymes, and cloning the resulting pieces using the polymerase chain reaction (PCR). This technique has been used by forensic scientists to compare repeated sequences in DNA for use in criminal investigations or for establishing family relationships (see Subtopic **3.5**).

Different techniques are used to identify genes that have not been recognised before. These are known as 'de novo' methods. Many of the new sequencing methods use indirect approaches based on tagging nucleotides with fluorescent markers. Others use 'nanopore' methods in which the strand of DNA to be analysed passes through a protein pore and data is recorded on a microchip. DNA can be analysed rapidly in this way and it may soon be possible to sequence a human genome in just a few hours. The field of DNA sequencing is advancing rapidly and some of the newest methods have the potential to be adapted to recognise other molecules such as drugs or other chemicals.

Databases have also been invaluable in genome research and are used for storing long sequences of data about proteins. One example of an institution that builds and maintains extensive biological databases is the European Bioinformatics Institute (EBI). This research institute is located on the Wellcome Trust Genome Campus near Cambridge in the UK, where large parts of the original Human Genome Project were carried out. It is part of the European Molecular Biology Laboratory (EMBL) and provides data to the scientific community and industry. The first completed genomes from viruses, phages and organelles were deposited in the EMBL database in the early 1980s. Since then hundreds of complete genome sequences have been added to the database.

There are many other databases, some of which are outlined here in Table 3.3. Most are freely available online to scientists and students so that all of us can use them to compare DNA and protein sequences – you can find them by typing the database name into any search engine.

Database	Description
ENA Genomes Server	overview of all complete genomes deposited in the European Nucleotide Archive
Ensembl	a joint project between EMBL–EBI and the Sanger Institute to produce and maintain automatic annotation of eukaryotic genomes
Ensembl Genomes	provides access to genomes of non-vertebrate species
GenBank (US National Center for Biotechnology Information)	a collection of all publicly available DNA sequences, providing the scientific community with up-to-date and comprehensive DNA sequence information

Table 3.3 Some databases that hold information on genomes.

Test yourself

1 The β-chain of the hemoglobin molecule contains 146 amino acids. State the number of nucleotides that are needed to code for this protein.
2 Define the following terms:
   a gene
   b allele
   c genome
   d gene mutation.
3 Outline the meaning of the term 'base substitution mutation'.

3.2 Chromosomes

Chromosome structure

Chromosomes are made of DNA molecules and carry the genetic code for each organism but prokaryotic and eukaryotic chromosomes are different in their structure. Prokaryotes have a much simpler chromosome than eukaryotes. Prokaryotes contain a circle of DNA that is often concentrated in one area of the cell, whereas eukaryotes have linear DNA that is associated with histone proteins. Some of these proteins are structural and others regulate the activities of the DNA. Prokaryotes have additional genetic material in the form of small circular structures known as plasmids. Prokaryotes are much simpler organisms and so require fewer genes to maintain themselves. The differences between prokaryotic and eukaryotic genetic material are summarised in Table **3.4**.

Prokaryotic DNA	Eukaryotic DNA
cell contains a circular chromosome, sometimes called a nucleoid	chromosomes are made of linear DNA molecules enclosed in a nucleus, bound by a double membrane
cell contains additional genetic material as small circular plasmids	no plasmids
DNA is 'naked' and is not associated with proteins	DNA is associated with histone proteins
cell contains just one circular chromosome	cell contains two or more chromosome types

Table 3.4 A comparison of prokaryotic and eukaryotic genetic material.

During the phase of the cell cycle known as interphase, eukaryotic chromosomes are in the form of long, very thin threads, which cannot be seen with a simple microscope. As the nucleus prepares to divide, these threads undergo repeated coiling and become much shorter and thicker (Figure **3.4**). When stained, they are clearly visible even at low microscope magnifications.

Chromosome structure is described in greater detail in Topic **7**.

Eukaryotic chromosomes

Eukaryotic species have two or more chromosome types. Their chromosomes form pairs, which are known as homologues. Homologous pairs are about the same length and carry the same sequence of genes at the same locations along their length. The form of the genes (alleles) on each of the pair are not necessarily the same because, in sexually reproducing organisms, one chromosome will have been inherited from each of the two parents. So a gene that determines flower colour in a plant would be at the same location on each chromosome in a homologous pair but the allele on the maternal chromosome might not be the same as that on the paternal chromosome.

Learning objectives

You should understand that:
- Prokaryotes have one chromosome, which is a circular DNA molecule.
- Some prokaryotes have plasmids as well as a chromosome but eukaryotes do not.
- Eukaryotes have chromosomes that are linear DNA molecules associated with histone proteins.
- Eukaryotes have different chromosomes that carry different genes.
- Each species has a characteristic number of chromosomes.
- Haploid nuclei have one of each type of chromosome.
- Diploid nuclei have two of each type of chromosome, called homologous pairs of chromosomes.
- Homologous chromosomes carry the same genes but not necessarily the same alleles.
- A karyogram shows an organism's chromosomes in homologous pairs arranged in order of decreasing length.
- Sex chromosomes determine the sex of an organism; autosomes are the chromosomes that do not determine sex.

Homologous chromosomes are found in the nuclei in the cells of **diploid** organisms. But if each chromosome exists alone with no partner, the cell is said to be **haploid**. Human **somatic cells** are diploid and contain 46 chromosomes in 23 homologous pairs; human gametes are haploid and contain only 23 chromosomes, one of each pair found in the body cells.

The number of chromosomes found in the cells of an organism is a characteristic feature of that organism. The diploid numbers of chromosomes in some well-studied species are shown in Table **3.5**.

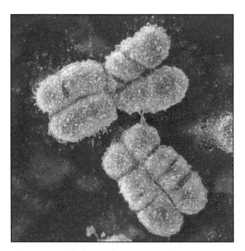

Figure 3.4 Coloured scanning electron micrograph of human chromosomes. They have replicated prior to cell division and so consist of two identical copies (chromatids) linked at the centromere (×7080).

Organism	Diploid number of chromosomes
Canis familiaris (domestic dog)	78
Pan troglodytes (chimpanzee)	48
Homo sapiens (human)	46
Mus muscularis (house mouse)	40
Oryza sativa (rice)	24
Drosophila melangosta (fruit fly)	8
Parascaris equorum (parasitic worm)	2

Table 3.5 The diploid number of different species.

Diploid a diploid nucleus contains two copies of each chromosome, in homologous pairs
Haploid a haploid nucleus contains one chromosome of each homologous pair
Homologous chromosomes a pair of chromosomes with the same genes but not necessarily the same alleles of those genes
Somatic cell a body cell that is not a gamete

In 2005, the International Rice Genome Sequencing Project, which began in 1998, succeeded in making rice the first crop plant whose genome had been mapped. Because rice is such an important staple food worldwide, an understanding of its genetic make-up is a key tool for researchers working on improved strains to feed a burgeoning human population. The sequencing of the rice genome involved cooperation between biologists in ten different countries – the project was led by scientists in Japan but involved teams from the United States, China, France, Taiwan, India, Thailand, Korea, Brazil and the UK.

Genome size

Genome size is the total number of nucleotide base pairs within one copy of a single genome. Measurements are often made in numbers of base pairs. It is interesting to note that an organism's complexity is not proportional to its genome size (Table **3.6**) – many organisms have more DNA than humans. Nor is variation in genome size proportional to the number of genes. This is due to the high proportion of non-coding DNA that is found in some organisms.

Karyotyping

Chromosomes have unique banding patterns that are revealed if they are stained with specific dyes. These patterns enable us to study the structure and type of chromosomes present in an organism. The technique has been used in prenatal diagnosis and in forensic science. Chromosomes are

Organism	Genome size in base pairs	Notes
T2 phage (a virus which infects bacteria)	3569	first RNA genome sequenced
E. coli (a bacterium)	4.6×10^6	
Drosophila melanogaster (fruit fly)	130×10^6	
Homo sapiens (human)	3200×10^6	
Paris japonica (Japanese native pale-petal)	$150\,000 \times 10^6$	largest known plant genome
Protopterus aethiopicus (marbled lungfish)	$130\,000 \times 10^6$	largest known vertebrate genome

Table 3.6 Genome sizes of some organisms.

prepared from the nuclei of dividing cells. If human cells are being studied, a sample of cells may be taken from amniotic fluid, in the case of a fetus, or from a blood sample. Division is halted at the end of prophase when chromosomes are condensed. A **karyogram** is a photograph or diagram of the stained chromosomes. Each chromosome is a characteristic length and each one has a homologous partner. The karyogram image is organised so that each chromosome is separated from the others and they are arranged in order of their size, as shown in Figure **3.5**. A karyogram shows the **karyotype** of the cell – that is, the number and types of chromosomes present in its nucleus. It indicates the sex of an individual because it shows the sex chromosomes, and in prenatal diagnosis it is possible to check for chromosome abnormalities such as those leading to Down syndrome. You can read more about prenatal diagnosis in Subtopic 3.3.

Karyogram a diagram or photograph of the chromosomes from an organism

Karyotype the number and type of chromosomes present in a nucleus

Sex chromosomes and autosomes

The **sex chromosomes** are the last two chromosomes shown in the human karyograms in Figure **3.5** (in the bottom right hand corner, in

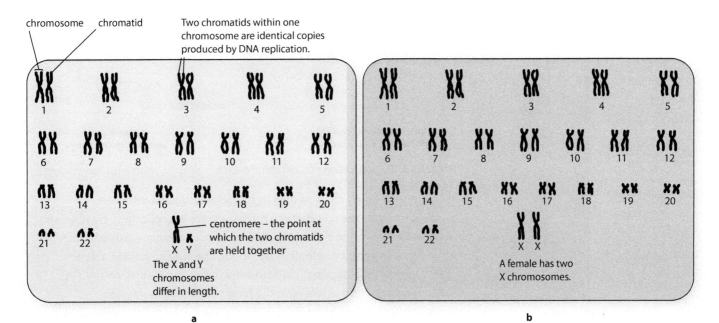

Figure 3.5 Karyograms for a human male **a** and a human female **b**.

each case). These chromosomes determine the sex of an individual – a female has two X chromosomes (XX) while a male has one X and one smaller Y chromosome (XY). The other 44 chromosomes in a human karyotype are known as autosomes and they determine the other characteristics of the individual.

Nature of science

Scientific advance follows technical innovation – Cairns' technique

Hugh John Forster Cairns (1922–) is a British molecular biologist who demonstrated by autoradiography that the bacterium *E. coli* contains a single, circular molecule of DNA.

A prediction of the Watson–Crick model of DNA replication was that there would be a replication fork formed as DNA copied itself. John Cairns tested this by allowing replicating DNA in *E. coli* bacteria to incorporate a radioactive marker in the form of tritiated thymidine. Cells cultured in tritiated thymidine take up the radioactive hydrogen isotope ^{3}H (tritium) and the thymine nucleotides become labelled. As cells undergo mitosis, the radioactive ^{3}H in the labelled thymine nucleotides is incorporated into the new cells that are produced. As ^{3}H decays, it emits a beta particle, which produces a dot on a photographic plate, so that the location of the radioactivity can be identified (Figure 3.6). The image produced on a photographic plate by radioactivity is called an autoradiograph.

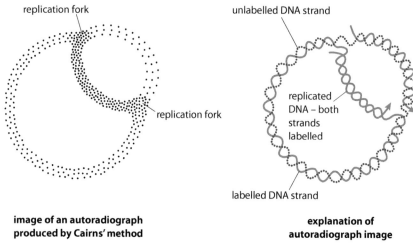

Figure 3.6 The replication of *E. coli* chromosome.

Cairns observed rings of dots in his autoradiograph after one replication cycle. As he observed further cycles of replication, he observed replication forks (the Y-shaped junctions where double-stranded DNA splits apart into two single strands) around the ring so he was able to conclude that the bacterial chromosome is circular. Cairns' techniques have also been used to measure the length of DNA molecules.

Since Cairns' experiments in 1963, further research has enabled scientists to observe replication of chromosomes and support his conclusions without the use of radioactivity.

How long is the DNA in a human body?

The human genome is the information contained in one set of human chromosomes, which are built up of about 3 billion base pairs (bp) of DNA in 46 chromosomes (22 autosome pairs + 2 sex chromosomes). The total length of DNA present in one adult human is calculated by multiplying:

(length of 1 bp) × (number of bp per cell)
    × (number of cells in the body)

So total length = $(0.34 \times 10^{-9} \text{ m}) \times (6 \times 10^9) \times 10^{13}$

$= 2.0 \times 10^{13}$ metres

That is about the distance of travelling 70 times from the Earth to the Sun and back.

Test yourself

4 State **two** differences between prokaryotic and eukaryotic chromosomes.
5 Define what is meant by a 'karyogram'.
6 Define the term 'autosome'.

Exam tip
Make sure you can define pairs of terms which are used in the same topics. For example: karyogram and karyotype, proteome and genome, chromosome and chromatid, diploid and haploid.

3.3 Meiosis

Learning objectives

You should understand that:
- When a diploid nucleus divides by meiosis, four haploid nuclei are produced.
- Halving the number of chromosomes in gametes allows for sexual reproduction in a lifecycle, because when two gametes fuse the diploid number is restored.
- DNA replication occurs before meiosis so that all chromosomes consist of two sister chromatids.
- Early in meiosis homologous chromosomes pair up and crossing over and condensation follow.
- Orientation of the pairs of homologous chromosomes before they separate is random.
- The chromosome number is halved in the first division of meiosis by separation of homologous chromosomes.
- Crossing over and random orientation lead to increased genetic variation.
- Genetic variation is also increased by the fusion of gametes from different parents.

There are a few organisms that are haploid – for example, the male honey bee, *Apis mellifera*. This means it produces gametes through mitosis. Female worker bees have 32 chromosomes in their body cells but male drones have only 16.

Meiosis is a reduction division

Meiosis is a type of cell division that produces gametes (sex cells). In any organism, each cell that is produced as a result of meiosis has half the number of chromosomes of other cells in the body.

Eukaryotic body cells have a **diploid** nucleus, which contains two copies of each chromosome, in homologous pairs (Subtopic **3.2**). As we have seen, humans have a diploid number of 46 chromosomes in 23 pairs, mangos and soybean both have 40 chromosomes in 20 pairs, and the camel has 70 chromosomes in 35 pairs.

During sexual reproduction, two gametes fuse together so, in order to keep the chromosome number correct in the offspring that are produced, each gamete must contain only one of each chromosome pair. That is, it must contain half the diploid number of chromosomes, which is called the **haploid** number. During gamete formation, meiosis reduces the diploid number to the haploid number – for this reason, meiosis is called a reduction division. At the moment of fertilisation, the normal diploid number is restored as the gametes fuse.

So, in the camel, the haploid sperm (35 chromosomes) and haploid egg (35 chromosomes) fuse at fertilisation to form the diploid zygote, with 70 chromosomes.

The process of meiosis

Meiosis occurs in a series of stages, as illustrated in Figure **3.7**, which result in the production of four cells.

Whereas mitosis, used to replace or repair cells, is achieved with one cell division, meiosis involves two divisions – the first reduces the number of chromosomes by half and the second produces four gametes each containing the haploid number of chromosomes. Exactly the same terms are used for the names of the stages, but since meiosis involves two divisions, the phases are numbered I and II.

The first division is very similar to mitosis and the second division is exactly the same as mitosis.

Prophase I

During interphase, before the start of prophase, chromosomes are replicated and consist of two identical sister chromatids joined at the centromere. In prophase I, these chromosomes now supercoil and the homologous pairs of chromosomes line up side by side.

Although the genes carried by each chromosome pair are identical, the alleles may not be. Exchange of genetic material between the pair can occur at this point. Chromatids may become entangled, break and re-join so that alleles are exchanged between homologous chromosomes during a process called crossing over. New combinations of alleles are formed and genetic variety among the resulting gametes increases.

The final step in prophase I is the formation of spindle microtubules and the breakdown of the nuclear envelope.

In each homologous pair, one chromosome is a maternal chromosome and the other a paternal chromosome. After crossing over, the chromatids recombine to produce new and unique combinations of alleles, different from both the maternal and the paternal arrangements. This is called recombination.

Metaphase I

Chromosomes line up on the equator at the centre of the cell. Each one attaches by a centromere to the spindle microtubules. The alignment of the chromosomes is random so that maternal and paternal chromosomes can appear on either side of one another on the equator. This means that either chromosome from a pair may move into each daughter cell during the first division at anaphase I, which results in increased genetic variety among the gametes.

Anaphase I

The microtubules now contract towards opposite poles. The pairs of sister chromatids remain together but the homologous pairs are separated. This is the reduction division where the chromosome number is halved from diploid to haploid.

Telophase I

Now spindles break down and a new nuclear envelope forms around each new nucleus. Cytokinesis follows and the cell splits into two cells, each containing only one chromosome of each homologous pair. Each chromosome, however, still consists of two sister chromatids at this point.

The second division of meiosis now follows to separate the two sister chromatids.

Prophase II

In each of the two cells resulting from meiosis I, new spindle microtubules start to form, the chromosomes re-coil and the nuclear envelope begins to break down.

Metaphase II

The nuclear envelope is broken down and individual chromosomes line up on the equator of each cell. Spindle fibres from opposite ends of the cell attach to each chromatid at the centromere.

Anaphase II

Sister chromatids are separated as the centromere splits and spindle fibres pull the chromatids to opposite ends of the cell.

Telophase II

Nuclear envelopes form around the four new haploid nuclei and the chromosomes now uncoil. A second cytokinesis occurs, resulting in four cells.

Meiosis I

1 Prophase I

- nuclear envelope breaks up as in mitosis
- crossing over of chromatids may occur
- homologous chromosomes pair up to form a bivalent

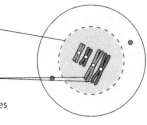

Bivalent showing crossing over that may occur:

 chromatids may break and may reconnect to another chromatid

- centromere
- chiasma – point where crossing over occurs (plural, chiasmata)
- one or more chiasmata may form, anywhere along length

At the end of prophase 1, a spindle is formed.

2 Metaphase I (showing crossing over of long chromatids)

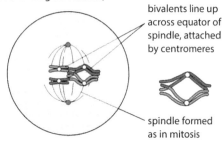

- bivalents line up across equator of spindle, attached by centromeres
- spindle formed as in mitosis

3 Anaphase I

Centromeres do not divide, unlike mitosis.

Whole chromosomes move towards opposite ends of spindle, centromeres first, pulled by microtubules.

4 Telophase I

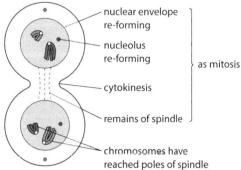

- nuclear envelope re-forming
- nucleolus re-forming
- cytokinesis
- remains of spindle
- chromosomes have reached poles of spindle

} as mitosis

Meiosis II

5 Prophase II

- nuclear envelope and nucleolus disperse
- centrioles replicate and move to opposite poles of the cell

6 Metaphase II

- chromosomes line up separately across equator of spindle

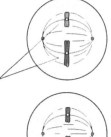

7 Anaphase II

- centromeres divide and spindle microtubules pull the chromatids to opposite poles

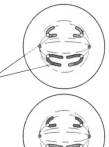

8 Telophase II

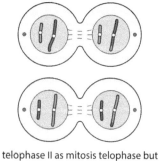

telophase II as mitosis telophase but four haploid daughter cells formed

Figure 3.7 The stages of meiosis in an animal cell. Note that the cells are shown with just two homologous pairs of chromosomes to make it easier to understand the process.

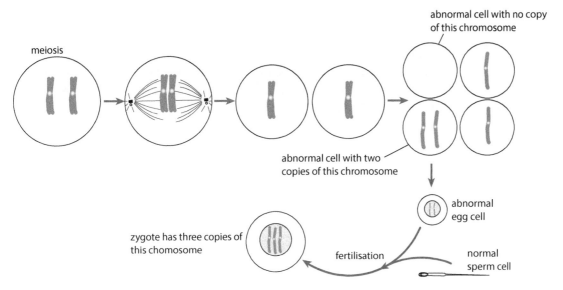

Figure 3.8 Non-disjunction at anaphase II of meiosis. Non-disjunction can also occur at anaphase I.

Non-disjunction

Non-disjunction is a failure of homologous pairs of chromosomes to separate properly during meiosis. It results in gametes that contain either one too few or one too many chromosomes. Those with too few seldom survive, but in some cases a gamete with an extra chromosome does survive and after fertilisation produces a zygote with three chromosomes of one type, as shown in Figure **3.8**. This is called a **trisomy**.

Trisomy in chromosome 21 results in the human condition known as **Down syndrome** (Figure **3.9**). A gamete, usually the female one, receives 24 chromosomes instead of 23 and a baby with 47 instead of the usual 46 chromosomes in each cell is born.

Karyotyping is used when there is concern about potential chromosome abnormalities. Cells from an unborn child are collected

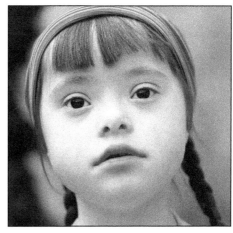

Figure 3.9 People with Down syndrome have characteristic physical features.

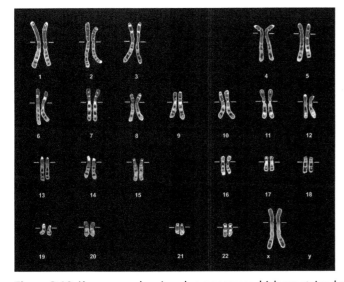
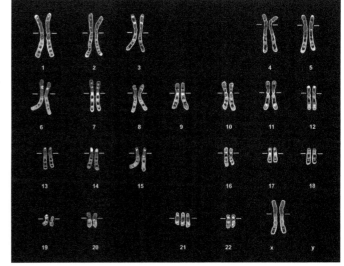

Figure 3.10 Karyogram showing chromosomes which are stained and photographed for a person with a normal chromosome complement and for a person with trisomy 21 (Down syndrome).

in one of two ways – chorionic villus sampling (CVS) or amniocentesis (Subtopic **11.4**). The cells are grown in the laboratory and a karyogram is prepared (Subtopic **3.2**). This is checked for extra or missing chromosomes (Figure **3.10**). The procedure is normally used if the mother is over the age of 35 years because Down syndrome is more common in babies of older mothers and can be detected using this method.

Meiosis and variation

Meiosis not only halves the chromosome number, it also promotes genetic variation in the gametes and individuals that are produced from them. There are several reasons for this:
- Each chromosome of the homologous pair carries different genetic information so that the gametes formed are genetically different.
- Different homologous pairs arrange themselves independently on the spindle and also separate independently so that gametes contain different combinations of each chromosome pair.
- Crossing over during prophase I means that genetic material is exchanged between homologous chromosomes producing entirely new combinations.
- At fertilisation, gametes from different parents fuse together promoting yet more genetic variation among the offspring produced.

Nature of science

Careful observation – discovery of meiosis by microscopic examination

The first observations of meiosis were made in 1876 by a German biologist, Oscar Hertwig (1849–1922), as he studied the eggs of sea urchins under an early microscope. He made careful, accurate drawing of his observations of gametes from several species including the salamander (Figure **3.11**). Later the Belgian biologist Eduord van Beneden (1846–1910) observed the behaviour of chromosomes during meiosis in the eggs of a parasitic worm. But it was not until 1890 that August Weismann (1834–1914) realised that two cell divisions were needed to produce haploid gametes from diploid cells so that the correct number of chromosomes would be maintained. The term 'meiosis' was first used in 1905.

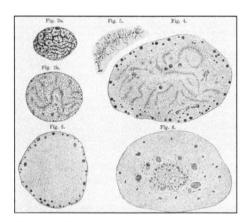

Figure 3.11 Drawings by Oscar Hertwig, showing the nuclei from the aquatic salamander previously known as Triton but now part of the genus *Eurycea*.

Prenatal screening

Obtaining fetal cells for karyotyping by **amniocentesis** involves taking a sample of amniotic fluid from the mother between weeks 14 and 16 of her pregnancy. **Chorionic villus sampling** (CVS) means taking a sample of cells from the chorionic villi, which are the fine projections of the placenta embedded in the lining of the uterus. This can be done 8–10 weeks into the pregnancy. Both methods carry a small risk of damaging the fetus or even causing a miscarriage. If the results of the test reveal abnormalities, the parents may be offered the option to terminate the pregnancy. The test may show an abnormality but it cannot give any indication about the likely severity of the condition.

Questions to consider

- Karyotyping is a procedure involving medical and ethical decisions. Who should make the decision to carry out the procedure – the parents or health care staff? How important are legal and religious arguments?
- Both procedures carry the risk of a miscarriage. How can this potential risk to the unborn child be balanced with the parents' desire for information? What safeguards should be in place when the karyotyping procedure is used?
- Does the information that can be obtained from the karyogram outweigh the risk to the unborn child?
- If the karyogram indicates a genetic abnormality, should the parents be permitted to consider a termination of the pregnancy?

Test yourself

7 State phase or phases of meiosis in which the following events occur:
   a homologous chromosomes separate
   b the nuclear envelope breaks down
   c the centromere splits
   d the sister chromatids line up on the equator
   e the chromosomes uncoil
   f reduction division occurs.
8 Define the term 'homologous chromosomes'.
9 Meiosis is a reduction division. Outline what this means.
10 State the number of cells formed when meiosis in one parent cell is completed.
11 Outline what you understand a trisomy to be.
12 State **one** example of non-disjunction in humans.
13 State **two** characteristics of chromosomes that are used to identify them when karyotyping.
14 The soybean has 40 chromosomes in its root cells. State the name given to this number of chromosomes.

Learning objectives

You should understand that:
- The principles of inheritance were determined experimentally by Mendel, who crossed large numbers of pea plants.
- Gametes are haploid and so contain only one allele of each gene. During meiosis, the two alleles of each gene in a diploid cell separate to different haploid daughter nuclei.
- Haploid gametes containing one allele of each gene fuse to produce diploid zygotes with two alleles, which may be the same or different.
- Dominant alleles mask the effects of recessive alleles. Codominant alleles have combined effects.
- Recessive alleles of autosomal genes account for many genetic disorders in humans.
- Dominant or codominant alleles account for some genetic disorders. Others disorders are sex linked.
- Due to their location on sex chromosomes, sex-linked genes show different inheritance patterns to autosomal genes.
- Most genetic disorders in humans are very rare but many are known.
- Mutation rate is increased by radiation and mutagenic chemicals, which can lead to cancer and genetic disease.

3.4 Inheritance

In the study of genetics, there are a number of specialist terms used. It will be helpful to understand and remember these key words.

Genotype the alleles possessed by an organism – each allele is represented by a letter; chromosomes come in pairs and so alleles come in pairs – a genotype is therefore represented by a pair of letters, such as **TT** or **Tt**

Phenotype the characteristics of an organism; a characteristic may be an external feature, such as the colour of flower petals, or internal, such as sickle-cell anemia

Dominant allele an allele that has the same effect on the phenotype when in either the homozygous or heterozygous state; the dominant allele is always given a capital letter – for example, **T**

Recessive allele an allele that only has an effect on the phenotype when in the homozygous state; a recessive allele is always given the lower case of the same letter given to the dominant allele – for example, **t**

Codominant alleles pairs of alleles that both affect the phenotype when present in the heterozygous state; these alleles are represented in a different way in genetics – a capital letter is chosen to represent the gene and then other (superscript) letters represent the alleles (for example, in human blood grouping, A and B are codominant alleles and are represented as I^A and I^B)

Locus the specific position of a gene on a homologous chromosome; a gene locus is fixed for a species – for example, the insulin gene is always found at the same position on chromosome 11 in humans

Homozygous having two identical alleles at a gene locus; the alleles may both be dominant or both recessive – for example, **TT** or **tt**

Heterozygous having two different alleles at a gene locus – for example, **Tt**

Test cross testing a dominant phenotype to determine if it is heterozygous or homozygous – for example, crossing either **TT** or **Tt** with **tt**; if there are any offspring with the recessive phenotype, then the parent with the dominant phenotype must be heterozygous (**Tt**)

Carrier an individual with one copy of a recessive allele that causes a genetic disease in individuals that are homozygous for this allele

Pure-breeding individuals of the same phenotype that, when crossed with each other, produce offspring which also all have that same phenotype

Discovering the principles of inheritance

The study of genetics aims to explain similarities and differences between parents and their offspring. Today we discuss genes, chromosomes and DNA but the study of inheritance began in the 19th century with the work of a monk, Gregor Mendel, who lived in what is now the Czech Republic. Mendel knew nothing of DNA or chromosomes but his studies of plant breeding are crucial to our understanding of inheritance. From about 1856 onward he carried out hundreds of painstaking experiments with pea plants, which he grew at his monastery. He chose plants with characteristics he wanted to study – such as those with short or long stems, or with wrinkled or smooth seeds – and he made crosses between them. He pollinated selected plants with characteristics of interest, using pollen from specially chosen plants, and grew new plants from the seeds that were produced. He observed, counted and recorded the different characteristics of all the plants that he grew.

Mendel's early studies involved the inheritance of just one pair of characteristics. He conducted what is now called a **monohybrid cross** using **pure-breeding** plants, which when crossed with each other always produce offspring that resemble their parents. Mendel crossed pure-breeding short pea plants with pure-breeding tall plants and grew seeds that were produced from the crosses (Figure **3.12**). In the new generation of peas (known as the first filial or F_1 generation) he noted that all the plants were tall.

Mendel self-pollinated these tall F_1 plants and grew seeds from the cross to produce a second filial or F_2 generation. He discovered that in this generation both tall and short plants were present (Figure **3.12**). Mendel recorded the results from 1064 plants. He found 787 tall plants and 277 short plants. Approximately three quarters of the plants were tall, or put another way, the ratio of tall : short plants was 3 : 1

Mendel carried out other monohybrid crosses with several other characteristics and obtained similar results. These are shown in Table **3.7**.

The results showed that inheritance did not produce a blending between the characteristics of the parent plants – so, for example, in the first cross there were no plants of medium height – and inheritance is described as being 'particulate'. We call the inherited particles genes or alleles but Mendel called them 'factors'. He understood that factors were transmitted to offspring in gametes.

Mendel also noted that short plants 'reappeared' in the F_2 generation despite the fact that there were no short plants in the F_1 generation.

Mendel used thousands of pea plants in his experiments. Without this amount of data he would have been unable to establish the ratios he observed with any certainty. If fewer replicates are used in any trial, the degree of uncertainty increases – the results are less **reliable**.

Characteristic	Cross made	Numbers produced in F_2 generation	Ratio calculated
height of stem	tall × short	787 tall, 277 short	2.84 : 1
petal colour	purple × white	704 purple, 244 white	3.15 : 1
seed shape	smooth × wrinkled	5474 smooth, 1859 wrinkled	2.96 : 1
seed colour	yellow × green	6022 yellow, 2001 green	3.01 : 1

Table 3.7 Mendel's experimental results.

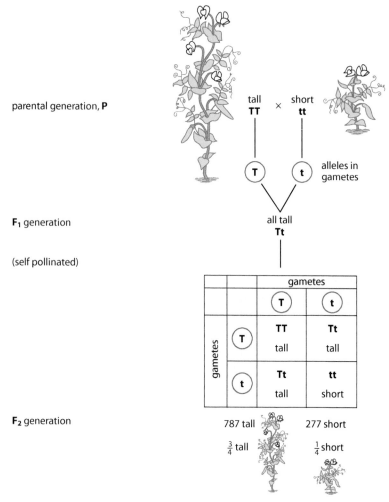

Figure 3.12 Diagrams to show Mendel's crosses of tall and short pea plants, with modern knowledge of alleles included.

Although all the F_1 plants are tall they must contain a 'factor' from their short parent that is 'masked' and does not reappear until the F_2 generation. Today, the factor for tallness is described as being **dominant** to the factor for shortness, which is said to be **recessive**. If a dominant and a recessive allele are present together the dominant allele masks the recessive allele.

Mendel proposed that factors were transmitted to offspring via gametes and came to the conclusion that gametes contain only one of the factors for height, for example, while the seeds and plants that grow from them contain both. Put another way, we would now say that gametes are haploid and seeds are diploid.

Without knowledge of genes and alleles, Mendel demonstrated that gametes contain one allele of each gene – nowadays we know that the two alleles of each gene separate into different haploid daughter nuclei during meiosis. Furthermore, his experiments showed that the fusion of gametes at fertilisation produces diploid zygotes (which in Mendel's experiments developed into seeds) with two alleles of each gene. The reappearance of short plants in the F_2 generation proves that the two alleles may be the same or different.

Determining allele combinations (genotypes) and characteristics (phenotypes) in genetic crosses

Using a Punnett grid

A genetic diagram called a Punnett grid can be used to work out all the possible combinations of alleles that can be present in the offspring of two parents whose genetic constitutions (genotypes) are known. Punnett grids show the combinations and also help to deduce the probabilities of each one occurring.

When working out a problem, it is helpful to follow a few simple steps.
1 Choose a letter to represent the gene. Choose one that has a distinctly different upper and lower case for the alleles – so for example O, P and W would not be good choices. It is useful to base the letter on the dominant phenotype – so for example R = red could be used for petal colour.
2 Represent the genotype of each parent with a pair of letters. Use a single letter surrounded by a circle to represent the genotype of each gamete.
3 Combine pairs of the letters representing the gametes to give all the possible genotypes of the offspring. A Punnett grid provides a clear way of doing this.
4 From the possible genotypes, work out the possible phenotypes of the offspring.

Worked examples **3.1** and **3.2** show how to tackle genetics problems using these steps.

Parental generation the original parent individuals in a series of experimental crosses
F_1 generation the offspring of the parental generation
F_2 generation the offspring of a cross between F_1 individuals

Worked examples

3.1 Suppose that fur colour in mice is determined by a single gene. Brown fur is dominant to white. A mouse **homozygous** for brown fur was crossed with a white mouse. Determine the possible genotypes and phenotypes of the offspring.

Step 1 Choose a letter. Brown is dominant so let **B** = brown fur and **b** = white fur.
Step 2 We are told the brown mouse is homozygous so its genotype must be **BB**. Since white is recessive, the genotype of the white mouse can only be **bb**. If a **B** were present, the mouse would have brown fur.
Step 3 Set out the diagram as shown.
Step 4 The Punnett grid shows that all the offspring will be phenotypically brown and their genotype will be **Bb**.

parental phenotypes: brown white
parental genotypes: **BB** **bb**
gametes: Ⓑ Ⓑ Ⓑ Ⓑ
Punnett grid for F_1:

	gametes from brown parent	
	Ⓑ	Ⓑ
gametes from white parent Ⓑ	Bb brown	Bb brown
Ⓑ	Bb brown	Bb brown

3.2 Seed shape in the pea plant is controlled by a single gene. Smooth shape is dominant to wrinkled shape. A plant that was heterozygous for smooth seeds was crossed with a plant that had wrinkled seeds. Determine the possible genotypes of the offspring and the phenotype ratio.

Step 1 Choose a letter. Smooth is the dominant trait but **S** and **s** are hard to distinguish so use another letter, such as **T**.

Step 2 We are told the smooth plant is heterozygous so its genotype must be **Tt**. Since 'wrinkled' is a recessive trait, the genotype of the wrinkled seed plant must be **tt**.

Step 3 Set out the diagram as shown, in exactly the same way as before. Notice that, in this case, the smooth-seeded parent produces two different types of gamete because it is heterozygous.

Step 4 Here the Punnett grid shows us that half of the offspring will have smooth seeds with the genotype **Tt** and half will have wrinkled seeds with the genotype **tt**.

parental phenotypes: smooth   wrinkled
parental genotypes: Tt   tt
gametes: T   t   t   t

Punnett grid for F$_1$:

	gametes from smooth-seed parent	
	T	t
gametes from wrinkled-seed parent — t	Tt smooth	tt wrinkled
t	Tt smooth	tt wrinkled

? Test yourself

15 Define each of the following terms.
  a genotype
  b phenotype
  c dominant allele
  d recessive allele
  e homozygous
  f heterozygous

16 If red **R** is dominant to yellow **r**, state the phenotype of each of the following genotypes.
  a RR   **b** Rr   **c** rr

17 State the gametes produced by a parent with each of the following genotypes.
  a RR   **b** rr   **c** Rr

18 Copy and complete the Punnett grid. Green seed colour **G** is dominant to purple seed colour **g**.

	gametes from green parent	
	G	g
gametes from green parent — G	GG green	
g		

114

Codominance and multiple alleles

In Worked examples **3.1** and **3.2**, one of the alleles completely dominates the other, so in a heterozygous genotype the phenotype is determined solely by the dominant allele. In **codominance**, both alleles have an effect on the phenotype.

The examples of the mouse coat colour and the smooth and wrinkled peas are both known as **monohybrid crosses** because they involve just one gene with two alleles: brown **B** and white **b**, or smooth **T** and wrinkled **t**. There are many other cases in which genes have more than two alleles. One example of this is human blood groups.

The ABO human blood grouping is an example of both codominance and multiple alleles. There are three alleles – I^A, I^B and **i**.

I^A and I^B are codominant and both are dominant to **i**. This results in four different phenotypes or blood groups.

A person's blood group depends on which combination of alleles he or she receives. Each person has only two of the three alleles and they are inherited just as though they are alternative alleles of a pair. Table **3.8** shows the possible combinations of alleles and the resulting phenotypes.

Genotype	Phenotype or blood group
$I^A I^A$	A
$I^A i$	A
$I^B I^B$	B
$I^B i$	B
$I^A I^B$	AB
ii	O

Table 3.8 Human blood groups and their genotypes.

Worked examples

3.3 Celia is blood group A and her husband Sanjeev is blood group B. Their daughter Sally is blood group O. Determine the genotypes of Celia and Sanjeev.

Step 1 The alleles are represented by I^A, I^B and **i**.

Step 2 To be blood group A, Celia could have genotype $I^A I^A$ or $I^A i$.
To be blood group B, Sanjeev could have genotype $I^B I^B$ or $I^B i$.
To be blood group O, Sally could **only** have the genotype **ii**.

Step 3 Each of Sally's two alleles have come from her parents, so she must have received one **i** from her mother and one **i** from her father, as shown in the Punnett grid.

	gametes from Sanjeev	
gametes from Celia	I^B	i
I^A		
i		ii Sally group O

3.4 Hair shape in humans is a codominant characteristic. Straight hair and curly hair are codominant alleles and the heterozygote has wavy hair. Daryll and Shaniqua both have wavy hair. Deduce the probabilities that their children will have straight hair, curly hair or wavy hair.

H^S = straight hair and H^C = curly hair.

Since both Daryll and Shaniqua have wavy hair, their genotypes must both be $H^S H^C$.

The Punnett grid shows that the probabilities of a child inheriting each hair type are:
- straight hair 25% ($\frac{1}{4}$)
- curly hair 25% ($\frac{1}{4}$)
- wavy hair 50% ($\frac{1}{2}$).

parental phenotypes:	wavy	wavy
parental genotypes:	$H^S H^C$	$H^S H^C$
gametes:	H^S H^C	H^S H^C

Punnett grid:

		gametes from Shaniqua	
		H^S	H^C
gametes from Daryll	H^S	$H^S H^S$ straight hair	$H^S H^C$ wavy hair
	H^C	$H^S H^C$ wavy hair	$H^C H^C$ curly hair

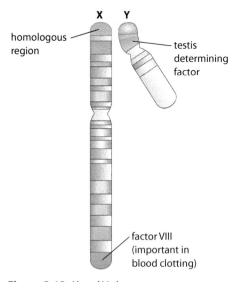

Figure 3.13 X and Y chromosomes.

Sex linkage the pattern of inheritance that is characteristic for genes located on the X chromosome

Sex chromosomes

Humans have one pair of chromosomes that determine whether the person is male or female. These chromosomes are called the **sex chromosomes**. Each person has one pair of sex chromosomes, either **XX** or **XY**, along with 22 other pairs known as **autosomes**. The **X** chromosome is longer than the **Y** and carries more genes. Human females have two **X** chromosomes and males have one **X** and one **Y**.

Sex chromosomes are inherited in the same way as other chromosomes. The ratio of phenotypes female : male is 1 : 1. This means, at fertilisation, there is always a 50% chance that a child will be a boy and 50% that it will be a girl.

Sex chromosomes and genes

The sex chromosomes not only carry the genes that control gender, the X chromosome also carries genes called sex-linked or X-linked genes. These genes occur only on the X chromosome and not on the Y chromosome, which is much shorter (Figure 3.13). The Y chromosome carries alleles that are mainly concerned with male structures and functions.

Sex linkage has a significant effect on genotypes. Females have two X chromosomes, so they have two alleles for each gene and may be homozygous or heterozygous. In a female, a single recessive allele will be masked by a dominant allele on her other X chromosome. Males only have one allele on their X chromosome with no corresponding allele on the Y chromosome, so a recessive allele will always be expressed in a male.

A female who is heterozygous for a sex-linked recessive characteristic that does not affect her phenotype is called a **carrier**.

Examples of sex-linked characteristics

Two examples of sex-linked human characteristics are hemophilia and red–green colour blindness.

Hemophilia is a condition in which the blood of an affected person does not clot normally. It is a sex-linked condition because the genes controlling the production of the blood-clotting protein factor VIII are on the X chromosome. A female who is $X^H X^h$ will be a carrier for hemophilia. A male who has the recessive allele $X^h Y$ will be a hemophiliac. Figure 3.14 is a pedigree chart showing how a sex-linked condition like hemophilia may be inherited. Notice that hemophilia seldom occurs in females, who would have to be homozygous for the recessive allele $X^h X^h$. This condition is usually fatal *in utero*, resulting in a miscarriage. Today, hemophilia is treated by giving the affected person the clotting factor they cannot produce.

A person with red–green colour blindness has difficulty distinguishing between red and green. Red–green colour blindness is inherited in a similar way to hemophilia. A female who is $X^B X^b$ is a carrier for colour blindness and a male with just one copy of the recessive allele will be colour blind. Remember that a man cannot be a carrier for a sex-linked gene.

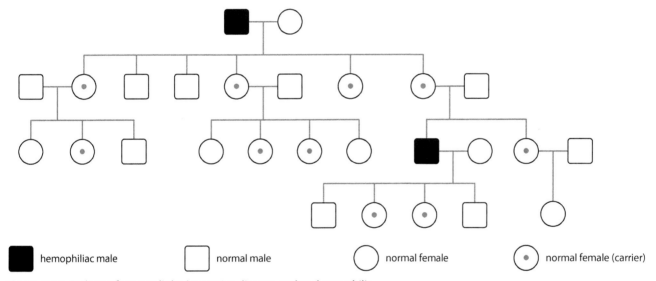

Figure 3.14 Pedigree for a sex-linked recessive disease, such as hemophilia.

Worked example

3.5 A woman who is homozygous for normal vision married a man who is red–green colour blind. Determine the possible types of vision inherited by their two children, one girl and one boy.

Step 1 Standard letters are used for these alleles – normal vision is X^B and colour blind is X^b. The X is always included.

Step 2 The woman is homozygous for normal vision so her genotype must be $X^B X^B$. Since the man is colour blind, his genotype must be $X^b Y$.

Step 3 Set out the diagram as shown.

Step 4 The Punnett grid shows that a daughter will have normal vision, but be a carrier for red–green colour blindness. A son will have normal vision.

parental phenotypes:	woman	man
parental genotypes:	$X^B X^B$	$X^b Y$
gametes:	X^B X^B	X^b Y

Punnett grid for F_1:

	gametes from man	
	X^b	Y
gametes from woman X^B	$X^B X^b$ girl, normal vision, carrier	$X^B Y$ boy, normal vision
X^B	$X^B X^b$ girl, normal vision, carrier	$X^B Y$ boy, normal vision

Pedigree charts

Pedigree charts, like the ones shown in Figures **3.14** and **3.15**, are a way of tracing the pattern of inheritance of a genetic condition through a family. Specific symbols are always used, and the chart is set out in a standard way. The horizontal lines linking the male and female in a generation indicate a marriage (or mating) and the vertical lines indicate their offspring. Offspring are shown in the order of their birth. For example, in the family shown in Figure **3.15**, the oldest individual affected with the genetic condition is II2, who is a male.

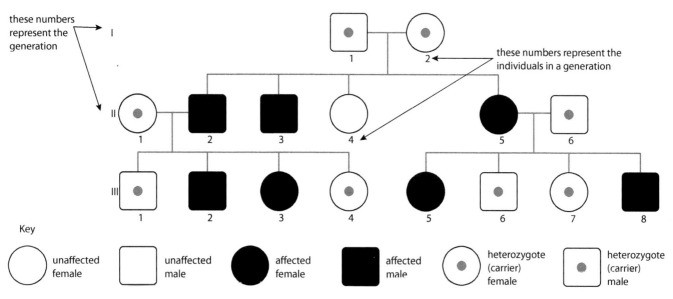

Figure 3.15 This pedigree chart shows the occurrence of a genetic condition known as brachydactyly (short fingers) in a family.

Genetic diseases

As well as the inherited genetic diseases that are carried on the sex chromosomes and inherited with the sex of an individual, there are many other inherited disorders that occur due to mutations in single genes on the autosomal chromosomes (numbers 1–22). Most are caused by the presence of two recessive alleles, though some – such as Huntington's disease – are caused by a dominant allele (Table **3.9**). Genetic 'diseases' are unlike other diseases such as a cold or flu. They cannot be 'caught' from another person but they may be passed through families from parent to child. More than 6000 physiological 'diseases', caused by mutations to single genes, are known. Most are very rare (Table **3.9**) and there is a wide variation in the occurrence of some genetic disorders between different racial types and geographic locations of people or their ancestors.

Genetic disease	Frequency	Cause	Gene location (chromosome number)
Cystic fibrosis (CF)	Variable. 1 in 25 caucasians are carriers of disease. In Europe, about 1 in 2000 babies are born with CF. In the USA, the incidence is 1 in 3500.	Autosomal recessive allele – more than 500 different mutations of the CF gene have been found.	7
β thalassemia	Most common in Asia, Middle East and Mediterranaean areas where malaria was or is endemic. 1% of people here may be affected. α and β thalassemias are the world's most common single-gene disorders.	Autosomal recessive allele of the gene that codes for the beta chain of hemoglobin.	11
Sickle-cell disease	Most common in people whose ancestors come from sub-Saharan Africa but also in South and Central America, Mediterranean countries and India. In the USA, 1 in 500 African-American babies and 1 in 1000 Hispanic-American babies are affected. 1 in 12 African-Americans and up to 4 in 10 West Africans are carriers of the allele. In West Africa, 1% of the population are affected.	Autosomal recessive mutation, which causes substitution of one amino acid in the beta chain of hemoglobin.	11
Huntington's disease (HD)	Rare disease affecting about 1 in 20 000 people. Males and females are equally affected and the disease occurs in all ethnic and racial groups.	Autosomal dominant mutation of the HD gene, which increases the length of a repeated sequence of CAG.	4

Table 3.9 Some information on four well-documented genetic diseases that affect human metabolism (data from World Health Organization).

Worked examples

3.6 Cystic fibrosis (CF) is a genetic disorder that causes the excessive production of thick sticky mucus. It is due to a recessive allele that is not sex linked. The pedigree chart shows two generations of a family. A filled-in symbol represents an individual who has cystic fibrosis.

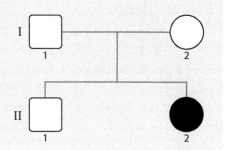

Deduce the genotypes of the parents I1 and I2.
Deduce the probability that II1 is heterozygous.

Step 1 Cystic fibrosis is a recessive disorder so the 'normal' condition, without cystic fibrosis, is dominant. It is useful to choose **N** to represent the normal allele.

Step 2 Neither of the parents, I1 and I2, have cystic fibrosis so both must have at least one normal allele **N**. Since cystic fibrosis is recessive and II2 has the condition, she must have the genotype **nn**.

II2 received one allele from each of her parents so both of them must have passed one **n** allele to her. Both parents must have one **n** but they do not have cystic fibrosis so their genotype must be heterozygous **Nn**.

The pedigree chart could now be redrawn, to show that the parents are heterozygous carriers.

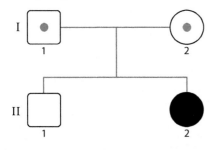

Step 3 Now that both parents are known to be heterozygous, a Punnett grid can be drawn.

Step 4 Person II1 does not have cystic fibrosis and so could have the allele combination shown by any of the shaded boxes. The probability of being heterozygous is 2 out of 3, or $\frac{2}{3}$, or 66%.

	gametes from I1	
	N	**n**
gametes from I2 **N**	NN normal	Nn normal
gametes from I2 **n**	Nn normal	nn II2 with CF

3.7 The pedigree chart shows the family history of a recessive human condition called woolly hair. A filled-in symbol indicates that the person has woolly hair. Deduce whether this condition is sex linked or not.

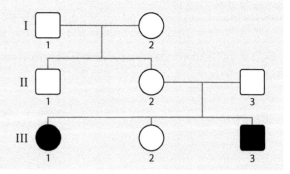

Step 1 Remember that in a sex-linked condition, the allele occurs only on the X chromosome and males only have one X chromosome.

Step 2 Using **N** to represent the condition, we can see that female III1 must be **nn** as she has the condition and thus has inherited one **n** from each parent.

Step 3 If woolly hair is not sex linked, both her parents would be **Nn** as they have normal hair.

Step 4 If it is sex linked, her mother (II2) would be $X^N X^n$ and her father (II3) would be $X^n Y$. This would mean he has the recessive allele and no dominant allele. If the condition is sex linked, he would have woolly hair, which he does not. This proves that it is not sex linked.

Ethics of identification of genotype

Although the sequencing of the bases of the human genome has been completed (Subtopic **3.1**), the task of identifying all of the genes is on going. As these genes are found, it may be possible for a person to be screened for particular alleles of genes that could affect them – for example, by increasing their susceptibility to cancer or the likelihood that they will develop Alzheimer's disease in later life. Alleles for genetic diseases such as Huntington's disease, whose onset is typically later in life, can already be identified.

Questions to consider

- Does simply knowing the sequence of the three billion base pairs of the human chromosomes tell us anything about what it means to be human?
- Should third parties such as health insurance companies have the right to see genetic test results or demand that a person is screened before offering insurance cover or setting the level of premiums?
- If treatment is unavailable, is it valuable to provide knowledge of a genetic condition that a person may carry?
- Knowledge of an individual's genome has implications for other members of their families. Should their rights be protected?

Worked example

3.8 The pedigree chart below shows the inheritance of a particular genetic condition in a family. A filled in circle or square means that the individual is affected – that is, shows the genetic condition. The condition is not sex linked. Deduce whether the characteristic is dominant or recessive.

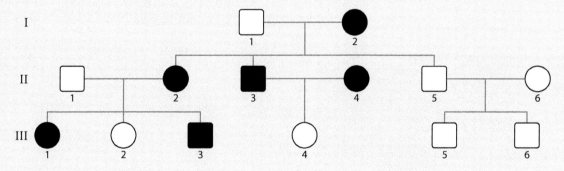

We can see that the genetic condition is dominant for the following reasons:
- affected individuals occur in every generation
- every affected individual has at least one affected parent.

Mutation rates

Mutations are spontaneous, permanent changes in the base sequence of DNA – they can occur at any time in any organism. Usually the natural rate of mutations is very low but some environmental factors can significantly increase the number and frequency of mutations. Ionising radiation from nuclear plants, ultraviolet light (UV), X-rays and mutagenic chemicals such as formaldehyde, benzene and tobacco tar are **mutagens**, which increase mutation rate.

Scientists tracking mutation rates after events such as the nuclear bombing of Hiroshima and Nagasaki in Japan during the Second World War, and nuclear accidents at Three Mile Island in the USA in 1979 and Chernobyl in Ukraine in 1986, noted significant increases in the rates of cancer among people who were affected by radioactive fallout (Figure **3.16**). Mutations may cause the onset of a cancer if cell division is disrupted by the mutation (Subtopic **1.6**).

Mutations caused by mutagens that affect body cells are not usually inherited – only mutations occurring in cells that produce gametes will be passed on to the next generation. This fact has been used by researchers investigating the inheritance of genes in species such as the fruit fly *Drosophila* which can be studied in the laboratory.

Nature of science

Experimental design – quantitative measurements with replicates lead to increased reliability

In the study of statistics, replication is defined as the repetition of an experiment in the same or similar conditions. Repeating or replicating experiments to generate larger amounts of data increases the **reliability** of the conclusions that can be drawn. The concept of reliability requires that the results must be more than a 'one-off' finding, and must be repeatable.

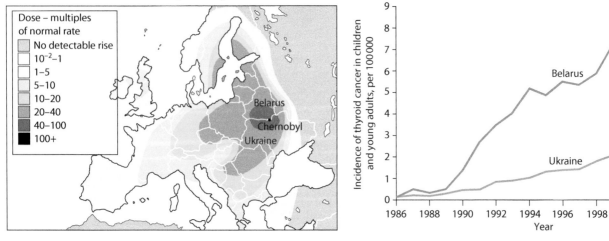

Figure 3.16 The explosion at the Chernobyl nuclear reactor in Ukraine in 1986 resulted in increased radiation doses across Europe. In the years that followed, the numbers of children in Ukraine and neighbouring Belarus suffering thyroid cancer increased significantly because of exposure to radioactive fallout.

Mendel made thousands of individual crosses (or replicates) as he carried out his pea plant experiments and obtained a large number of numerical results. Other researchers who have performed similar crosses under the same conditions have obtained similar results. Replicating an experiment in this way reinforces the findings and increases the likelihood of the wider scientific community accepting any hypothesis that is proposed.

Genetics pioneers rejected and ignored

Gregor Mendel (1822–1884) was an Augustinian monk in the Abbey of St Thomas in Brno, a town in what is now the Czech Republic. Over a period of 7 years, he cultivated and tested different pea plants and studied their visible characteristics. In 1866, he published a paper on the inheritance of characteristics in pea plants, which he called 'Experiments on plant hybridization', in *The Proceedings of the Natural History Society of Brünn*. In it, he set out his two laws of inheritance – the 'law of segregation' and the 'law of independent assortment of characteristics'. Although Mendel sent copies to well-known biologists, his ideas were rejected. For the next 35 years, his paper was effectively ignored yet, as scientists later discovered, it contained the entire basis of modern genetics.

In 1951, 2 years before the structure of DNA was determined, Dr Barbara McClintock presented a paper on 'jumping genes' at a symposium at the Cold Spring Harbor research laboratories in the USA. She explained how genes could be transferred from one chromosome to another. Her work was rejected by fellow scientists and ignored for 30 years until rediscovered. Now she is recognised as a pioneer in the field and her work is acknowledged as a cornerstone of modern genetics. In 1983, Dr McClintock received a Nobel Prize for her work.

Questions to consider

- Why is the work of some scientists ignored while that of others becomes readily accepted?
- How important is it for a new theory to be presented in a famous journal or at a well-attended meeting?
- Do you think Mendel's work was ignored because it was not widely published, or because he was not a well-known scientist, or for some other reason?
- Barbara McClintock was a young woman when she presented her work. Her work represented a completely different view of how genes might behave. How significant is this type of paradigm shift in gaining acceptance for a new theory or discovery?

Test yourself

19 A single gene controls the synthesis of the protease enzyme pepsin in the camel. State how many alleles controlling pepsin synthesis will be present in the following cell types in the camel:
  a a stomach cell   **b** a brain cell   **c** an egg cell.
20 State which phenotype (blood group) in the ABO human blood grouping system always has a homozygous genotype.
21 State **two** sex-linked characteristics in humans.
22 State the standard symbol in a pedigree chart for a woman who shows a particular genetic condition.
23 State the term used to describe alleles if they both have an effect on the phenotype in a heterozygous genotype.
24 State the name given to an individual who has one copy of a recessive allele that would cause a genetic disease when homozygous.

Learning objectives

You should understand that:
- In gel electrophoresis, fragments of DNA or proteins move in an electric field and are separated according to their size.
- The polymerase chain reaction (PCR) is used to copy and amplify minute quantities of DNA.
- DNA profiling involves comparison of DNA from different individuals.
- Genetic modification (GM) involves the transfer of genes between species.
- A clone is defined as a group of genetically identical organisms derived from a single original parent cell.
- Natural methods of cloning are found in many plant species and some animal species.
- Animals can be cloned from embryos that are broken up into more than one group of cells.
- Adult animals can be cloned from differentiated cells in some cases.

3.5 Genetic modification and biotechnology

Genetic modification and **biotechnology** have opened up new opportunities in forensic science, agriculture, medicine and food technology. As knowledge has grown, science has enabled people to manipulate the unique genetic identity of organisms. Gene transfer, cloning and stem cell research have raised questions about the safety and ethics of techniques that have been unknown to previous generations.

DNA profiling

At a crime scene, forensic scientists check for fingerprints because a person's fingerprint is unique and can be used to identify them. Forensic scientists also collect samples of hair, skin, blood and other body fluids left at a crime scene because they all contain a person's DNA and that too is a unique record of their presence.

Matching the DNA from a sample to a known individual is called **DNA profiling**. In forensic science, DNA profiles from crime scenes can be used to establish the possibility of guilt or prove a suspect innocent (Figure 3.17). DNA profiling can also be used to determine paternity. For example, a woman might believe that a particular man is the father of her child. By comparing DNA samples from all three individuals – the woman, the man and the child – paternity can be established.

The polymerase chain reaction (PCR)

DNA profiles can only be done if there is sufficient DNA to complete the procedure. Sometimes, at a crime scene or when a body is found after a very long time, only a minute amount can be collected. The **polymerase chain reaction (PCR)** is a simple method that makes millions of copies of tiny amounts of DNA so there is sufficient to produce a profile. This is done at high temperature using a special type of DNA polymerase enzyme. Samples of DNA are treated at 95 °C so that their two strands are separated. After cooling to 60 °C, DNA primers are attached to the strands and free nucleotides are added, together with DNA polymerase. Complementary strands of DNA are synthesised on the two strands so that after one cycle of PCR two copies of the original DNA are produced. The process is then repeated more than 20 times until sufficient DNA has accumulated. Technicians must take great care when handling the original sample so that it is not contaminated with their own or other DNA.

Gel electrophoresis

Gel **electrophoresis** is a method used to separate fragments of DNA on the basis of size and the electric charge they carry. It can identify natural variations found in every individual's DNA.

Any DNA sample usually contains long molecules that are too large to be used for profiling. Enzymes, called **restriction enzymes**, are used to cut DNA into fragments at very precise points in the base sequences.

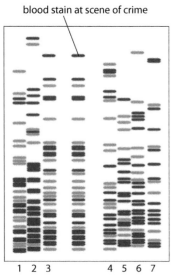

Figure 3.17 DNA profile of a blood stain found at the scene of a crime compared with profiles from seven suspects. Which suspect was at the scene of the crime? What is the evidence to support your answer?

Since each individual has a unique DNA sequence, the positions of these cutting sites will vary, giving a mixture of different fragment sizes. DNA profiling often examines repetitive sequences of so-called 'satellite' DNA that vary in their degree of repetitiveness from person to person. These are called variable number tandem repeats (VNTRs) and short tandem repeats (STRs). These regions have repeated sequences of DNA, which are very similar in close relatives but so variable in unrelated people that non-relatives are extremely unlikely to have the same repeated sequences.

The DNA fragments are placed in a well in a plate of gel (a jelly-like material) and an electric field is applied. Each DNA fragment has a small negative charge and so will move in the electric field, through the gel. The distance a fragment can move depends on its size – smaller fragments move most easily through the gel matrix and travel further, while larger fragments are left behind close to their starting point. After the fragments have been separated in the gel, they are stained and produce a unique pattern of bands called a **DNA profile** (Figures **3.17** and **3.18**).

Short tandem repeats

Each STR is shared by between 5 and 20% of unrelated people but forensic scientists can examine many STRs at the same time by using sequence-specific fragments of DNA primers to find the repeated sections. The pattern of repeats can identify an individual to a high degree of accuracy so STR analysis is an excellent forensic tool. The more STR regions that are examined the more accurate the test becomes.

Gene technology

Gene technology, which is also called genetic modification (GM) or **genetic engineering**, involves the transfer of genes from one species to another in order to produce new varieties of organisms with useful or desirable characteristics.

Selective plant and animal breeding has been carried out by humans for thousands of years as people tried to develop cattle that produced high milk yields or crops with better resistance to disease, for example. In these cases, animals or plants of the same species were chosen for breeding because of their particular characteristics. Over many generations of selection, the desired characteristics increase in frequency in the population.

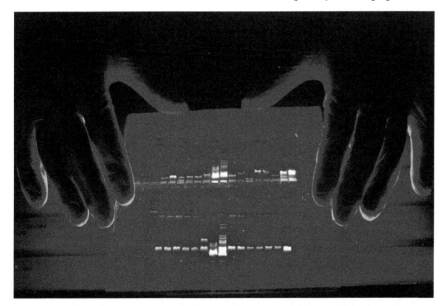

Figure 3.18 Scientist examining an agarose electrophoresis gel used to prepare a DNA profile. The sample of DNA is marked with a radioactive substance, so the DNA banding pattern appears pink under ultraviolet light. The pattern is preserved by applying radiographic film to the gel.

DNA profile databases

In the USA, the FBI has a national database of DNA profiles from convicted criminals, suspects, missing persons and crime scenes. The data that is held may be used in current investigations and to solve unsolved crimes. There are many commercial laboratories that carry out DNA profiling analysis on behalf of law enforcement agencies. Many of them check 13 key STR sequences in DNA samples, which vary considerably between individuals. The FBI has recommended that these should be used because they provide odds of one in a thousand million that two people will have the same results.

CODIS is a computer software program that operates the national database of DNA profiles. Every American State has a statutory right to establish a DNA database that holds DNA profiles from offenders convicted of particular crimes. CODIS software enables laboratories to compare DNA profiles electronically, linking serial crimes to each other and identifying suspects from profiles of convicted offenders. CODIS has contributed to thousands of cases that have been solved by matching crime scene evidence to known convicted offenders.

Questions to consider

- DNA profiles do not show individual base sequences but only identify repeated sequences. How much confidence should be placed on DNA evidence?
- How secure is DNA profiling?
- What are the implications for society if the authorities were to hold a DNA profile for every person?
- What safeguards should be in place to protect the rights of individuals whose DNA profiles have been placed on a database but who have not been convicted of a crime?
- Is it right to convict a person on DNA evidence alone?

GM sheep benefit humans

Factor XI was first recognised in 1953. Factor XI is part of the cascade of blood clotting factors that form the chain leading to a protective clot. The incidence of factor XI deficiency is estimated at 1 in 100 000. In some people, the symptoms are similar to hemophilia, but it is not a sex-linked condition and it affects men and women equally. In many countries, treatment involves obtaining factor XI from fresh-frozen plasma and since factor XI is not concentrated in plasma, considerable amounts of it are needed. It is hoped that the production of clotting factors in GM sheep's milk will lead to new and better treatments.

Gene technology gives us the new ability to transfer genes from one species to another completely different species in just one generation. For example, bacterial genes have been transferred to plants, human genes transferred to bacteria and spider genes transferred to a goat.

Gene transfer is possible because the genetic code is universal. No matter what the species, the genetic code in the DNA spells out the same information and produces an amino acid sequence in one species that is exactly the same in any other species.

Usually, in gene transfer, only single genes are used – for example, the gene for producing human blood-clotting factor XI has been transferred to sheep, which then produce the factor in their milk.

The technique of gene transfer

One of the first important uses of gene transfer was to produce insulin for diabetic patients whose own bodies do not produce insulin properly. Many years ago, insulin was obtained from cow or pig pancreases but the process was difficult and the insulin was likely to be contaminated. Today, diabetics inject themselves with human insulin that has been made by modified *E. coli* bacteria (Figure **3.19**).

There are key three steps in the process:
- obtaining the desired human insulin gene in the form of a piece of DNA
- attaching this DNA to a **vector**, which will carry it into the **host** cell (*E. coli*) – the vector used is the **plasmid** found inside the bacterium
- culturing *E. coli* bacteria so that they translate the DNA and make insulin, which is collected.

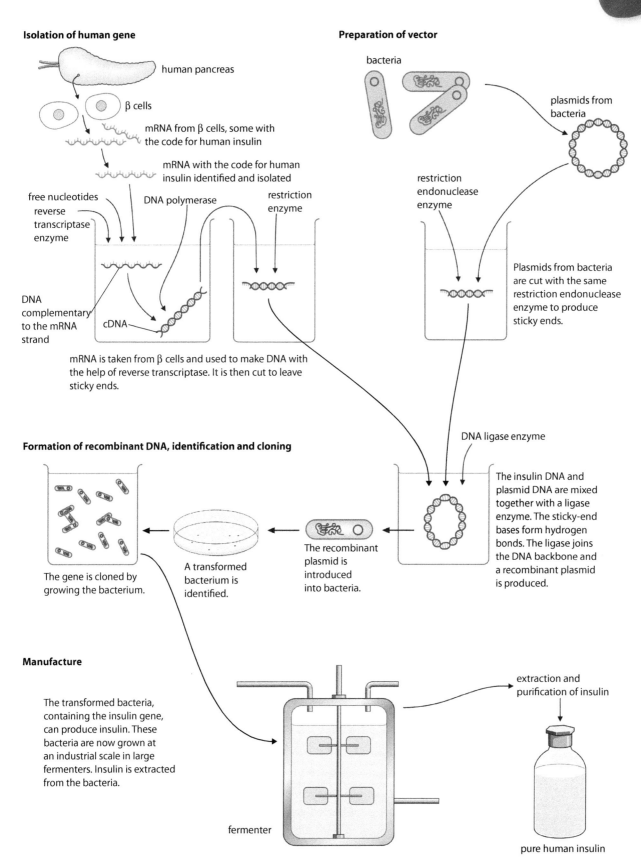

Figure 3.19 Stages in producing a transgenic bacterium.

The first GM plant was produced in 1982. It was a tobacco plant containing bacterial genes that gave it antibiotic resistance.

New developments in genetic modification

While strawberries, pineapples, sweet peppers and bananas have all been genetically modified to remain fresh for longer, a new variety of golden-coloured rice has been genetically modified so that it contains high levels of a substance called beta-carotene. Beta-carotene gives carrots their orange colour but, more importantly, the body converts it to vitamin A, which is essential for the development of pigments in the retina of the eye. Three genes had to be introduced into rice so it could produce beta-carotene. Two of these came from daffodils and the third was taken from a bacterium.

Enriched Golden Rice™ is a valuable dietary supplement for people whose diet is low in vitamin A and who might otherwise suffer vision problems or blindness. Since Golden Rice™ was developed in the 1990s, research has continued into its use and help in enriching diets. Some interesting and controversial issues have arisen about the ownership of the rights to the seeds, the flavour of the rice and the publication of research data from experimental studies.

Genetically modified organisms (GMOs)

More than 100 plant species have been genetically modified and many trials have taken place to assess their usefulness. In comparison, there are very few examples of genetically modified animal species. Most genetic engineering has involved commercial crops such as maize, potatoes, tomatoes and cotton. Plants have been modified to make them resistant to pests and disease and tolerant to herbicides. Genetically engineered animals, on the other hand, have mainly been farmed for the products of the inserted genes – most common are proteins such as factor XI, used to treat hemophilia in humans, and α1 antitrypsin, a shortage of which can lead to lung and liver problems.

Herbicide tolerance

Herbicides are used to kill weeds in crop fields but they are expensive and can affect local ecosystems as well as cultivated areas. One commonly sprayed and very powerful herbicide is glyphosate, which is rapidly broken down by soil bacteria. For maximum crop protection, farmers needed to spray several times a year. But now, the genes from soil bacteria have been successfully transferred into both maize and soy bean plants making them resistant to the herbicide. Farmers can now plant the modified seeds, which germinate along with the competing weeds. Spraying once with glyphosate kills the weeds and leaves the crops unaffected. The maize and soy bean then grow and out-compete any weeds that grow later when the glyphosate has broken down in the soil. Yields are improved and less herbicide has to be used.

Reducing pollution

Pigs fed on grains and soybean meals produce a lot of phosphate in their manure. Phosphate causes pollution and **eutrophication** in the environment. Genetically modified pigs have been developed with a gene from the bacterium *E. coli*. The bacteria make an enzyme, phytase, which releases the digestible phosphorus found in grains and soybeans. Genetically modified pigs produce this enzyme in their saliva and so digest their food better. More phosphorus becomes available to them and less goes undigested. The pigs absorb the nutrients into their blood, so they grow better, and much less phosphate is released in their manure.

How do scientists assess the risks associated with genetically modified organisms?

Genetic modification (GM) of plants and animals is potentially enormously helpful to the human race but it raises ethical and social questions, which are the source of heated debate. Assessing the risks posed by genetically modified organisms is a difficult task because scientists at the cutting edge of research cannot have full knowledge of the potential of the organisms they produce. All genetically modified plants and animals are subjected to safety testing but that is not to say that all possible risks can be assessed. Some of the possible benefits for the future are listed below.

- As our population increases and more people need feeding, modifying plants and animals to increase yield or to be able to grow in places where they previously could not, will provide more food. Plants can be made tolerant to drought or salt water so that food can be grown in difficult areas.
- Crop plants that are disease resistant not only increase yields but also reduce the need for applying potentially harmful pesticides.
- Many substances, such as human growth hormone, a blood-clotting factor, antibodies and vitamins, are already being made by genetically modified organisms to improve human health.

On the other hand, there are those who are greatly concerned by the use of genetically modified plants and animals.

- They argue that animals could be harmed by having these genes inserted.
- There is concern that people consuming genetically modified plants and animals could be harmed.
- The long-term effects of genetically modified crops in the environment are not known. Plants or animals could 'escape' into the environment and their genes might become incorporated into wild populations, with unknown effects.
- Human food crops could become controlled by a small number of biotechnology companies.
- GM seeds/plants may be more expensive, preventing poorer farmers from buying them. Wealth might become concentrated in a smaller percentage of the population, which might damage the local economy.
- More genetically modified organisms might lead to a reduction in natural biodiversity.

Ethics and genetic modification

There is much discussion in the media about genetic modification. On the one hand, some see it as 'the next green revolution, capable of saving the world from starvation', while others are extremely concerned about 'unleashed genes having catastrophic effects on the environment'.

Questions to consider

- Keeping in mind that the genetic code is universal, what do you consider to be the differences between genetic modification and standard plant and animal breeding?
- Read about the precautionary principle, which is explained in Subtopic **4.4**. How should the precautionary principle be applied to genetic modification of farmed plants and animals?
- At what level should gene technology be controlled – laboratory, government or international?

Clone a group of genetically identical organisms or a group of cells derived from a single parent cell

Cloning

Cloning happens naturally – identical twins or triplets are a **clone**. Cloning is also very widespread in agriculture and horticulture and has been used for many years to propagate new plants from cuttings (Figure **3.20**), taken from roots, stems or leaves. Many plant species have natural methods of cloning themselves in order to reproduce. Strawberry plants put out runners from which new plants grow, and English elm trees reproduce using suckers, which produce new trees that are a clone of the original plant. These are shown in Figures **3.21** and **3.22**.

1 Cut a shoot from the parent plant using a sharp knife.

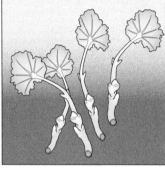
2 Dip the cut end of each shoot into rooting powder, to help new roots form.

3 Place each shoot into a pot of potting compost.

4 Each shoot will soon develop roots, and become a new plant.

Figure 3.20 Diagrams to show the stages of taking cuttings from a geranium plant.

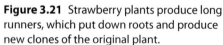

Figure 3.21 Strawberry plants produce long runners, which put down roots and produce new clones of the original plant.

Roots from the parent tree spread out near the surface of the soil.

Stems (suckers) grow from these roots. In time, these can form new trees.

Figure 3.22 English Elm trees do not produce seeds but produce suckers. Clones of the parent tree grow from these.

Animal clones can be produced from embryos after *in vitro* fertilisation. The ball of cells formed as the zygote begins to divide can be separated into several parts in a Petri dish and each part can go on to form a genetically identical embryo. This type of reproductive cloning produces more individual animals with desirable characteristics for farmers and animal breeders.

Clones from differentiated animal cells

Cells from a newly fertilised egg are not differentiated and have not specialised into the different cells they will become. Until recently, cloning an animal from another animal was impossible because the cells in an animal's body had already become nerves, skin, muscles and so on. Differentiated cells have many of their genes switched off, so before they can be used for cloning, the genes have to be switched back on.

The first successful clone made from adult, differentiated animal cells was Dolly the sheep (Figure **3.23**). The breakthrough was made in 1997 by Sir Ian Wilmut and his team at a laboratory at the Roslin Institute in Edinburgh, Scotland, where Dolly was created. Unlike previous clones, Dolly was created from the fusion of an ovum with a nucleus from a mammary gland cell of an adult sheep, creating a genetic replica of the original adult animal. The procedure is known as somatic-cell nuclear transfer because the nucleus used comes from a body, or somatic, cell – in this case a mammary gland cell.

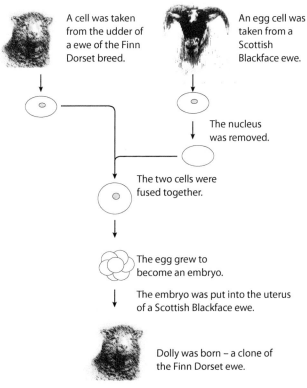

Figure 3.23 Stages in the creation of Dolly, the first successful clone.

Nature of science

Assessing risk in science – GM maize and the monarch butterfly

The monarch butterfly is native to the USA (Figures **3.24** and **3.25**). In the winter, it hibernates along the coast of southern California or in the forests of central Mexico. As spring arrives, thousands of butterflies begin to head northwards in one of the most spectacular insect migrations in the world. By the time the butterflies reach their breeding ground, in the American corn belt, they may have flown as far as 4800 kilometres (3000 miles).

One commonly grown GM crop in this area is maize (corn), which contains a gene from the soil bacterium *Bacillus thuringiensis*. Maize carrying the Bt gene produces its own toxin to kill insect pests so that fewer pesticides need to be sprayed on it to keep it pest free. Controversy about the crop and its possible threat to monarch butterfly populations began in 1999 when a paper in the journal *Nature* suggested that the crop might be killing monarch butterfly caterpillars. Entomologist John Losey suggested that dustings of pollen from Bt maize which fell on milkweed plants (the main food source of the butterflies) growing in a maize field might harm the butterflies. Researchers at Iowa State University said they had found monarch butterfly caterpillars were seven times more likely to die when they ate milkweed plants dusted with pollen from Bt-corn rather than unmodified maize in the laboratory.

This study was discredited in 2001 in a study published in the *Proceedings of the National Academy of Sciences*, which investigated the effects of GM maize on wild butterflies, and demonstrated what happens in the real environment. The team included not only industry-sponsored scientists but also independent researchers. For butterflies to be at risk, they would have to be present in the maize fields at the time that pollen was shed and they would also need to be exposed to enough pollen for it to be harmful. The team found that this was not the case – significant numbers of butterflies were not present in the maize fields at the time when pollen was shed, and therefore massive exposure did not occur in nature.

Monarch butterfly populations are declining but face many threats to their survival, including natural events. For example, in 1995 a snowstorm killed more than 5 million butterflies as they travelled north. Estimates show that in an average year less than 5% of the caterpillars that hatch survive to adulthood. Human activities such as deforestation and development destroy their habitat and pesticides used in maize fields destroy milkweeds, the sole food source for the caterpillars. Nevertheless, headlines in the popular press still claim that 'GM Crops Kill Butterflies'. In fact, butterflies may be safer in a Bt field than they are in a conventional field, where they may be exposed to chemical pesticides that kill not just caterpillars but most other insects as well.

Figure 3.24 Monarch butterfly caterpillar feeding on milkweeds.

Figure 3.25 Monarch butterfly.

Questions to consider

- Why is it important that laboratory-based experiments are supported by observations in the natural world?
- What is the significance of including both independent and industry-sponsored scientists in research into the monarch butterfly?
- How important are stories published in the press to our understanding of scientific issues?

Test yourself

25 Explain the purpose of the polymerase chain reaction.
26 State the characteristic of DNA fragments that causes them to separate during electrophoresis.
27 State **two** uses of DNA profiling.
28 State the characteristic of the genetic code that allows a gene to be transferred between species.
29 State the name of the rings of DNA found in prokaryotic cells that can act as gene vectors.
30 State **two** enzymes used to make a recombinant plasmid.
31 State **two** cell types used to make Dolly the sheep.

Exam-style questions

1   Which of the following statements is correct?

    A Both men and women can be carriers of a sex-linked allele.
    B Mitosis reduces the number of chromosomes in a cell from the diploid to the haploid number.
    C All of the genetic information in an organism is called a genome.
    D The polymerase chain reaction separates fragments of DNA in a gel. [1]

2   The fruit fly, *Drosophila*, has sex chromosomes X and Y, the same as humans. 'White eye' is caused by a sex-linked allele, which is recessive. In the pedigree chart shown, which **two** individuals must be carriers of the white allele?

    A I4 and II3
    B I4 and II2
    C II1 and II2
    D II2 and III2

[1]

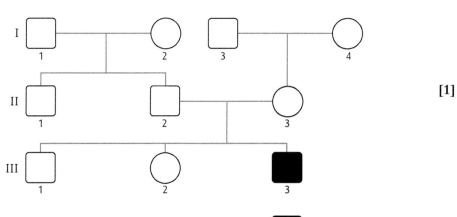

this fly has white eye

3 The karyogram shows:

   A Down syndrome
   B a sex chromosome trisomy
   C a normal human female
   D a normal human male   [1]

4 Which of the following describes the behaviour of chromosomes during prophase I and metaphase II of meiosis?

	Prophase I	Metaphase II
A	Chromosomes undergo supercoiling.	Sister chromatids are separated.
B	Homologous chromosomes pair up together.	Homologous pairs of chromosomes line up on the equator.
C	Homologous chromosomes pair up together.	Chromosomes line up on the equator.
D	The nuclear envelope reforms.	Chromosomes line up on the equator.

[1]

5 The diagram shows a DNA electrophoresis gel. The results are from a single probe showing a DNA profile for a man, a woman and their four children.

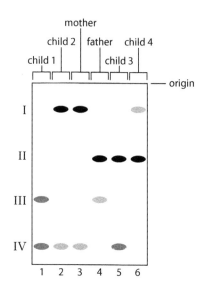

Which child is least likely to be the biological offspring of the father?

   A Child 1
   B Child 2
   C Child 3
   D Child 4   [1]

6 The following is a sequence of bases in a DNA molecule that has been transcribed into an RNA molecule.

CGGTAAGCCTA

Which is the correct sequence of bases in the RNA molecule?

    A CGGTAAGCCTA
    B GCCAUUCGGAU
    C CGGUAAGCCUA
    D GCCUTTCGGAU   [1]

7 a In the normal allele for the β-chain of hemoglobin, there is a triplet on the coding strand of the DNA that is C–T–C. In people suffering from sickle-cell anemia, what has this triplet mutated to? [1]

   b Because of this mutation, one amino acid in the polypeptide chain of the β-subunits in hemoglobin is abnormal. Name the normal amino acid and also the amino acid resulting from the sickle-cell mutation. [1]

   c Explain why this mutation leads to sickle-cell anemia. [3]

8 Ludovica is blood group AB and is expecting a baby with her husband Mikhail who is blood group A. Mikhail's mother was group O. Deduce the possible genotypes and phenotypes of their baby using a Punnett grid. [4]

9 Genetic modification involves the transfer of DNA from one species to another. Discuss the potential benefits and possible harmful effects of genetic modification. Include a named example. [8]

10 Outline how a DNA profile is obtained, including one way in which it has been used. [5]

11 Karyotyping involves arranging the chromosomes of an individual into pairs. Describe one application of this process, including the way in which the chromosomes are obtained. [5]

4 Ecology

Introduction

Almost the entire surface of the Earth – the land, rivers, lakes, seas and oceans – is home to organisms of one kind or another. It has been estimated that there are as many as 10 million different species on Earth and understanding where and how they live and interact is a branch of biology known as ecology. Humans are not the most numerous species on Earth (there are many more bacteria and insects, for example), but humankind is having a disproportionate effect on the world's ecosystems as damage is caused by pollution, rainforest destruction and global warming.

4.1 Species, communities and ecosystems

Learning objectives

You should understand that:
- A species is a group of organisms that have the potential to interbreed to produce fertile offspring.
- If members of a species are in different, separated populations, they may not be able to interbreed.
- Species are defined as autotrophs or heterotrophs depending on their source of food. (Some species are both.)
- Consumers are defined as heterotrophs that feed on living organisms by ingestion.
- Detritivores are defined as heterotrophs that feed on dead organic matter (detritus) by internal digestion.
- Saprotrophs are heterotrophs that obtain nutrients from dead organic matter by external digestion.
- A community is a group of populations living and interacting with each other in a specific area.
- A community interacting with its abiotic environment forms an ecosystem.
- The abiotic environment is the non-living part of an ecosystem, which supplies inorganic materials to autotrophs.
- Nutrient cycling in an ecosystem maintains the supply of inorganic nutrients.
- Ecosystems have the potential to be indefinitely sustainable.

'A true conservationist is a man who knows that the world is not given by his father but borrowed from his children.'

John James Audubon (1785–1851)

The concept of an ecosystem

The idea of the ecosystem provides a convenient way for us to consider and describe the interactions between groups of organisms in their natural environment: an ecosystem is defined as all the groups of organisms living in an area plus the non-living environment with which they interact. One of the most important and recognisable groups in any ecosystem is a species. We define a species as a group of organisms that are able to interbreed and produce fertile offspring. So, lions are a species because they breed with other lions and produce more lions. But if lions are separated from one another by long distances and exist in isolated populations, two individual lions may not actually interbreed even though they remain members of the same species and retain the potential do so.

If species are separated in different populations for long periods of time, over many generations the two groups may evolve and become different from one another. You can find out more about this process on Topic **5**.

Populations of different species that live together in the same ecosystem form a larger group known as a community. A community comprises all the organisms in the ecosystem. So in a pond ecosystem, the community includes populations of fish, frogs, pond snails, water beetles, pond algae and so on.

The community is affected by the abiotic (non-living) factors in the environment. In a pond, these might include the pH of the water, the nature of the underlying rocks or the amount of sunlight. In a terrestrial ecosystem, climate and physical conditions such as availability of water and type of soil are important. These non-living factors in any ecosystem affect the survival and growth of living things found there, as well as their ability to reproduce and their distribution in the ecosystem (Figure **4.1**).

Autotrophs and heterotrophs

Species are divided into types depending on their method of obtaining food. Autotrophs are species that are able to make their own food from basic inorganic materials and this group includes all plants that can photosynthesise. Autotrophs (which means 'self feeding') use light energy to synthesise sugars, amino acids, lipids and vitamins, using simple inorganic substances such as water, carbon dioxide and minerals. Heterotrophs are consumers – species that obtain their food from organic matter. This group includes herbivorous and carnivorous animals, which feed on living organisms. A few organisms, such as *Euglena*, are capable of feeding both autotrophically and heterotrophically (see *Nature of science*, page 140).

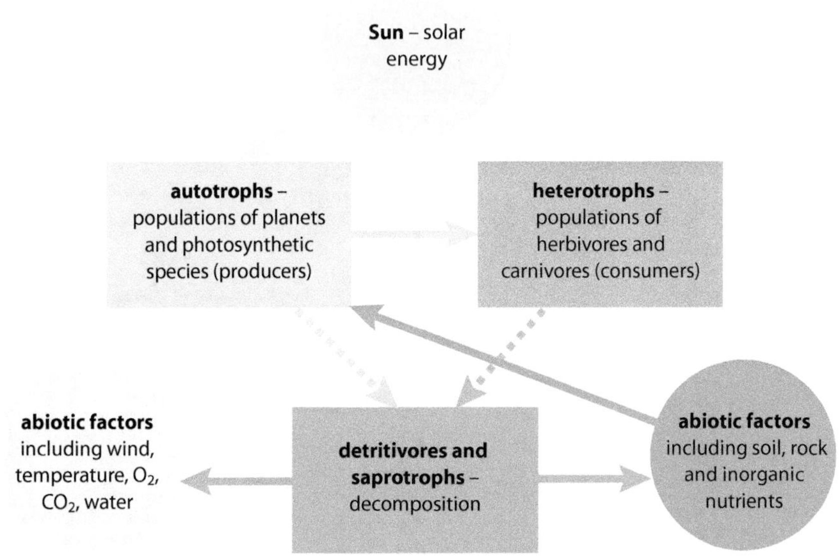

In order to understand the topics in this chapter, it is important to remember the following terms, which are used in the study of ecology.

Species a group of organisms that can interbreed and produce fertile offspring

Population a group of organisms of the same species who live in the same area at the same time

Community a group of populations living and interacting with each other in an area

Ecosystem a community and its abiotic environment

Ecology the study of relationships between living organisms and between organisms and their environment

Trophic level the position of an organism in a food chain

Figure 4.1 The components of an ecosystem.

Mesocosms

A mesocosm is a small-scale, self-sustaining natural system that can be set up in a laboratory to mirror conditions that may occur on a larger scale. Scientists can use a mesocosm to study part of an ecosystem under controlled conditions and draw inferences about how the ecosystem works in the natural environment. Indoor mesocosms such as growth chambers provide experimenters with a way to control variables such as temperature and light so that the effect of environmental change on species and communities can be predicted. Mesocosms have been used to study the feeding habits of fish when in the presence of different proportions of plankton and competitors and also to observe the effect of climate change on carbon dioxide in shallow ponds simulated in cylinder-shaped vessels.

Two important groups of heterotrophs are detritivores and saprotrophs, which feed on dead organic matter. These organisms are vital to the well-being of any ecosystem because of their recycling role. When an organism dies, the remains of its body provide nutrients for detritivores and saprotrophs, which feed on them in different ways. **Detritivores** are organisms such as earthworms, woodlice and millipedes that ingest dead organic matter such as fallen leaves or the bodies of dead animals (Figure 4.2). On the other hand, **saprotrophs** – which include bacteria and fungi – secrete digestive enzymes onto organic matter and then absorb their nutrients in a digested form (Figure 4.3). Saprotrophs are therefore responsible for the decomposition of organic matter and are often referred to as **decomposers**. Saprotrophic bacteria and fungi are the most important decomposers for most ecosystems and are crucial to the recycling of inorganic nutrients such as nitrogen compounds. Recycled inorganic compounds can be re-used over and over again by autotrophs, which in turn continue to grow and provide food for heterotrophic consumers.

If there is enough sunlight to provide energy for the autotrophs to photosynthesise, and the community itself can maintain the recycling of inorganic materials within the abiotic environment, an ecosystem like this has the potential to remain stable and self-sustaining for a long period of time – as long as there are no adverse interferences from outside. Catastrophic natural events and human interference are two factors that can disrupt otherwise stable ecosystems. It is important to remember that sustainable, stable systems are vital for the continued survival of all species, including our own.

Figure 4.2 Millipedes feed on dead leaves on the forest floor and recycle the nutrients they contain via their feces.

Figure 4.3 Fungi secrete enzymes on to the dead material and absorb digested material into their cells.

Sustainable development – international measures to conserve fish stocks

In recent years, there have been several alarming reports on declining fish populations worldwide (Figure **4.4**). In 2003, 29% of open-sea fisheries were in a state of collapse, defined as a decline to less than 10% of their original yield.

To make fishing sustainable in the future, fish populations must be monitored and quotas or closed seasons agreed internationally to reduce fishing during the breeding seasons so that fish numbers can recover after an area has been fished. Some agreements are already in place to control net sizes so that smaller immature fish can be left in the water to breed.

In many parts of the world, bigger vessels, bigger nets and new technology for locating fish are no longer improving catches, simply because there are fewer fish to catch. But where fishing is banned or regulated by international treaties, biodiversity can improve and fish populations may be restored. International cooperation is essential if measures like these are to be successful because neither fish nor trawlers are limited to the jurisdiction of just one country.

The European Union (EU) has enacted several measures to try to ensure a sustainable fishing industry in the North Sea. These include restricting sizes of nets, a ban on drift nets (which catch many different species together) and the imposition of quotas for different fish. Recently, following a public outcry, new rules have been implemented to prevent the return of excess fish to the sea. Until now, fishing boats often had to throw away perfectly good fish if they had exceeded their legally allowed quotas or had caught prohibited species. Sometimes, they also threw back fish that were small, or not popular for human consumption, and therefore of limited value. Some boats discarded as much as two-thirds of their catch. But once bony fish are brought to the surface, their swim bladders are damaged and the fish die – throwing them back does not conserve fish stocks. Fish stocks can only be restored by changing the way the quota rules work and are monitored – so that the number of fish caught and killed is limited, rather than the number brought to shore to sell – and by encouraging selective methods of fishing (for example, by restricting the sizes of nets and banning drift nets). International agreements to support such efforts, such as the new EU rules, are the only way to save fish stocks and maintain sustainable fishing industries for the future.

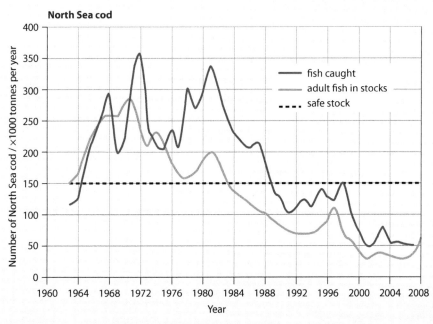

Figure 4.4 Stocks and fishing mortality in North Sea cod between 1963 and 2008. The population of North Sea cod has fallen greatly since the end of the 1960s, almost certainly as a direct result of over-fishing. The horizontal line shows the minimum stock size that has been calculated will allow the cod population to be maintained at a viable level.

Nature of science

Looking for trends and discrepancies – are all plants and algae autotrophic?

Autotrophs or 'self-feeders' are organisms that produce complex organic compounds from simple substances in their environment. In the majority of cases, autotrophs are plants and algae, which use light from the Sun as a source of energy.

One organism that is neither plant nor alga but which was once classified with the plants because it can photosynthesise is *Euglena*. *Euglena* is a unicellular organism with chloroplasts within its cell that enable it to feed autotrophically, like a plant. But *Euglena* can also feed heterotrophically, taking in organic materials as animals do. When *Euglena* was first discovered, organisms were classified into just two kingdoms – the plant kingdom and the animal kingdom – and *Euglena* was impossible to place. Then in the 19th century, Ernst Haeckel added a third kingdom, which he called the Protista, to accommodate organisms like *Euglena* that display characteristics of both plants and animals.

Some plants exist that do not photosynthesise for their entire life cycles. So far about 3000 have been recognised and all are vascular flowering plants (angiosperms). Some of these plants were once classified as saprophytes because they have no chlorophyll but it is now known that they exist in symbiotic relationships with certain fungi, which supply them with nutrients. This relationship is known as **myco-heterotrophy**. Plants such as the bird's nest orchid *(Neottia nidus-avis)* and the ghost pipe *(Monotropa uniflora)* are obligate myco-heterotrophs for part of their life cycle and photosynthetic for the rest. The bird's nest orchid is leafless and grows on the roots of trees, taking nutrients from decaying organic material with the help of fungal mycorrhizae. Its thick fleshy roots resemble a bird's nest and contain the fungus that helps the plant to feed.

A second group of non-photosynthetic plants are called holoparasites. These plants take water, sugars and other nutrients from a host plant through direct contact with it. One example is the dodder (*Cuscata* sp.). Dodder has thin stems with leaves that are reduced to tiny scales. Climbing over other plants, it attaches itself to their stems by extensions called haustoria that penetrate the vascular bundles of the host (Figure 4.5). Once it is established, the root of the dodder dies and it is entirely dependent on its host.

Perhaps these plants should lead us to the conclusion that not all plants are always autotrophs and that there are exceptions to the general rule.

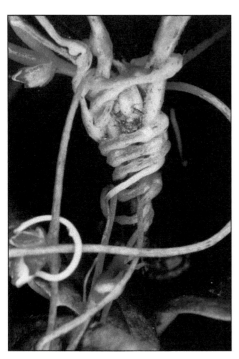

Figure 4.5 Haustoria of dodder, or strangleweed, winding around the host plant.

Test yourself

1 Define the following terms.
   a species
   b ecology
   c population
   d community
   e ecosystem.
2 Distinguish between an autotroph and a heterotroph.

3 Distinguish between consumers, detritivores and saprotrophs.
4 State the name of the feeding group that these organisms belong to:
   a a Peruvian condor feeding on dead sea lions
   b an American alligator feeding on a fish
   c an Australia koala feeding on eucalyptus leaves.

4.2 Energy flow

Food chains

Every organism needs food to survive but eventually it too is eaten. In any ecosystem, there is a hierarchy of feeding relationships that influences how nutrients and energy pass through it. The sequence of organisms that provide food for one another is known as a food chain.

Green plants are autotrophs, which start food chains because they are able to capture light energy from the Sun. Plants are called producers because they 'produce' organic compounds by photosynthesis (Subtopic 2.9). These organic compounds contain chemical energy that has been converted from light energy by the process of photosynthesis. Every other organism in a food chain gets its organic compounds from its food and so is called a heterotroph or consumer.

Consider a hyena eating a cheetah. The cheetah could have eaten an antelope, which in turn had eaten leaves from a plant. Thus, the four organisms are interrelated by their food and form a food chain:

plant → antelope → cheetah → hyena

Every organism fits somewhere in a food chain, and although the organisms that make up the food chain will vary from place to place, almost every food chain starts with a green plant. It may be any part of the plant – the leaves, roots, stems, fruits, flowers or nectar.

Figure **4.6** shows three examples of food chains from different ecosystems. Notice that the arrows in a food chain always point in the direction in which the energy and nutrients flow.

When an organism dies, it provides nutrients for a series of detritivores and saprotrophs, which play a vital role in returning inorganic nutrients to the abiotic environment, from which they are re-used by autotrophic organisms.

Trophic levels

Every ecosystem has a structure that divides organisms into trophic levels on the basis of their food sources. Trophic means 'feeding' and every organism in a food chain is on a particular feeding level.

Green plants are producers and are at the lowest trophic level. Above them come all the consumer levels. The first consumers, or primary consumers, are always herbivores. Any organism above the herbivores will

Learning objectives

You should understand that:
- Most ecosystems rely on light energy from the Sun.
- Light energy is converted to chemical energy in carbon compounds during photosynthesis.
- Chemical energy in carbon compounds passes through food chains as organisms feed.
- Energy released during respiration is converted to heat.
- Heat cannot be converted to any other form of energy by living organisms.
- Heat is lost from ecosystems.
- Energy losses at each trophic level restrict the biomass of higher trophic levels and also the length of food chains.

be a carnivore and these can be listed as secondary consumer, tertiary consumer and so on. A food chain can therefore be summarised as:

producer → primary consumer → secondary consumer → tertiary consumer

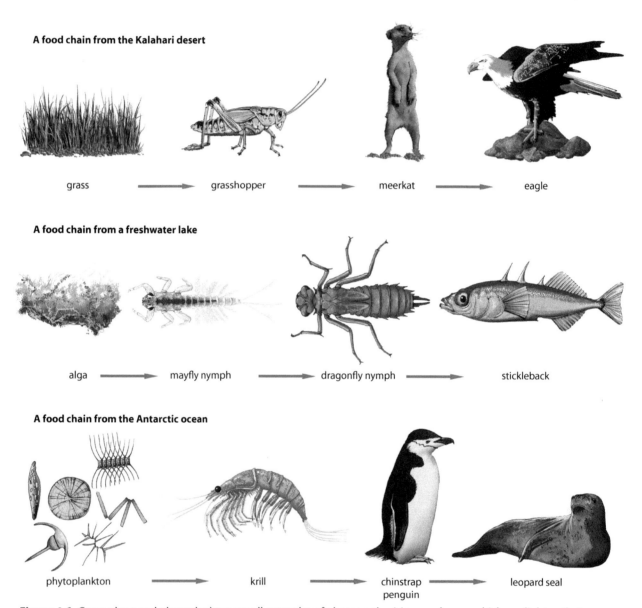

Figure 4.6 Grass, algae and phytoplankton are all examples of photosynthesising producers, which use light as their source of energy. Almost all food chains start with light as the initial source of energy.

Food webs

Few consumers feed on only one source of food. For example, this food chain describes one set of feeding relationships:

grass → beetle → tree creeper → sparrowhawk

But beetles eat a wide range of plants, tree creepers eat other types of insect and sparrowhawks eat other birds. So this food chain could be interlinked with many others. A food web like the one shown in Figure 4.7 shows

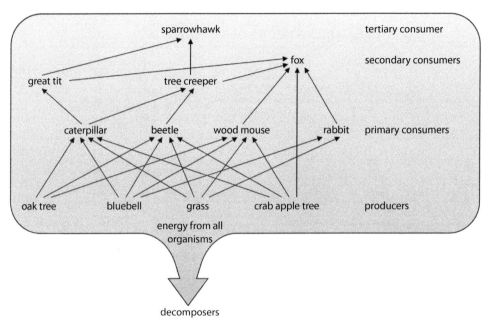

Figure 4.7 A food web in oak woodland.

a much more realistic picture of the feeding relations of the organisms in a habitat. Notice how organisms change trophic levels depending on what they are eating at any particular time. In Figure **4.7**, for example, the fox is a primary consumer when it is eating a crab apple but a secondary consumer when it is eating a woodmouse.

Test yourself

5 In the food web in Figure **4.7**, state the trophic level of:
  a the tree creeper when eating a caterpillar
  b the fox when eating a great tit.

Energy and food chains

Arrows in a food chain show the direction of flow of both the energy and nutrients that keep organisms alive. Energy flow through an ecosystem can be quantified and analysed. These studies reveal that, at each step in the food chain, energy is lost from the chain in various ways. Some is not consumed, some leaves the food chain as waste or when an organism dies, and some is used by living organisms as they respire (Figure **4.8**). In all three cases, the lost energy cannot be passed to the next trophic level.

Consider an area of African savannah where grass, antelopes and cheetahs form a simple food chain.

- **Energy loss 1 – not consumed.** The grass stores energy from photosynthesis but the antelopes only eat some parts of the grass, so they do not consume all the energy it has stored.
- **Energy loss 2 – not assimilated, or lost through death.** The grass that is eaten passes through the digestive system of the antelope but not

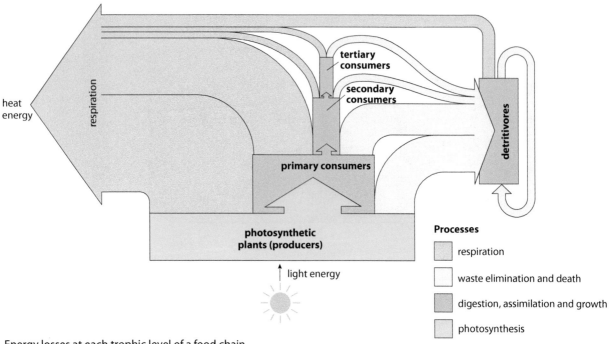

Figure 4.8 Energy losses at each trophic level of a food chain.

all of it is digested and absorbed, so some passes out in the feces. If an antelope dies and it not eaten by a predator, but decayed by detritivores and saprotrophs, the energy in its body is lost to this food chain.
- **Energy loss 3 – cell respiration.** The antelope uses energy to move and to keep its body temperature constant. As a result, some energy is lost to the environment as heat.

The assimilated energy remaining after respiration goes into building the antelope's body and this energy becomes available to the cheetah when it eats the antelope.

Ecologists represent the transfer of energy between trophic levels in diagrams called **energy pyramids**. The width of each of the layers in the pyramid is proportional to the amount of energy it represents. So the antelope → cheetah energy transfer would appear as in Figure 4.9. This section of an energy pyramid shows that only about 10% of energy from the antelope passes to the cheetah and about 90% has been lost.

Energy losses occur at every step in a food chain, as the energy pyramid in Figure 4.10 illustrates. One consequence of this is that the quantity of biomass at each trophic level decreases due to the loss of waste products at each transfer of energy. Every link in the chain results in losses, so that eventually there will be insufficient energy to support any further trophic levels. Most food chains commonly contain between three and five species, and seldom more than six. The energy that enters an ecosystem as light is converted to stored chemical energy and finally lost as heat.

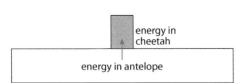

Figure 4.9 A simple energy pyramid for a single energy transfer.

Energy flow and nutrient recycling

All the organic matter from an organism, including everything from living or dead material to waste, is eventually consumed by other organisms.

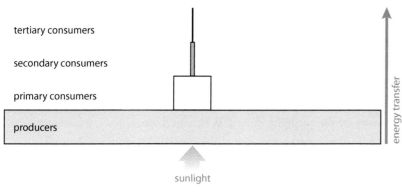

Figure 4.10 A generalised energy pyramid.

All these organisms respire and release energy as heat. All the energy that enters ecosystems as light energy, and is trapped by photosynthesis, will eventually be converted to heat and become unavailable to be used again by living things.

Nutrients, on the other hand, are continually **recycled**. A nitrogen atom may be absorbed as nitrate by a plant root and used to make an amino acid. The amino acid may pass into an animal when the plant material is eaten, and then pass out of the animal's body during excretion. Soil bacteria may convert urea in the excreted material back into nitrate and the cycle begins again.

Decomposers in the soil, the saprotrophic bacteria and fungi, are essential for the recycling of nutrients. You can find out more about nutrient cycles in Subtopic **4.3**.

Feeding the world

Different types of farming and food production can be compared with respect to their trophic levels and efficiency of energy conservation. As arable crops are grown or animals produced, a certain amount of energy is used in the growth of biomass and production of offspring. In a terrestrial ecosystem food is usually harvested at trophic levels 1 and 2 (producers and herbivores). From an energy point of view, farming systems that produce crops are much more efficient than those that produce animals because the energy has passed through only one trophic level.

One hectare of land can produce approximately 7.5 tonnes of wheat (equivalent to 11 000 loaves of bread) or 0.3 tonnes of beef (1200 steaks). Many more people can be provided with food from the wheat rather than the beef.

A steadily rising world population has increased the demand for food and led to environmental problems including loss of forests and soil degradation and, as newly emerging economies worldwide have seen incomes rise, the demand for meat has also increased.

Questions to consider

- Why is arable farming more efficient than farming animals?
- How important are other aspects of animal farms – for example, milk production, wool and leather?
- Do you think there may be a limit to the amount of meat that farmers can produce?

4 ECOLOGY

Nature of science

Using theories to explain natural phenomena – the concept of energy flow and food chains

Plants are the primary source of energy in nearly all ecosystems, but energy capture and photosynthesis are not efficient processes so not all of the energy of the sunlight is used. When herbivores feed, the energy transferred from plant to herbivore is also not 100% efficient. Not all of the plant material is eaten and absorbed, and some energy is lost in movement and respiration. The same is true for carnivores eating prey animals. Only about 10% of the energy in producers is passed to herbivores and a similar low percentage of energy is passed from herbivores to carnivores (Figure **4.11**).

All along a food chain or food web, energy is lost at each trophic level through respiration and waste. This is why ecosystems rarely contain more than four or five trophic levels. There is simply not enough energy to support another level and understanding the concept of energy flow explains why food chains are limited in length.

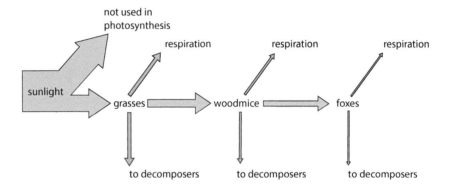

Figure 4.11 Energy losses in a food chain (not to scale). Energy is measured in kJ m^{-2} y^{-1}. Only a small percentage of the energy in each level is transferred to the next.

Test yourself

6 Explain what the arrows represent in a food chain.
7 Describe what is meant by a food web.
8 State the initial energy source for most food chains.
9 Define 'trophic level'.
10 List the **three** ways energy is lost when moving from one trophic level to the next.
11 The leaves of a tree store 20 000 J m^{-2} y^{-1} of energy. Estimate the amount of energy stored by the caterpillars that feed on the leaves.

4.3 Carbon recycling

The carbon cycle

Carbon is one of the most important elements that are recycled in an ecosystem (Figure **4.12**). Inorganic carbon dioxide in the atmosphere is trapped or 'fixed' as organic carbon compounds during photosynthesis. Carbon dioxide gas needed for photosynthesis passes by diffusion from the atmosphere into land-dwelling autotrophs and dissolved carbon dioxide diffuses from water into aquatic organisms. Some of this carbon is soon returned to the atmosphere or the water as the plants respire. The other steps in the carbon cycle follow the same path as food chains. As herbivores eat plants, and carnivores eat herbivores, the carbon compounds move from plants to animals. Respiration by any organism in this sequence returns carbon to the environment as carbon dioxide. When a plant or animal dies, carbon compounds in their bodies provide nutrition for detritivores and saprotrophs and may also be respired, returning carbon dioxide to the atmosphere.

Methane, peat and fossil fuels

In some conditions, plants and animals do not decay completely when they die and organic carbon in their bodies may not be released directly as carbon dioxide as they decompose. In wetlands, where the soil is waterlogged and the concentration of oxygen is very low, organic material is broken down by anaerobic **methanogenic** bacteria (Achaeans), which produce **methane** as a byproduct. Peat bogs represent the largest natural source of atmospheric methane, a greenhouse gas many times more harmful than carbon dioxide. A large proportion of this methane is recycled by methane-eating bacteria (methanotrophs) in the peat bog, which form a symbiotic relationship with peat-bog moss. The bacteria supply additional carbon dioxide to the moss, which in turn provides the bacteria with a habitat.

Methane is oxidised to carbon dioxide and water released into the atmosphere. Thus, eventually, carbon dioxide rejoins the carbon cycle and contributes to the carbon dioxide in the atmosphere, although some remains in the ground.

Learning objectives

You should understand that:
- In the atmosphere, carbon is present as carbon dioxide gas, while in aquatic environments, carbon is present as dissolved carbon dioxide and hydrogen carbonate ions.
- Carbon dioxide is converted into carbohydrates and other carbon compounds by autotrophs.
- Carbon dioxide enters the cells of autotrophs from the air or water by the process of diffusion.
- Carbon dioxide produced during respiration diffuses out of organisms into the surrounding air or water.
- In anaerobic conditions, methane is produced from organic matter by methanogenic bacteria (Achaeans), and diffuses into the atmosphere or accumulates in the ground.
- In the atmosphere, methane is oxidised to carbon dioxide and water.
- When organic matter is not fully decomposed because of acidic or anaerobic conditions in waterlogged soils, peat forms.
- Oil, gas and coal are formed from partially decomposed organic matter from past geological eras. Oil and gas accumulate in porous rocks.
- Combustion of biomass and fossilised organic matter produces carbon dioxide.
- Some animals, such as corals and molluscs, have parts composed of calcium carbonate and they can become fossilised in limestone.

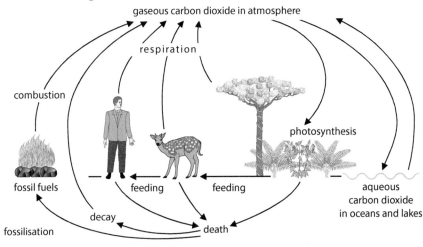

Figure 4.12 The carbon cycle.

4 ECOLOGY

Methane as a greenhouse gas

Methane has a greater potential to cause global warming than carbon dioxide (Subtopic **4.4**) but it is present in much smaller amounts in the atmosphere. It is produced naturally by termites as they digest their food and significant amounts are also released from the oceans. Human activity has increased the release of methane during the extraction of gas, oil and coal, and in farming. Cattle and other ruminants produce methane as part of their digestive process while waste treatment plants and rice fields are also additional sources of methane.

Peat is also produced in wetlands where partly decayed vegetation accumulates and flooding prevents oxygen flow in the soil, so that anaerobic and sometimes acidic conditions persist. For peat to form, the production of biomass must be greater than its chemical breakdown. The first stage in peat formation takes place when water levels are low and aerobic micro-organisms act on decaying vegetation in surface layers. As this layer becomes covered by further layers of vegetation and thus subjected to anaerobic conditions in wetter, deeper layers, it becomes preserved and changes very little with time. Depending on the local conditions, the types of peat from different areas can be quite different in their degree of decomposition. Areas where peat is found also have specific kinds of plants, such as heather, sphagnum moss and sedges. Since organic matter accumulates over thousands of years, peat deposits also provide records of past vegetation and climates.

Peat is harvested as a fuel in some parts of the world as it can be compressed into bricks and unlike coal, burns without producing smoke (Figure **4.13**). But peat is not classified as a renewable energy resource because its rate of extraction is far greater than its very slow rate of formation which can take between 1000 and 5000 years.

Figure 4.13 Peat is cut to be used as a fuel. Some farmers use it as an additive to improve soil structure.

Some partially decomposed organic matter from past geological eras has become compressed and fossilised in a process that has taken millions of years. Conditions on Earth at the time prevented decomposers from feeding on this material and over time it was fossilised to become **fossil fuels**. Vast coal, oil and natural gas deposits have been formed deep under the ground where they contain reserves of carbon that are locked in and excluded from the carbon cycle for very long periods of time. The carbon trapped in these fuels cannot return to the atmosphere unless the fuels are burned, when carbon in them combines with oxygen in the air to form carbon dioxide. Over a very long period of geological time, fossil fuel formation has gradually lowered the carbon dioxide level of the Earth's atmosphere, but in more recent times this balance has been upset. As people burn wood, peat, coal, oil or gas, the carbon molecules locked up in them for thousands or millions of years are released back into the atmosphere as carbon dioxide. As carbon dioxide re-enters the carbon cycle it has the potential to cause global warming (Subtopic **4.4**).

Carbon flux

Carbon flux is defined as the flow of carbon from one 'carbon pool' to another. It is the net difference between the carbon removal and the carbon addition. For the Earth's atmosphere, carbon is removed by plant growth, mineral formation and dissolving in the ocean, while it is added through respiration, fossil-fuel burning and volcanic activity. Scientists monitor carbon flux in order to build up a picture of changes and disturbances in the balance of the carbon cycle (Figure **4.14**).

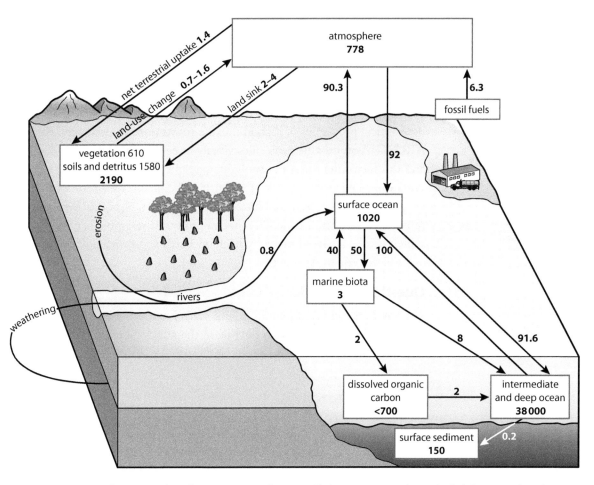

Figure 4.14 Diagram showing carbon fluxes in a natural system. Carbon stores are shown in Gt (gigatonnes) and carbon fluxes are shown in gigatonnes per year (Gt C yr^{-1}).

Nature of science

Obtaining evidence – how is reliable data on carbon dioxide and methane concentrations obtained?

The level of carbon dioxide in the atmosphere has been increasing since the end of the 19th century. The first measurements of the gas were made by Charles Keeling (1928–2005) who took careful measurement of samples from Mauna Loa in Hawaii and the Antarctica in 1957 and 1958. Since then records of the gas have been kept continuously and the

Fossil formation

Fossils are the remains of organisms that died millions of years ago but which have not decayed. The formation of coal began when ancient forests died and became compressed in anaerobic conditions. Oil and gas were formed from the bodies of dead marine organisms that sank to the bottom of the ancient oceans, where they were heated and compressed deep below the surface for millions of years. Limestone is formed from the remains of marine organisms such as corals and molluscs. It too provides a reserve of carbon in the form of calcium carbonate. Calcium-containing shells, bones and hard parts of these organisms were trapped in sediments and have become fossilised in limestone.

For more information on the Orbiting Carbon Observatory, visit www.nasa.gov/oco.

Keeling curve (named after Charles Keeling) and additions to it are well recognised (Figure **4.15**). Today, many different types of data are collected about the Earth's climate (Figure **4.16**) and the concentrations of both carbon dioxide and methane are recorded.

Carbon dioxide has been measured using a weather satellite instrument called AIRS (Atmospheric Infrared Sounder) developed by a NASA scientist Moustafa Chahine (1935–2011). AIRS works by measuring the infrared light emitted by carbon dioxide molecules. Gas molecules absorb infrared rays emitted by the Earth's surface and then re-radiate them at a slightly lower energy level. The exact frequencies of the emissions depend on temperatures. About 3 million measurements are taken in this way every day and a computer processes the information that is collected. AIRS focuses on a section of the atmosphere known as the middle troposphere but a second satellite, NASA's Orbiting Carbon Observatory (OCO), collects data from the entire atmospheric column. The OCO uses optical properties of carbon dioxide to measure its presence using an AIRS. Carbon dioxide molecules vibrate at certain frequencies of light and this instrument uses the sunlight as a light source to monitor the vibrations. The satellite circles the Earth and collects light that has passed through the atmosphere and into three spectrometers, two of which respond to carbon dioxide. The spectra produced resemble bar codes, which can be analysed precisely back on Earth and provide data on areas as small as a square kilometre.

Question to consider

- Look at the curve shown in Figure **4.15**. The dotted line shows the trend of increasing carbon dioxide in the atmosphere but what do the peaks and troughs along the line indicate?

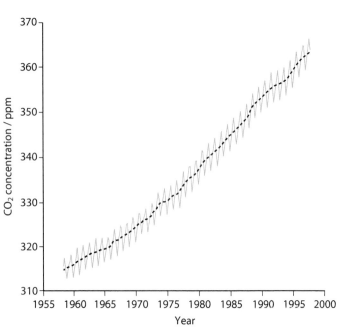

Figure 4.15 Atmospheric carbon dioxide concentration measured at monthly intervals in Hawaii.

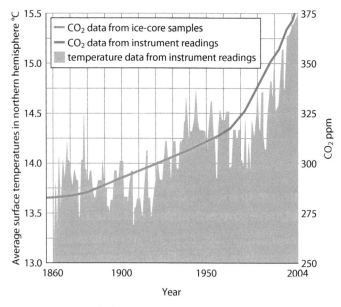

Figure 4.16 Graph showing CO_2 data from different sources and evidence of global warming.

Test yourself

12 State the difference between energy movement and nutrient movement in an ecosystem.
13 State **two** ways in which the formation of peat differs from the formation of coal.

4.4 Climate change

How the greenhouse effect works

Certain gases, the most important of which are carbon dioxide and water vapour, enable the atmosphere to retain heat. Without these gases in the atmosphere, the Earth's temperature would be too low to support life. The warming effect of these gases is known as the **greenhouse effect** because it is caused in a similar way to the warming of a greenhouse.

A greenhouse is made of glass, which allows shortwave radiation from the Sun to pass through it. As the sunlight passes through the glass, the radiation is absorbed, changed into heat – which has a longer wavelength – and re-radiated. Glass is less transparent to these long wavelengths and heat is trapped in the greenhouse, making it warmer inside. So-called 'greenhouse gases' in the Earth's atmosphere (such as carbon dioxide, methane and water vapour) act in a similar way to the greenhouse glass. They trap heat that is radiated from the Earth's surface and keep the Earth at a comfortable temperature for life to exist (Figure **4.17**).

Learning objectives

You should understand that:
- The two most significant greenhouse gases are carbon dioxide and water vapour.
- Methane and oxides of nitrogen are also greenhouse gases but they have less impact.
- Longer wavelength radiation is emitted by the warmed Earth.
- Greenhouse gases re-absorb longer wavelength radiation, retaining heat in the atmosphere and thus reducing heat loss from Earth.
- The impact of a greenhouse gas on global temperatures and climate patterns depends on how well it absorbs long wavelength radiation and on its concentration in the atmosphere.
- Concentrations of carbon dioxide in the atmosphere have been increasing since the start of the Industrial Revolution 200 years ago. Average global temperatures have also been rising over the same period.
- Increases in carbon dioxide concentration in recent years are mainly due to increases in the combustion of fossil fuels.

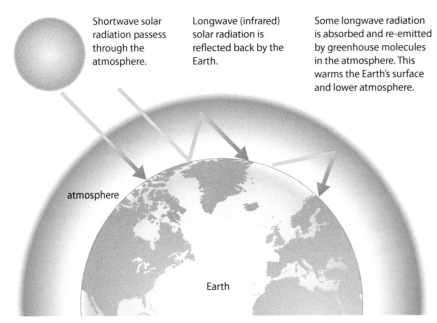

Figure 4.17 Greenhouse gases trap heat, warming the atmosphere.

HFCs as greenhouse gases

Chlorofluorocarbons (CFCs) were used in aerosols and as refrigerants but were found to damage the ozone layer when released into the atmosphere. They are being replaced by hydrofluorcarbons (HFCs), but this is leading to an additional problem because HFCs are greenhouse gases. Note that damage to the ozone layer is *not* a cause of the greenhouse effect or increased global temperatures.

Greenhouse gases, global warming and climate patterns

Carbon dioxide currently forms only 0.04% of the atmospheric gases but it plays a significant part in the greenhouse effect. Other greenhouse gases include water vapour, methane, oxides of nitrogen and fluorocarbons (FCs).

The human population has increased dramatically in recent history, with a consequent increase in demand for energy in industry, transport and homes. Most of this energy demand has been met by burning fossil fuels, mainly oil, coal and gas. Burning fossil fuels releases both carbon dioxide and oxides of nitrogen. This activity has raised the concentration of carbon dioxide in the Earth's atmosphere significantly since the mid-1800s, a period which has coincided with increasing industrialisation (Figure **4.18**), and by more than 20% since 1959 (Figure **4.15** and Table **4.1**). In the tropical regions of the world huge rainforests trap carbon dioxide through photosynthesis and have been important in maintaining the low level of atmospheric carbon dioxide. Humans have upset this balance by deforesting vast areas of forest for agriculture and timber production. Forest destruction has multiple effects, but the most important for the atmosphere are the loss of carbon dioxide uptake by photosynthesis and the increase in carbon dioxide released from the rotting or burnt vegetation.

Methane is another important greenhouse gas. It is produced by human activity when organic waste decomposes in waste tips. It also comes from rice paddies and from cattle farming. More rice is being planted as the human population increases and more cattle are being farmed for meat. Cattle release methane from their digestive systems as they process their food.

The influence of increased concentrations of greenhouse gases has produced changes in global temperatures and climate patterns. Rising levels of greenhouse gases are believed to be causing an enhancement

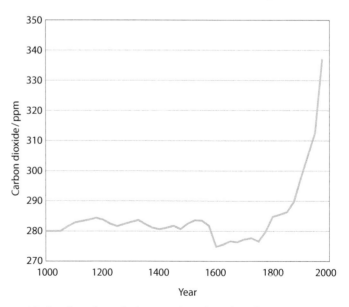

Figure 4.18 Graph to show the increase in carbon dioxide concentration in the atmosphere since the Industrial Revolution.

Year	Carbon dioxide concentration / parts per million (ppm)	% increase from the previous value	% increase from 1959
1959	316		
1965	320	1.3	1.3
1970	326	1.9	3.2
1975	331	1.5	4.7
1980	339	2.4	7.3
1985	346	2.1	9.5
1990	354	2.3	12.0
1995	361	2.0	14.2
2000	369	2.2	16.8
2002	373	1.1	18.0
2004	378	1.3	19.6
2006	382	1.1	20.9
2008	386	1.0	22.2

Table 4.1 The atmospheric carbon dioxide levels monitored at the Mauna Loa laboratory in Hawaii since 1959. Although the percentage increase from the previous year tends to show a downward trend from 2000, the percentage increase from 1959 continues to rise (source: NOAA Earth System Research Laboratory).

of the natural greenhouse effect. Scientists have shown that the Earth is experiencing a rise in average global temperature, known as **global warming** (Figure **4.16**) which is thought to be happening because of this enhanced greenhouse effect. Climatologists are concerned that, as a result of all this activity, humans are adversely affecting our atmosphere.

Some possible results of global warming might be:
- melting of ice caps and glaciers
- a rise in sea levels, causing flooding to low-lying areas
- changes in the pattern of the climate and winds – climate change – leading to changes in ecosystems and the distributions of plants and animals
- increases in photosynthesis as plants receive more carbon dioxide.

Some may argue, for political or other reasons, that it is hard to justify action to reduce the impact of the release of greenhouse gases, but there are many compelling arguments that individuals and societies should consider.

- Global warming has consequences for the entire human race and an international solution is needed to tackle the problems. It is not always the case that those who produce the most greenhouse gases suffer the greatest harm so it is essential that measures to reduce emissions are taken with full international cooperation.
- If industries and farmers in one area invest money to reduce their greenhouse gases while those in other areas do not, an economic imbalance may be created in favour of the more polluting enterprises, who can offer services more cheaply.
- Consumers can be encouraged to use more environmentally friendly goods and services.

Climate change

The term 'global warming' is somewhat simplistic because some areas of the world will be colder as a result of an increase in greenhouse gases. A better term is 'climate change'. Climatologists try to make predictions about changes in weather patterns using computer-generated climate models.

- Scientists can argue that it is better to invest in a sustainable future and prevent further harm.
- It is unethical for one generation to cause harm to future generations by not taking action to address the problem of greenhouse gases.

Increasing carbon dioxide levels and fragile ecosystems

Coral reefs

Coral reefs are fragile ecosystems, which are very sensitive to rising carbon dioxide levels. They are sometimes known as the 'rainforests of the sea' because of their extensive biodiversity. Coral polyps extract calcium from seawater and use it to build the elaborate limestone skeletons that make up a coral reef. As well as large numbers of fish and other species, coloured species of algae live within a reef and give it its colourful appearance. Corals respond to the stress of higher sea temperatures which result from global warming by expelling these algae – an event known as coral bleaching (Figure **4.19**). Some coral can recover from bleaching but often the coral dies, and the entire reef ecosystem disappears.

Ocean acidification occurs when higher CO_2 levels in the atmosphere cause the oceans to absorb more carbon dioxide. Acidic oceans may inhibit the growth of producers and so affect food chains generally, but it is also a direct threat to corals. In a more acidic ocean, corals cannot form their skeletons properly and their growth slows down. It has been estimated that a doubling of atmospheric carbon dioxide will reduce calcification in some corals by as much as 50%. Predicted rises in sea level caused by melting sea ice could also affect some reefs by making the water too deep to allow adequate sunlight to reach them.

Figure 4.19 Bleached coral.

The Arctic

The ecosystems of the Arctic include the tundra, permafrost and the sea ice, a huge floating ice mass surrounding the North Pole (Figure **4.20**). Scientists studying these regions have recorded considerable changes in recent years. Average annual temperatures in the Arctic have increased by approximately double the increase in global average temperatures and the direct impacts of this include melting of sea ice and glaciers.

Trouble in Greenland

The snow in Greenland is not all pristine white. There are patches of brown and black dotted all over the landscape. This is cryconite, a mixture of desert sand, volcanic ash and soot from burnt fossil fuels carried by the wind to Greenland from hundreds or thousands of kilometres away. The more soot there is, the blacker the snow becomes. Black objects absorb the Sun's heat more quickly than white ones do, so the sooty covering makes the underlying snow melt faster. As it melts, the previous year's layer of cryconite is exposed, which makes the layer even blacker. This positive feedback is causing rapid melting of the Greenland ice sheets.

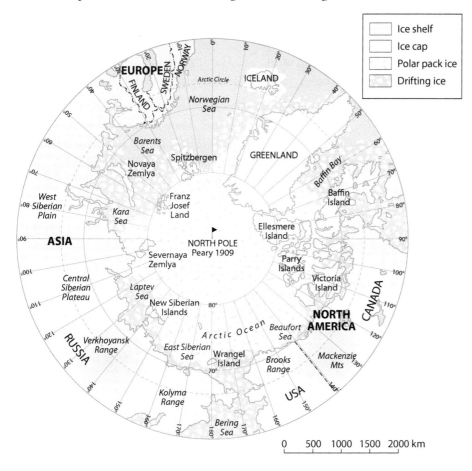

Figure 4.20 Extent of the Arctic region.

Melting sea ice affects many species. Algae, which are important producers in Arctic food chains, are found just beneath the sea ice. As the ice disappears, so do the algae and this affects numerous other organisms that use them as food. Populations of marine mammals, caribou and polar bears are also affected and have already been forced to adapt to changes in their habitats. Polar bear numbers are decreasing at an alarming rate.

Forest and tundra ecosystems are also important features of the Arctic environment. In Alaska, substantial changes in forest life, including increases in insect pests, have been observed as some species from temperate climates are extending their ranges to the north. Detritus trapped in frozen tundra is also released as the ground thaws. The detritus decomposes and the carbon dioxide and methane produced are released into the atmosphere, contributing to rising greenhouse gas levels.

Nature of science

Assessing risk in science – the precautionary principle

The precautionary principle suggests that if the effect of a change caused by humans may be harmful to the environment, action should be taken to prevent it, even though there may not be sufficient data to prove that the activity will cause harm. The precautionary principle is most often applied to the impact of human actions on the environment and human health, as both are complex and the consequences of actions may be unpredictable.

One of the cornerstones of the precautionary principle, and a globally accepted definition, comes from the Earth Summit held in Rio in 1992.

'In order to protect the environment, the precautionary approach shall be widely applied by States according to their capabilities. Where there are threats of serious or irreversible damage, lack of full scientific certainty shall not be used as a reason for postponing cost-effective measures to prevent environmental degradation.'

There are many warning signs to indicate that climate change will have a serious effect on ecosystems across the world. It is clear that global temperatures are increasing and there is a significant probability that this is caused by human activities. It is also likely that weather patterns will alter and cause changes in sea levels and the availability of land for farming. The challenge to governments, industries and consumers is to take action without waiting for definitive scientific proof to be forthcoming.

Question to consider

- Are there any reasons to doubt that climate change is occurring?

Test yourself

14 List **three** groups of organisms in the carbon cycle that transfer carbon dioxide to the atmosphere.
15 State the process in the carbon cycle by which carbon dioxide is fixed as organic carbon-containing compounds.
16 List **three** greenhouse gases.
17 List **five** human activities that are causing an increase in the levels of greenhouse gases in the atmosphere.
18 State **two** ways in which a rise in average global temperature is having an effect on coral ecosystems.

Exam-style questions

1. Which of the following statements is correct?

    A A community is the place where several different species live.
    B Heterotrophs are organisms that feed off organic matter.
    C Decomposer bacteria are detritivores.
    D A habitat is a community and its abiotic environment. [1]

2. Which of the following statements is correct?

    A The precautionary principle states that if the effects of a human-induced change would be very large, those responsible for the change must prove that it will not be harmful before proceeding.
    B The greenhouse gases methane and oxides of nitrogen have increased global temperatures by converting shortwave radiation to longwave radiation.
    C In the carbon cycle, the process of complete decomposition of plant remains stores carbon in the form of fossil fuels.
    D Coral can grow faster in oceans which have more carbon dioxide dissolved in them because they are warmer. [1]

3. Which substances increase the greenhouse effect by the greatest amount?

    A nitrogen and methane
    B sulfur dioxide and nitrogen
    C methane and HFCs
    D oxygen and HFCs [1]

4. The diagram below shows part of a carbon cycle involving only plants. What are the processes labelled 1 and 2?

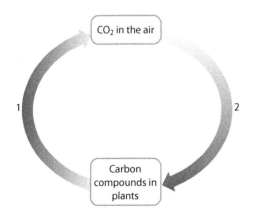

	1	2
A	photosynthesis	combustion
B	decomposition	respiration
C	respiration	decomposition
D	respiration	photosynthesis

5. The total energy in an area of grassland was analysed and found to be $400 \, kJ \, m^{-2} \, y^{-1}$. Construct a labelled pyramid of energy for this grassland for the first three trophic levels, assuming an energy loss of 90% at each level. It is not necessary to draw it to scale. [3]

6 According to data given by Kleiber (1967), a cow may weigh as much as 300 rabbits. The following data are also relevant:

	1 cow	300 rabbits
Total body mass/kg	600	600
Daily food intake (kg hay)	7.5	30.0
Daily heat loss/kJ	82 000	329 000
Daily mass increase/kg	0.9	3.6

   a   i   Calculate the increase in mass per tonne of hay consumed for both the cow and the rabbits (tonne =1000 kg). Show your working. [4]
       ii Explain why this increase in mass is different from the daily increase in mass shown in the table. [2]
   b   Suggest a reason why the rabbits have a higher daily heat loss than the cow. [2]

7 An island off the coast of South America is the breeding ground for sea lions (*Otaria flavescens*), which give birth on the island. The island has a plentiful supply of algae, which supplies food for fish. There are large numbers of animals present some of which occupy precarious positions on the rocks. After giving birth adult sea lions return to the sea to feed on fish.

   a   Draw a food chain to represent the feeding relationships described. [3]
   b   Label the trophic level of each organism you have included. [2]

Abandoned and dead sea lion pups are found on a beach on the mainland. Here, there are many vultures, which feed on the dead pups.

   c   State the feeding group of a vulture which feeds on a dead sea lion pup. [1]

8 Charles Darwin wrote: 'It is interesting to contemplate an entangled bank at the side of a meadow, clothed with plants of many kinds with birds singing on the bushes, various insects flying about and worms crawling through the damp earth ...'

With reference to the 'entangled bank at the side of a meadow':

   a   i   State whether the 'bank' is an ecosystem or a community. [1]
       ii Give a reason for your choice. [1]
   b   State two ways in which birds and plants may interact with one another. [2]
   c   Name one organism that is mentioned which is a secondary consumer. [1]
   d   Outline how nutrients are recycled in this area. [3]

Evolution and biodiversity 5

Introduction

Over long periods of time and many generations, the genetic make-up of species may change as they become adapted to new surroundings or altered conditions. One result of these changes may be the evolution of new varieties and species. There is strong evidence for the evolution of life of Earth, both from the fossil record and from organisms that are alive today. Natural selection provides an explanation of how evolution might have occurred. Classification of organisms helps us to understand their ancestry and our observations of biodiversity, as well as providing further evidence for the mechanism of evolution.

5.1 Evidence for evolution

What is evolution?

Life on Earth is always changing. Just by looking at any group of individuals of any species – whether humans, cats or sunflowers – you can see that individuals are not all the same. For example, the people in Figure **5.1** vary in height, hair colour, skin tone and in many other ways. How do these differences within a species occur? How do different species arise?

Variation within a species is a result of both genetic and environmental factors. We say that selection pressures act on individuals and because of variation, some may be better suited to their environment than others. These are likely to survive longer and have more offspring.

The characteristics of a species are inherited and passed on to succeeding generations. The cumulative change in these heritable characteristics over generations is called evolution. If we go back in time, then existing species must have evolved by divergence from pre-existing ones. All life forms can therefore be said to be linked in one vast family tree with a common origin.

What evidence is there for evolution?

The fossil record

Fossils, such as the one shown in Figure **5.2**, are the preserved remains of organisms that lived a long time ago. They are often formed from the hard parts of organisms, such as shell, bone or wood. Minerals seep into these tissues and become hardened over time. As the living tissue decays, the minerals form a replica that remains behind. Soft tissue can sometimes be preserved in the same way, as can footprints and animal droppings. Most fossils become damaged over time or are crushed through land or sea movement, but some are discovered remarkably well preserved. The earliest

Learning objectives

You should understand that:
- Evolution takes place when heritable characteristics change, over many generations.
- Evidence for evolution comes from the fossil record.
- Artificial selection can result in evolution, as the selective breeding of domesticated animals shows.
- Evolution of homologous structures by adaptive radiation explains similarities in structure with differences in function.
- Populations of a species can diverge and this can eventually lead to the evolution of separate species.
- The continuous variation observed across the geographical range of related populations supports the idea of gradual divergence.

Evolution cumulative change in the heritable characteristics of a population

Figure 5.1 Most of the variation between humans is continuous variation, and is influenced by the environment as well as genes.

Remains of organisms discovered preserved in ice, tar and tree sap have also yielded important information about the evolution of species.

Figure 5.2 A fossil of Archaeopteryx, which is seen as an evolutionary link between reptiles and birds. It looked like a small dinosaur, but had feathers and could fly.

fossils found date from over 3 billion years ago, so the time scale of the fossil record is immense. Most fossils are of species that died out long ago, because they did not adapt to new environmental conditions.

The study of fossils is called palaeontology. Palaeontologists have been collecting and classifying fossils for over 200 years, but they have only been able to date them since the 1940s. Scientists do this by studying the amount of radioactivity in a fossil specimen. Over time the amount of radioactivity decreases because radioactive elements decay. The rate of decay is fixed for each element, so it is possible to date a fossil by measuring the amount of radioactivity present in it. Carbon-14 is used to study material up to 60 000 years old. For older material, other elements are used.

Although the fossil record is incomplete and fossils are very rare, it is possible to show how modern plants and animals might have evolved from previous species that existed hundreds or thousands of millions of years ago. For example, fossil sequences suggest how modern horses may have evolved from earlier species (Figure 5.3). It is important to recognise that we can never say that '**this** species evolved into **that** species', based on a fossil sequence – even when we have many fossils. All that we can say is that they appear to be related – that they probably share a common ancestor. Other species could well have existed too, for which no fossils have ever been found.

A few organisms seem to have changed very little. The horseshoe crab we see today is very similar to fossil specimens a million years old. This would seem to suggest that there has been little selection pressure on these crabs.

Observations of fossils provide evidence that life on Earth changes and that many of the changes occur over millions of years.

Selective breeding

Further evidence for the way evolution might occur comes from observations of selective breeding. People have altered certain domesticated species by breeding selected individuals in a process called **artificial selection**. Plants or animals with favourable characteristics are chosen and bred together, to increase the number of offspring in the next generation that have the favourable characteristics. Those individuals that do not have the desired features are not allowed to breed. People have domesticated and bred plants and animals in this way for thousands of years and, over many generations, this has resulted in the evolution of numerous breeds and varieties, which differ from each other and from the original wild ancestors.

Modern varieties of wheat, barley, rice and potatoes produce higher yields and are more resistant to pests and disease than ever before. Wheat and rice plants are shorter and stronger than varieties of a hundred years ago, so that they are less likely to be damaged by wind and rain and are easier to harvest. The plants of a hundred years ago were also very different from the original grasses that wheat was bred from 10 000 years ago. Many plants are also bred for their appearance, and ornamental varieties have

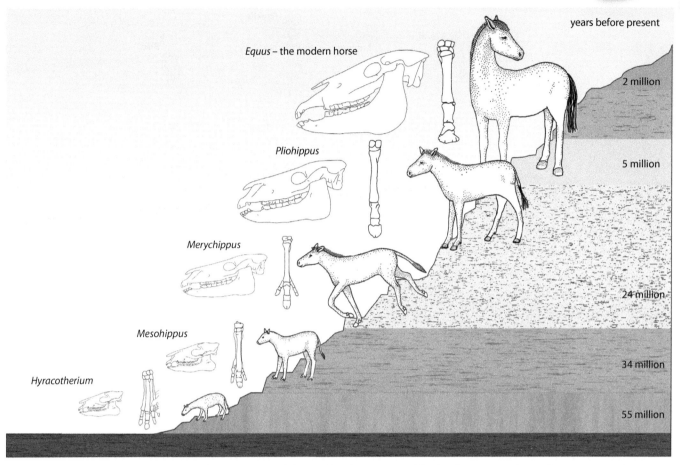

Figure 5.3 Some of the many species of fossil horses, and the modern horse, *Equus*. The fossil sequence shows that, over time, horses have developed single-toed hooves, longer legs and longer faces with larger teeth for grazing.

different petal shapes and colours from the original parent stock.

Animals are chosen and bred by farmers and animal breeders for special characteristics such as high milk yield in a cow, or good-quality wool in a sheep. Individuals with these characteristics are selected to breed, so that more of the next generation have these useful features than if the parents had not been artificially selected (Figure 5.4).

Although the driving force for artificial selection is human intervention, which is quite different from natural evolution, selective or artificial breeding does show that species can change over generations.

Homologous structures

The existence of **homologous structures** provides another strand of evidence for evolution. Homologous structures are anatomical features showing similarities in shape, though not necessarily in function, in different organisms. Their presence suggests that the species possessing them are closely related and derived from a common ancestor. A good example is the vertebrate **pentadactyl** limb. This is found in a large range of animals including bats, whales and humans, as shown in Figure 5.5. In each group, limbs have the same general structure and arrangement of bones but each one is adapted for different uses. Bird wings and reptile

Figure 5.4 Selective breeding of cows over many centuries has produced many breeds including the Guernsey, bred for the production of large quantities of fat-rich milk. Other breeds have been produced with flat backs to facilitate birthing, and longer legs for easier milking.

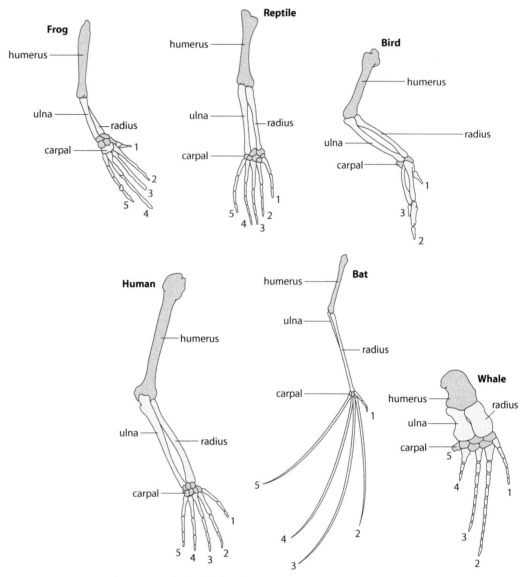

Figure 5.5 The forelimbs of animals with pentadactyl limbs all have a clearly visible humerus, radius, ulna and carpals.

Adaptive radiation is a term used to explain how organisms diverge into a range of new forms from a single common ancestor. It can occur if the environment changes and new sources of food or new habitats become available. The pentadactyl limb demonstrates adaptive radiation in the vertebrates, and Darwin's finches (Subtopic **5.2**) are an example of how one species adapts to exploit new resources.

limbs are also homologous structures. Even though a bird uses its wings for flying and reptiles use their limbs for walking, they share a common arrangement of bones.

Variation and divergence

For almost every species, there is genetic variation between individuals in a population and between populations in different areas further differences may be visible. For example, some groups of birds may have brighter plumage than others, some troupes of howler monkeys may be able to call more loudly than others and some groups of mice may have slightly darker fur than others giving them better camouflage. This **continuous variation** between populations, some very slight and some very obvious, provides support for the proposal that *populations* have the *potential* to diverge from one another and evolve into separate species. If this does happen, there is also the possibility that new species could be formed (Subtopic **5.2**).

Beyond reasonable doubt?

Evolutionary history is a difficult area of science. It is not possible to go back in time or conduct experiments to establish past events or what may have caused them. Nevertheless, in some cases, modern scientific methods can help establish beyond reasonable doubt what has taken place. Science is used to demonstrate the existence of phenomena we cannot observe directly. Many crucial scientific discoveries – such as atoms, electrons, viruses, bacteria, genes and DNA – have been made and accepted using indirect observation and the scientific method of 'inference'. The theory of evolution has been considered using multiple lines of research, including studies of anatomy, biochemistry and palaeontology, which provide empirical data consistent with the theory.

New methodologies that have been used include carbon dating of fossils, the study of DNA, comparison of exons and introns (Subtopic **7.1**) from existing organisms and their fossil ancestors, and the establishment of endosymbiosis with the discovery of rRNA sequences using PCR (Subtopics **1.5** and **2.7**). All of the data from these techniques have provided evidence of relationships between species that are consistent with the relationships suggested by the theory of evolution.

Questions to consider

- Do modern methods establish the theory of evolution beyond reasonable doubt?
- Will it ever be possible to 'prove' that evolution occurred?
- How do the techniques used in science differ from those that a historian might use to establish what has taken place in the past?

Test yourself

1 Define the term 'evolution'.
2 Outline what is meant by a homologous structure.
3 Outline the evidence for natural selection provided by selective breeding.

Learning objectives

You should understand that:
- Variation between individuals is required for natural selection to occur.
- Variation can be caused by mutation, meiosis and sexual reproduction.
- Adaptations are features that make an individual better suited to its environment and way of life.
- More offspring tend to be produced than the limited resources in the environment can support, which means there is a struggle for survival between individuals.
- Better adapted individuals tend to survive and produce more offspring while less well-adapted individuals tend to die or produce fewer offspring.
- When individuals reproduce, they pass on their heritable characteristics to their offspring.
- Natural selection increases the frequency of well-adapted individuals in a population and decreases the frequency of individuals without the adaptations, and thus leads to changes in the species.

Characteristics that an organism acquires during its lifetime, such as large muscles or special skills resulting from training, immunity to a disease or scars on its skin, cannot be passed on to its offspring. Only heritable characteristics, which are determined by genes, can be passed on to the next generation via the parents' gametes.

5.2 Natural selection

A mechanism for evolution

The theory of evolution by means of natural selection was proposed by Charles Darwin and Alfred Wallace. Darwin explained his ideas in a book called *On the Origin of Species by Means of Natural Selection*, published in 1859. The explanation remains a theory because it can never be completely proved but there is an abundance of evidence to support the key ideas, which are based on the following observations and deductions. Some terms we use now were not used by Darwin, who had no knowledge of genes or alleles. However, the fundamental basis of his argument was the same as outlined here.

1 Populations are generally stable despite large numbers of offspring

Organisms are potentially capable of producing large numbers of offspring and far more than the environment can support. Trees can produce thousands of seeds and fish hundreds of eggs. Yet few of these survive to maturity and we rarely see population explosions in an ecosystem.

2 Better adapted individuals have a competitive advantage

Both plants and animals in a growing population will compete for resources. These may be food, territory or even the opportunity to find a mate. In addition, predators and disease will take their toll. This competition will bring about a struggle for survival between the members of a population. Organisms that are well adapted to the conditions will be good at competing and will tend to survive to reproduce, passing on heritable traits to their offspring, while others die.

3 There is heritable variation within species

Different members of the same species are all slightly different and this variation is due to the mechanism of sexual reproduction. The process of meiosis produces haploid gametes and furthermore the genes in the gametes an individual produces may be present in different forms or alleles. When an egg is fertilised, the zygote contains a unique combination of genetic material from its two parents. Sexual reproduction gives an enormous source of genetic diversity, which gives rise to a wide variation among the individuals of a species.

4 Advantageous heritable traits become more frequent over generations

As a result of variation, some members of a population may be better suited (better adapted) to their surroundings than others. They may have keener eyesight, or have better camouflage to avoid predators. These individuals will out-compete others; they will survive better, live longer, and pass on the genes for their advantageous characteristics to more offspring. Gradually, as the process is repeated generation after generation, the proportion of these genes in the population as a whole increases. This is called natural selection, and it occurs as the 'fittest' (best adapted) survive to reproduce.

Natural selection and evolution

Once species have evolved to become well adapted to conditions in a stable environment, natural selection tends to keep things much the same. However, if the environment changes, a population will only survive if some individuals have heritable traits that suit them to the new conditions, and these then become more frequent in the population, because of natural selection. Three examples of how this can happen in a relatively short period of time are the beak adaptations of Galapagos finches after a change in food availability, the response of a moth population to pollution, and the emergence of new strains of bacteria following the introduction of antibiotics.

Darwin's finches

The finches living on the Galapagos Islands (Figure **5.6**), about 900 km off the coast of Ecuador, were important in shaping Darwin's ideas about natural selection. Studies of the birds, now known as Darwin's finches, continue to this day and modern DNA analysis indicates that all 13 species now found on the islands probably evolved from a small flock of about 30 birds that became established there around 1 million years ago.

When the birds first arrived, the Galapagos Islands were probably free of predators and initially the resources were sufficient for all the individuals. As the population grew, the finches started to adapt their feeding habits to avoid competition and as each group selected different

Sexual reproduction promotes variation

Mutations in genes cause new variations to arise, but sexual reproduction also increases variation in a population by forming new combinations of alleles.
- During meiosis, crossing over at prophase I and random assortment in metaphase I produce genetically different gametes (Subtopic **3.3**).
- Different alleles are also brought together at fertilisation, promoting more variation.

In species that reproduce asexually, variation can arise only by mutation.

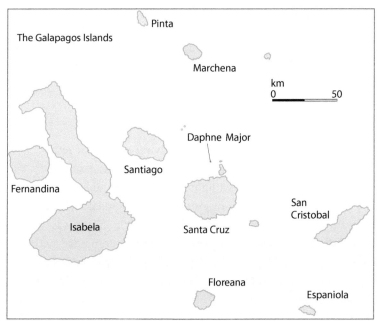

Figure 5.6 The Galapagos Islands.

foods, they developed differently until eventually a number of separate species were established. Today we recognise 13 different species including the cactus finch which has a long beak that reaches into blossoms, the ground finch with a short stubby beak adapted for eating seeds buried under the soil, and the tree finch with a parrot-shaped beak suited for stripping bark to find insects (Figure 5.7). Darwin's finches provide one of the best-known examples of adaptive radiation.

One island that has been extensively studied is Daphne Major, a tiny volcanic island just north of Santa Cruz. Biologists Peter and Rosemary Grant have collected data on the medium ground finch (*Geospiza fortis*) on this island for more than 30 years. In 1977, there was a serious drought on the island and the small seeds that the birds feed on were in short supply. Some of the birds, which had slightly larger beaks, were able to open larger seeds and were able to survive but birds with small beaks died. The following year, the Grants measured the offspring of the survivors and found that their beaks were about 3% larger, on average, than those of earlier pre-drought generations they had studied. They concluded that natural selection had acted on the population and larger-beaked birds had survived and reproduced more successfully. Females also chose their mates based on the size of their beaks so these two factors – ability to find food and ability to attract a mate – seem to be key influences in the eventual development of a new species. The Grants have continued to study the finches of Daphne Major and gathered more data on other factors to provide further evidence of natural selection.

Figure 5.7 Photographs of some of Darwin's finches from the Galapagos Islands: the tree finch, ground finch and cactus finch.

Industrial melanism

The peppered moth (*Biston betularia*) is a night-flying moth that rests during the day on the bark of trees, particularly on branches that are covered with grey–green lichen. It is a light speckled grey, and relies on camouflage against the tree branches to protect it from predatory birds.

In Britain in the mid-19th century, a black form of the moth was noticed (Figure 5.8). The appearance of this new colour coincided with the period of the Industrial Revolution when many factories were built and contributed to growing pollution in the atmosphere. This pollution killed the lichens that grow on the bark of trees, which became blackened with particles of soot.

Figure 5.8 Light and melanic forms of peppered moths on light and dark tree bark.

The colour of the moth is due to a single gene, which can be present in two forms. The common recessive form gives rise to a light speckled colour. The much less common dominant form gives rise to the black, melanic moth.

In the polluted areas, the speckled form was no longer camouflaged on the blackened tree bark, and was easily seen by birds that ate speckled moths. The black moths were better suited to the changed environment as they were camouflaged. Black moths survived and bred and the proportion of black moths with the dominant allele grew in the population.

In 1956, the Clean Air Act became law in Britain and restricted air pollution. Lichen grew back on trees and their bark became lighter. As a consequence, the speckled form of the peppered moth has increased in numbers again in many areas, and the black form has become less frequent.

Antibiotic resistance

Antibiotics are drugs that kill or inhibit bacterial growth. Usually, treating a bacterial infection with an antibiotic kills every invading cell. But, because of variation within the population, there may be a few bacterial cells that can resist the antibiotic. These individuals will survive and reproduce. Because they reproduce asexually, all offspring of a resistant bacterium are also resistant, and will survive in the presence of the antibiotic. In these conditions, the resistant bacteria have enormous selective advantage over the normal susceptible strain, and quickly out-compete them.

Treating a disease caused by resistant strains of bacteria becomes very difficult. Doctors may have to prescribe stronger doses of antibiotic or try different antibiotics to kill the resistant bacteria.

The problem of antibiotic resistance is made more complex because bacteria frequently contain additional genetic information in the form of plasmids, which they can transfer or exchange with other bacteria, even those from different species. Genes for enzymes that can inactivate antibiotics are often found on plasmids, so potentially dangerous bacteria can become resistant to antibiotics by receiving a plasmid from a relatively harmless species. Many bacteria are now resistant to several antibiotics

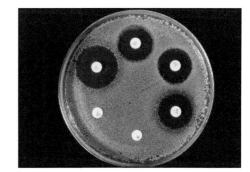

Figure 5.9 The grey–green areas on the agar jelly in this Petri dish are colonies of the bacterium *Escherichia coli*. The white card discs are impregnated with different antibiotics. This strain of *E. coli* is resistant to the antibiotics at the bottom left and has been able to grow right up to the discs, while the other discs have a 'zone of inhibition' around them.

(Figure **5.9**), so pharmaceutical companies are constantly trying to develop new antibiotics to treat infections caused by these multiple resistance forms of bacteria.

So-called 'superbugs', such as MRSA (methicillin-resistant *Staphylococcus aureus*) and *Clostridium difficile,* are bacteria that are resistant to many antibiotics. They have arisen partly as a result of overuse of antibiotics. Antibiotics used incorrectly, or too frequently, help to 'select' the resistant individuals, which then increase in numbers. Patients failing to take a complete course of medication can also encourage the survival of slightly resistant bacteria that might have been killed if the antibiotic had been taken properly.

Scientific theories and evidence

Natural selection is a **theory**. In science, the term 'theory' has a very specific meaning. Scientific theories require a hypothesis that can be tested by gathering evidence. If any piece of evidence does not fit in with the theory, a new hypothesis must be put forward and more scientific evidence gathered. It is important to recognise the difference between a theory and a **dogma**, which is a statement of beliefs that are not subject to scientific tests. Since Darwin's time evidence has been collected to support his theory. With new technologies such as DNA profiling and carbon dating, further evidence continues to accumulate.

Questions to consider
- How much evidence is needed to support a theory?
- What kind of evidence is needed to refute a theory?
- Will it ever be possible to prove that evolution has taken place?

Nature of science

Using theories to explain natural phenomena – the evolution of antibiotic resistance in bacteria

Read the information in the preceding paragraphs and use it to formulate an answer to this question:
- How important is it for scientists to develop theories to explain their proposals to a wider audience?

Test yourself

4 Explain why sexual reproduction is important for evolution.
5 Individuals in a population are often said to be 'struggling for survival'. State the key fact that causes this struggle.
6 If an environment changes, individuals with particular combinations of genes are more likely to survive. State the name given to this phenomenon.
7 Describe **two** examples of evolution in response to environmental change.

5.3 Classification of biodiversity

Learning objectives

You should understand that:
- The binomial system for naming species is used universally by biologists and has been agreed at many scientific meetings.
- Each new species that is discovered is given a two-part name using the binomial system.
- Taxonomists classify species using a hierarchy of eight taxa: domain, kingdom, phylum, class, order, family, genus and species.
- All organisms are classified into one of three domains: Archaea, Eubacteria and Eukarya.
- In a natural classification, each genus and higher taxon contains all the species evolved from one common ancestor.
- Taxonomists reclassify groups of species when evidence shows that species have evolved from different ancestors.
- Natural classification helps to identify new species and predict characteristics shared by species within a group.

Figure 5.10 Carolus Linnaeus, also known as Carl Linnaeus, was a Swedish botanist, physician and zoologist, who laid the foundations for the modern scheme of binomial nomenclature.

Natural (biological) classification attempts to arrange living organisms into groups that enable them to be identified easily and that show evolutionary links between them. The system of classification we use today has its origins in a method devised by the Swedish scientist Carolus Linnaeus (1707–1778).

The binomial system of classification

The classification of living organisms is simply a method of organising them into groups to show similarities and differences between them. More than 2000 years ago, the Greek philosopher Aristotle (384–322 BCE) classified organisms into two groups – plants and animals. This was useful as a starting point, but as the two main groups were sub-divided, problems started to appear. At that time, organisms were seen to be unchanging, so there was no understanding of evolutionary relationships. Many organisms discovered later did not fit into the scheme very well.

Birds were separated into a group defined as 'Feathered animals that can fly' so no place could be found for the flightless cormorant, a bird that does not fly. Bacteria, which were unknown at the time, were not included at all.

In 1735, Carolus Linnaeus (Figure **5.10**) adapted Aristotle's work, and his system forms the foundation of modern taxonomy. Taxonomy is the science of identifying, naming and grouping organisms.

Linnaeus gave each organism two Latin names – the first part of the name is a genus name and the second part a species name. Thus the binomial, or two-part, name for the American grizzly bear is *Ursus americanus* whereas a polar bear is *Ursus maritimus*. Linnaeus used Latin for

By convention, the genus name starts with a capital, while the species does not. Both are written in italic or underlined. Once an organism has been referred to by its full Latin name in a piece of text, further references abbreviate the genus to the first letter only – for example, *U. maritimus* (Table **5.1**).

his names. Latin has long been the language of medicine and science, and it is unchanging. If *Ursus maritimus* is mentioned anywhere in the world, scientists will know that polar bears are being discussed.

- The **genus** part of the name indicates a group of species that are very closely related and share a common ancestor.
- The **species** is usually defined as a group of individuals that are capable of interbreeding to produce fertile offspring.

Linnaeus developed structure in his classification system – for example, he grouped birds into birds of prey, wading birds and perching birds. Although it is possible to group living things in many different ways, over the last 200 years a hierarchical classification system has emerged, through agreement at many scientific meetings, which is now used by biologists everywhere. Modern taxonomists all over the world classify species using a hierarchy of groups called taxa (singular: taxon). There are eight levels to the hierarchy:

- domain
- kingdom
- phylum (plural: phyla)
- class
- order
- family
- genus (plural: genera)
- species.

Two examples of how species are classified are shown in Table 5.1.

Aristotle's original grouping of organisms into just two kingdoms has also been refined. Today the most widely accepted method of classification uses three domains – the Archaea, Eubacteria and Eukarya – divided into six kingdoms.

- Archaea:
  Kingdom Archaebacteria (ancient bacteria) – methanogens, halophiles and thermoacidophiles
- Eubacteria:
  Kingdom Eubacteria (true bacteria) – bacteria and cyanobacteria
- Eukarya:
  Kingdom Plantae – plants
  Kingdom Animalia – animals
  Kingdom Fungi – fungi
  Kingdom Protista – red algae and dinoflagellates

You will notice that viruses are not included in the classification of living things. They are not considered to be living because they cannot reproduce independently and need to take over a host cell to do so. Viruses differ widely in shape and appearance and are usually divided into groups and described by the shape of their protein coat.

	Polar bear	Lemon tree
Domain	Eukarya	Eukarya
Kingdom	Animalia	Plantae
Phylum	Chordata	Angiospermata
Class	Mammalia	Dicoyledoneae
Order	Carnivora	Geraniales
Family	Ursidae	Rutaceae
Genus	*Ursus*	*Citrus*
Species	*maritimus*	*limonia*

Table 5.1 The taxonomic hierarchy for a plant species and for an animal species.

Thermophilic bacteria inhabit hot sulfur springs and hydrothermal vents and survive at temperatures in excess of 70 °C and up to 100 °C in some cases.

Halophilic bacteria live in very salty environments such as the Dead Sea where the Sun has evaporated much of the water. They are also found in salt mines.

Methanogenic bacteria are anaerobes found in the gut of ruminants, as well as in waste landfills, paddy fields and marshland. They produce methane as a waste product of respiration.

Why is classification important?

Natural classifications group together organisms with many of the same characteristics and are predictive, so that by studying the characteristics of an organism it is possible to predict the natural group it belongs to.

Phylogenetic classifications are natural classifications that attempt to identify the evolutionary history of species. The role of taxonomy is to group species that are related by common ancestry. You can find out more about this in Subtopic **5.4**.

A natural classification such as the one devised by Linnaeus is based on identification of homologous structures that indicate a common evolutionary history. If these characteristics are shared by organisms then it is likely that those organisms are related. So the binomial system can be both a natural and a phylogenetic classification.

In summary, then, there are four main reasons why organisms need to be classified:

1. to impose order and organisation on our knowledge
2. to give each species a unique and universal name, because common names vary from place to place around the world
3. to identify evolutionary relationships – if two organisms share particular characteristics then it is likely that they are related to each other, and the more characteristics they share then the closer the relationship
4. to predict characteristics – if members of a particular group share characteristics then it is likely that other newly discovered members of that group will have at least some of those same characteristics.

Classifications change

New species are being found and new discoveries are being made about existing species all the time. Such discoveries may force us to rethink the way we classify living things. For example, in the past, the name 'bacteria' was given to all microscopic single-celled prokaryotes. But recent molecular studies have shown that prokaryotes can be divided into two separate domains, the Eubacteria and the Archaea, which evolved independently from a common ancestor. Molecular biology, genetics and studies of cell ultrastructure have shown that the Archaea and Eukarya (eukaryotes) are in fact more closely related to one another than either is to the Eubacteria. Similar principles are applied to all levels of classification. Taxonomists reclassify organisms when new evidence shows that they have evolved from different ancestral species.

The Kingdom Protista, which is not accepted as a true taxon by all taxonomists, includes eukaryotes that are either unicellular or, where they do form colonies, are not differentiated into tissues. The taxonomy of this group is likely to change because the group is paraphyletic, which means that all of its members may not share a common ancestor. A paraphyletic group is defined by what it does not have and may consist of an ancestor but not all of its descendants.

In 2013, previously unknown bacteria were found living in a deep lake nearly 500 m below the ice in Antarctica. These organisms have new characteristics unlike archaeans and other bacteria. Discoveries like this may mean that further changes will have to be made to our current system of classification in the future.

Carl Woese

Carl Woese introduced the three domains of life – the Archaea, Eubacteria and Eukarya – in 1977. In proposing these new classifications, he separated the prokaryotes into two groups – the Eubacteria and the Archaea – based on discoveries about the structure of their ribosomal RNA. His classification reflects the current view of the ancestry of the three major groups of organisms, but it was not accepted and used widely until the 1990s, when scientists' knowledge about RNA structure increased. (Note that the name Bacteria is now often used for the domain comprising the true bacteria, rather than Woese's original name, Eubacteria.)

Acceptance of a new paradigm

Several respected scientists objected to Woese's division of the prokaryotes into two domains, including microbiologist Salvador Luria (1912–1991) and evolutionary biologist Ernst Mayr (1904–2005).

Questions to consider
- To what extent is conservatism in science desirable?
- How much evidence must be presented before a new theory can be accepted?

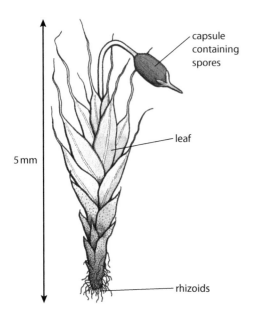

Figure 5.12 A moss, *Grimmia pulvinata*.

The main phyla of the plant kingdom

Members of the plant kingdom are eukaryotic, have cellulose cell walls and carry out photosynthesis. The kingdom is divided into several different phyla based on other similarities (Figure **5.11**).

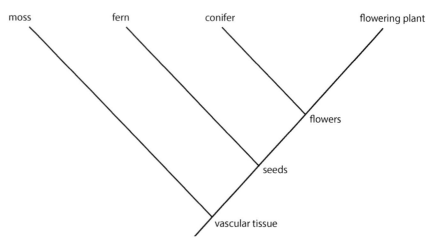

Figure 5.11 A cladogram showing the relationship between the main plant phyla. You can find out more about cladograms in Subtopic **5.4**.

Bryophyta

Plants in this phylum include the mosses (Figure **5.12**) and liverworts. These are the simplest land plants and are probably similar to the first plants to colonise the land some 400 million years ago.
- Bryophyta are usually small and grow in damp places because they have no vascular system to carry water.
- They reproduce by way of spores and these are contained in capsules on small stalks held above the plants.
- They have no roots, just thin filamentous outgrowths called rhizoids. They have no cuticle and absorb water across their whole surface.
- Liverworts have a flattened structure called a thallus but mosses have small simple leaves.

Filicinophyta

This group includes the club mosses, horsetails and ferns (Figure **5.13**).
- Filicinophyta have roots, stems and leaves and possess internal structures.
- Because of the support from woody tissue, some tree ferns grow to over 5 m in height.

Figure 5.13 Club moss hanging from a maple tree in Washington, USA.

- Some have fibrous roots, while others produce an underground stem called a rhizome.
- Like the bryophyta, they also reproduce by producing spores. In the ferns, these are found in clusters called sori on the undersides of the leaves.

Coniferophyta

Coniferophyta include shrubs and trees, such as pine trees, fir and cedar, which are often large and evergreen (Figure **5.14**). Some of the world's largest forests are comprised of conifers.

- Coniferophyta produce pollen rather than spores, often in huge amounts, as conifers are wind-pollinated plants.
- They produce seeds, which are found in cones on the branches.
- Most have needle-like leaves to reduce water loss.

Angiospermophyta

This group includes all the flowering plants, which are pollinated by wind or animals (Figure **5.15**). They range from small low-lying plants to large trees. Many of them are important crop plants.

- Angiospermophyta all have flowers, which produce pollen.
- They all produce seeds, which are associated with a fruit or nut.

Figure 5.14 A western white pine tree (*Pinus monticola*) in the Sierra Nevada, USA.

Some phyla of the animal kingdom

Organisms in the animal kingdom are characterised by being able to move and getting their nutrition by eating plants, other animals or both. Animals are divided into two groups – those that have a backbone (vertebrates) and those that do not (invertebrates).

Porifera

This group contains the sponges (Figure **5.16**). They have different types of cell, but no real organisation into tissues and no clear symmetry. All

Figure 5.16 A giant barrel sponge (*Xestospongia testudinaria*) in the Sulawesi Sea of Indonesia is around 10–20 cm in diameter and 10–20 cm tall.

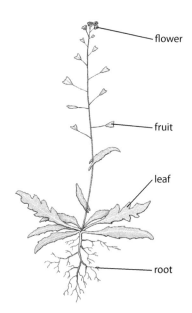

Figure 5.15 Shepherd's purse, *Capsella bursa-pastoris* – an example of a flowering plant.

sponges are aquatic and many produce a skeleton of calcium carbonate or silicon. They pump water through the numerous pores in the body wall and filter out food. Sponges have no nerves or muscular tissue.

Cnidaria

These are the sea anemones, corals and jellyfish (Figure **5.17**). Almost all are marine and have cells organised into tissues in two body layers. They feed on other animals by stinging them with special cells called nematocysts and trapping them in their tentacles. They have a mouth to take in food and use the same opening to get rid of waste. The box jellyfish and the Portuguese man-of-war are two of the most venomous animals on Earth.

Figure 5.17 *Dendrophyllia* coral polyps in the Red Sea. These polyps are 2–4 cm in diameter.

Platyhelminthes

These have three layers of cells and have a body cavity with a mouth and an anus. Some are free living in water while others are parasites, living inside other organisms. They have a flattened appearance, hence their common name 'flatworms' (Figure **5.18**). Most flatworms are small, but the tapeworm found in the intestines of animals may grow to several metres long.

Annelida

This group, known as the 'segmented worms', contains lugworms, earthworms and leeches. Some are aquatic, living in rivers, estuaries and mud, and others inhabit soil. All annelids have bodies that are divided into sections called segments. All of them have a simple gut with a mouth at one end and an anus at the other. Earthworms are important in agriculture because their burrowing aerates the soil and brings down organic matter from the surface, which helps to fertilise it.

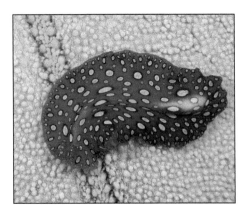

Figure 5.18 The Hawaiian spotted flatworm (*Pseudobiceros* sp.) is about 5 cm long.

Mollusca

This is the second largest animal phylum, containing over 80 000 species. The group includes small organisms like slugs and snails, as well as large marine creatures like the giant squid and octopuses (Figure **5.19**). Many produce an outer shell of calcium carbonate for protection.

Arthropoda

The arthropods comprise the largest animal phylum, and include aquatic animals such as the crustaceans (the crabs and lobsters), and terrestrial animals such as insects, spiders and scorpions. All have an **exoskeleton** made of chitin. They have segmented bodies and jointed limbs for walking, swimming, feeding or sensing. An exoskeleton places a restriction on their size: arthropods are never very big because they must shed their exoskeleton and produce a new, larger one in order to grow. The largest arthropod is the Japanese spider crab, which can be 4 m long. These crabs are marine so water gives their bodies buoyancy, enabling them to move. Well over 1 million arthropods are known and it is estimated that there may be at least as many more that have not yet been identified.

Figure 5.19 The giant octopus (*Enteroctopus* sp.) is one of the largest invertebrates and can be up to 5 m long.

Chordata

The phylum Chordata includes humans and other vertebrates, but not all chordates are vertebrates. The features shared by all chordates are:

- a dorsal nerve cord, which is a bundle of nerve fibres connecting the brain to the organs and muscles
- a notochord, which is a rod of cartilage supporting the nerve cord
- a post-anal tail
- pharyngeal slits, which are openings connecting the inside of the pharynx (throat) to the outside. These may be used as gills.

All chordates have these four features at some stage in their lives but, in the case of most vertebrates, they may only be seen in the embryo. The phylum includes the vertebrates (mammals, birds, amphibians, reptiles and fish) as wells the tunicates (sea squirts and salps) and the cephalochordates (lancelets). About half the phylum is made up of the bony fish (Class Osteichthyes).

Vertebrates

Vertebrate groups can be quickly identified from their external features if their skin and method of breathing are examined (Figure **5.20**). Another important feature is their method of reproduction. The young of mammals develop inside the body of the mother and are fed with milk from mammary glands after they are born. Birds lay hard-shelled eggs, while the eggs of reptiles are covered by a leathery shell. Bird and reptile eggs contain nutrients for the development of their embryos. Both fish and amphibians must reproduce in water – their eggs are released into the water and fertilised outside the female's body.

Designing a dichotomous key

A **dichotomous key** is a series of steps, each involving a decision, which can be used to identify unknown organisms. The key prompts us to decide, through careful observation, whether or not a specimen displays particular visible features, and allows us to distinguish between specimens on this basis.

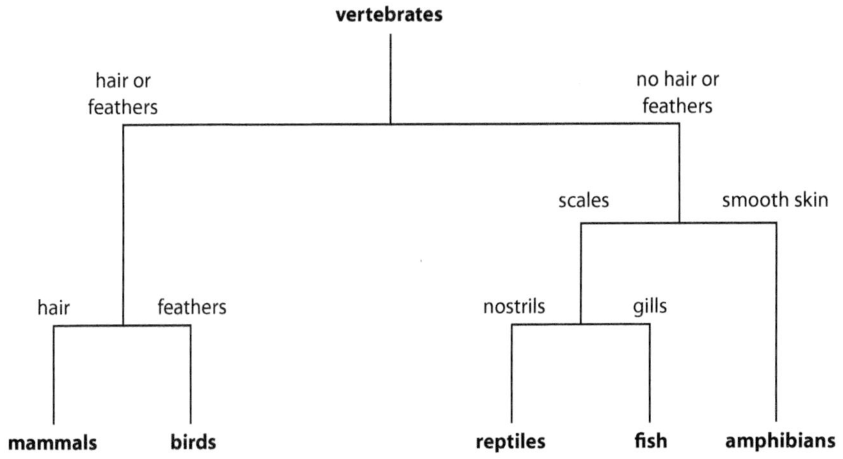

The Platyhelminthes, Annelida, Mollusca, Arthropoda and Chordata are animals that are 'bilaterally symmetrical'. They have a definite front and back end, and a dorsal (back) and ventral (belly) side.

All vertebrates have cartilage in addition to, or instead of, bone. Cartilage can either be flexible, like the cartilage in the human nose and ears, or hard and firm, like the cartilage in the larynx (voicebox). Cartilage also covers surfaces of bones at joints to ease movement. Calcified cartilage makes up the vertebrae and teeth of sharks. This is not true bone as it is dead material, whereas bone is a living tissue.

Figure 5.20 A simple key like this can be used to identify the vertebrate groups.

Constructing a dichotomous key

When constructing a key to identify organisms such as those shown in Figure **5.21**, first examine each specimen in the set carefully, and choose a characteristic that is present in about half of the individuals and absent in the others.

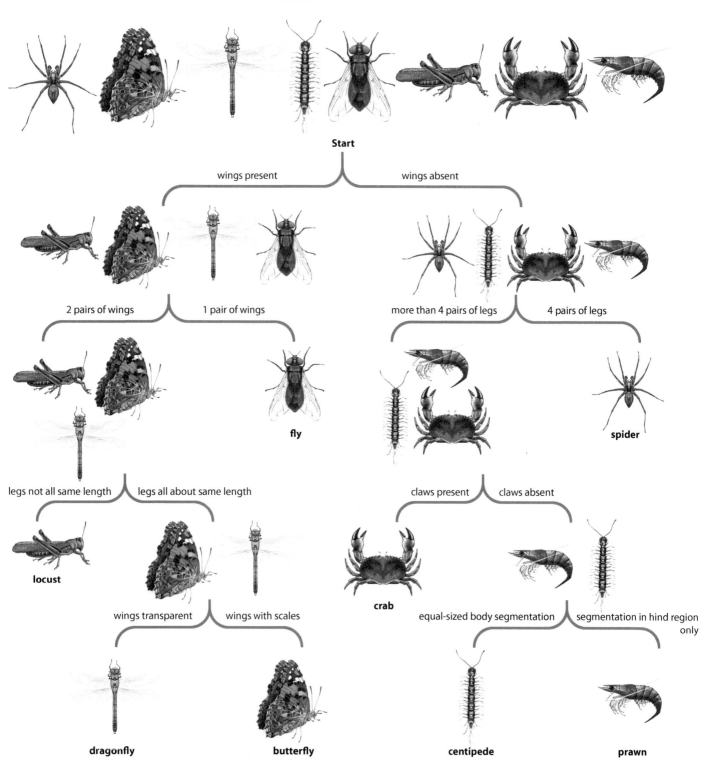

Figure 5.21 A dichotomous tree diagram distinguishing eight organisms.

For example, the presence of wings could be this first distinguishing characteristic, which effectively divides the specimens into two smaller groups.

Now for each group, another diagnostic feature must be chosen whose presence or absence divides the specimens into two further groups. A branching tree diagram can be constructed, as shown in Figure **5.21**, progressively dividing the specimens into smaller and smaller groups, until at the end of each branch a single individual is identified.

Finally, the tree diagram is 'translated' into a written key, in which the branch points are expressed as alternative statements. Each alternative either names the identified specimen or leads the user to a subsequent pair of statements, until an identification is reached. A well-written key is composed of a series of questions or steps, such that an organism that is being studied can only be placed in one of two groups. The style of the questions is therefore very important in the design of a good key.

So, for example, the dichotomous key arising from the tree diagram in Figure **5.21** would be as follows.

1	Wings present	go to 2
	No wings	go to 5
2	Two pairs of wings	go to 3
	One pair of wings	fly
3	Legs all approximately the same length	go to 4
	Hind pair of legs much longer than front two pairs	locust
4	Wings covered in scales	butterfly
	Wings transparent, not covered in scales	dragonfly
5	Four pairs of legs	spider
	More than four pairs of legs	go to 6
6	Pair of claws present	crab
	No claws	go to 7
7	Body clearly divided into equal-sized segments	centipede
	Body in two regions, segments only clear on hind region	prawn

Nature of science

Cooperation and collaboration – an international naming system

Local names for different species can cause confusion. What do you think is being described here: armadillo bug, cafner, wood bug, butchy boy, gamersow, chiggley pig, sow bug, chuggypig and pill bug?

All these terms are local names for the woodlouse *Porcellio scaber* or its relative *Armadillidium vulgare* and are used in different parts of Europe and North America.

Cooperation and collaboration between international scientists provided an agreed binomial name for the woodlouse so that wherever these organisms are studied, information about them can be attributed to the correct species.

Test yourself

8 List in order the levels in the hierarchy of taxa.
9 State the **two** names from the hierarchy of taxa that are used in the binomial system.
10 Identify the group of plants that is characterised by producing pollen and having seeds in cones.
11 State which group of plants is characterised by producing spores, having no root system and no cuticle.
12 State which group of animals is characterised by having jointed limbs and an exoskeleton.
13 State which group of animals is characterised by having segmented bodies with a mouth at one end and an anus at the other.
14 If you were making a dichotomous key to identify leaves, explain why the question 'Is the leaf large?' would not be useful.

Learning objectives

You should understand that:
- A group of organisms that have evolved from a common ancestor is called a clade.
- Evidence used to place a species in a clade can come from base sequences of genes or corresponding amino acid sequences in proteins.
- Differences in base sequences accumulate gradually so there is a positive correlation between the number of differences between the sequence of a gene in two species and the time since they diverged from a common ancestor.
- Characteristics (traits) can be analogous or homologous.
- A cladogram is a tree diagram that shows the most likely sequence of divergence of clades.
- Evidence from cladistics has shown that the classification of some groups based on their structure does not correspond with their evolutionary origins.

5.4 Cladistics

The universality of DNA and protein structures

Despite the incredible complexity of life, the building components of living organisms are not only simple in structure but are also universal.
- All living organisms use DNA built from the same four bases to store their genetic information and most use the same triplet code during translation. The few exceptions include mitochondria, chloroplasts and a group of bacteria.
- Proteins are built up from amino acids and living organisms make use of the same 20. In most cases, if a gene from one organism is transferred into another, it will produce the same polypeptide (if the introns have been removed from it – Subtopic **7.1**).

These facts indicate a common origin of life and provide evidence to support the view that all organisms have evolved from a common ancestor. Study of the genetic code and amino acids of an organism can provide evidence that links it to its close relatives and enables us to build up diagrams called **cladograms**, which show how species are related to one another in **clades**.

Clades and cladistics

Cladistics is a method of classification that groups organisms together according to the characteristics that have evolved most recently. Diagrams called **cladograms** divide groups into separate branches known as **clades** (Figure **5.22** and **5.28**).

One branch ends in a group that has characteristics the other group does not share. A clade contains the most recent common ancestor of the group and its descendants, so a clade contains all the organisms that have evolved from a common ancestor.

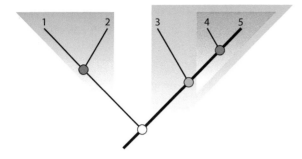

Figure 5.22 A cladogram with four clades.

Clade a group of organisms, both living and extinct, that includes an ancestor and all the descendants of that ancestor

Cladistics a method of classifying organisms using cladograms to analyse a range of their characteristics

Cladogram a diagram that shows species' evolutionary relationships to one another

Figure 5.22 shows five organisms forming part of an evolutionary tree.
- Organisms 1, 2, 3, 4 and 5 belong to the yellow clade.
- Organisms 1 and 2 belong to the blue clade.
- Organisms 3, 4 and 5 belong to the green clade.
- Organisms 4 and 5 belong to the red clade.
- The common ancestor for each clade is shown by the coloured spot at the branch point, or **node**.

Why do biologists need cladistics?

There are three important reasons for using cladistics to organise and discuss organisms.
- It is useful for creating systems of classification so that biologists can communicate their ideas about species and the history of life.
- Cladograms are used to predict the properties of organisms. A cladogram is a model that not only describes what has been observed but also predicts what might not yet have been observed.
- Cladistics can help to explain and clarify mechanisms of evolution by looking at similarities between the DNA and proteins of different species.

Finding evidence for clades and constructing cladograms

Phylogenetics is the study of how closely related organisms are and it is used to establish clades and construct cladograms. The modern approach is to use molecular phylogenetics, which examines the sequences of DNA bases or of amino acids in the polypeptides of different species to establish the evolutionary history of a group of organisms. Species that are the most genetically similar are likely to be more closely related. Genetic changes are brought about by mutation and, provided a mutation is not harmful, it will be retained within the genome. Differences in DNA accumulate over time at an approximately even rate so that the number of differences between genomes (or the polypeptides that they specify) can be used as an approximate evolutionary clock. This information can tell us how far back in time species split from their common ancestor. A greater number of differences in a polypeptide indicates that there has been more time for DNA mutations to accumulate than if the number is smaller. There is a positive correlation between the number of differences between two species and the time they evolved from a common ancestor.

Evidence from amino acids

We can expect that related organisms will have the same molecules carrying out particular functions and that these molecules will have similar structures. So by comparing proteins in different groups of organisms and checking them for similarities in amino acid sequences, it is possible to trace their ancestry. Chlorophyll, hemoglobin, insulin and cytochrome c, which are found in many different species, have all been studied in this way. Cytochrome c is found in the electron transport chain in mitochondria, where it plays a vital role in cell respiration. Its primary structure contains between 100 and 111 amino acids and the sequence has been determined for a great many plants and animals.

Below are the amino acid sequences of corresponding parts of the cytochrome c molecule from five animal species. Each letter represents one amino acid. Humans and chimpanzees have identical molecules, indicating that the two species are closely related. There is only one difference (shown in red) between the human cytochrome c and that of a rhesus monkey but rabbits and mice have nine differences when compared with humans, which indicates they are less closely related. This biochemical evidence supports the classification of the animals that has been made from morphological observations.

Human:

mgdvekgkki fimkcsqcht vekggkhktg pnlhglfgrk tgqapgysyt aanknkgiiw gedtlmeyle npkkyipgtk mifvgikkke eradliaylk katne

Chimpanzee:

mgdvekgkki fimkcsqcht vekggkhktg pnlhglfgrk tgqapgysyt aanknkgiiw gedtlmeyle npkkyipgtk mifvgikkke eradliaylk katne

Rhesus monkey:

mgdvekgkki fimkcsqcht vekggkhktg pnlhglfgrk tgqapgysyt aanknkgitw gedtlmeyle npkkyipgtk mifvgikkke eradliaylk katne

Rabbit:

mgdvekgkki fvqkcaqcht vekggkhktg pnlhglfgrk tgqavgfsyt danknkgitw gedtlmeyle npkkyipgtk mifagikkkd eradliaylk katne

Mouse:

mgdvekgkki fvqkcaqcht vekggkhktg pnlhglfgrk tgqaagfsyt danknkgitw gedtlmeyle npkkyipgtk mifagikkkg eradliaylk katne

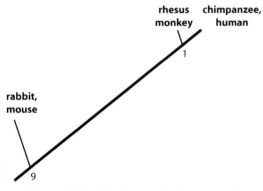

Figure 5.23 A cladogram for five mammal species.

The differences in the amino acid sequences in cytochrome c represented by the letters shown above can be tabulated as shown in Table 5.2. From these data, it is possible to construct a cladogram showing the relationships between these five organisms, as shown in Figure 5.23. There are no differences between rabbit and mouse so they have to be drawn together at the end of a branch, and the same applies to the chimpanzee and human. Rhesus monkey differs from chimpanzee and human by only one amino acid and so the branch point must be one unit from the end. Rabbit and mouse differ by nine amino acids and so the branch point must be nine units further down. Biochemical analysis of other molecules or comparison of DNA sequences would be needed to complete the separation of rabbit from mouse and human from chimpanzee.

Organism	Number of amino acid differences in cytochrome c compared with human
human	–
chimpanzee	0
rhesus monkey	1
rabbit	9
mouse	9

Table 5.2 Table comparing cytochrome c in five species.

Evidence from DNA

DNA molecules can be compared in a technique known as hybridisation. If a specific DNA sequence from an insect, a reptile and a mammal are compared in this way, the number of differences between the mammal and the reptile might be found to be 40, and between the insect and the mammal 72. This provides evidence that the reptile is more closely related to the mammal than the insect is and we can construct a cladogram that shows the relationship between these three species (Figure **5.24**).

The diagram in Figure **5.25** has been constructed in a similar way from DNA analysis of the Canidae (dog and wolf family). It shows that the domestic dog and the grey wolf are very closely related, but the grey wolf and Ethiopian wolf are more distantly related. The black-backed jackal and golden jackal are also very distantly related. Complex diagrams like this are usually constructed using specially designed computer software.

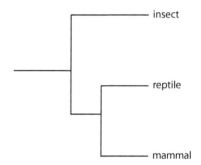

Figure 5.24 A cladogram showing the phylogenetic relationship between insects, reptiles and mammals.

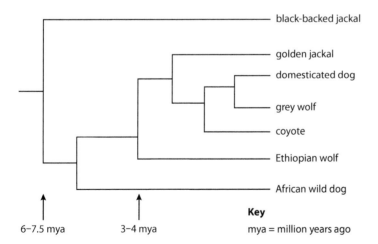

Figure 5.25 A cladogram showing the phylogenetic relationships in the Canidae family.

Worked example

5.1 Which apes are the closest living relatives of *Homo sapiens*?

Gibbons, orangutans, gorillas and chimpanzees have many physical similarities to human beings. For example human beings, chimpanzees and gorillas all have a cavity in the skull just above the eyes known as a frontal sinus. Gibbons, orangutans and other primates do not have this, so the physical evidence suggests that chimpanzees and gorillas are more closely related to human beings than gibbons and orangutans are. Evidence from the analysis of blood proteins also suggests that orangutans are more closely related to humans than gibbons are. This evidence can be shown as in Figure 5.26.

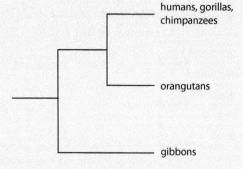

Figure 5.26 A cladogram showing the relationship between five apes.

Chimpanzees and gorillas are more closely related to humans than other living animals are but which are our closest living relatives?

To sort out the relationships between human beings, chimpanzees and gorillas, we must assess the evidence and check which features are shared. We can construct a table to summarise the evidence (Table **5.3**). Consider which of the three cladograms shown in Figure **5.27** is supported by most evidence.

Characteristic	Other primates such as baboons	Gorillas	Human beings	Chimpanzees	Cladogram supported by evidence
DNA evidence:					
number of chromosomes	42 or more	48	46	48	C
structure of chromosomes 5 and 12		different from other primates	like other primates	different from other primates	C
chromosome Y and 13		same as human beings	same as gorillas	like other primates	A
% genetic difference from humans	orangutan 3.1% rhesus monkey 7%	1.6%	–	1.2%	B
Molecular evidence:					
alpha hemoglobin compared with human	several differences	one amino acid difference	–	identical to humans	B
protein factor in blood	not variable	not variable	same variability as chimpanzees	same variability as humans	B
amino acid sequence in myoglobin		like chimpanzees	like other primates	same variability as humans	C

Table 5.3 Summary of molecular and DNA evidence for the relatedness of primates.

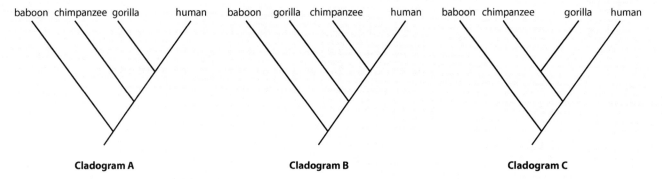

Figure 5.27 Three possible cladograms to show the relationship between human beings, chimpanzees and gorillas.

None of the cladograms can be proved to be correct from this evidence but cladogram B is the best supported based on the data and is therefore hypothesised to best reflect our current understanding of the evolutionary relationships of human beings. If further evidence is collected in future the hypothesis may be changed.

The shapes of cladograms

Cladograms can be drawn in one of two ways, as shown in Figure 5.28, which shows two formats for a cladogram of living vertebrate animals. By looking at the upper diagram we can see that the organism with the greatest number of differences from mammals branches off first. These organisms are the least closely related to mammals.

The relationship between reptiles and birds is the subject of much debate amongst scientists. Some reptiles (e.g. crocodiles and dinosaurs) are more closely related to birds than other reptiles (e.g. lizards and turtles). Cladists have suggested 'reptiles' is not a clade and a better grouping would be Archosaurs (crocodiles, dinosaurs and birds) Lepidosaurs (snakes and lizards) and Testudines (turtles and tortoises).

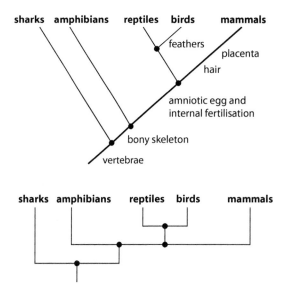

Figure 5.28 A cladogram shown in two different formats.

Cladograms and the classification of living organisms

Cladograms have been produced using information from fossils, morphology, physiology and molecular data. In most cases, they agree with each other, or the differences are small. When there is uncertainty, a second cladogram, built up using a different feature, can be constructed for comparison. Each cladogram can be thought of as a hypothesis about the relationships of the organisms it contains.

Homologous and analogous characteristics

Homologous structures are those that have evolved from a common ancestral form. The pentadactyl limb (Figure **5.5**) is an example of a homologous structure. As new species have evolved, exploiting new environmental conditions, the features of the basic structure of the limb have become adapted to each species' particular habitat.

The wing of a bat and the wing of an insect are structurally very different even though they perform a similar function. The two wings are said to be analogous characteristics and are not useful in establishing relationships between the two species (Figure **5.29**).

Occasionally, a cladogram based on molecular data indicates an evolutionary relationship that disagrees with other or older classifications. But where the different methods support each other, the evidence is more robust and where there is disagreement, it provokes scientific debate and the search for further evidence to support either one argument or the other. Cladistics is an objective method of classification and is a good example of the scientific method of studying organisms. Using cladistics, it is possible to determine whether our understanding of relationships is correct or whether it should be abandoned in favour of a competing hypothesis. For example, it was thought for a long time that orb-weaving spiders, which build intricate, orderly webs, had evolved from cobweb-weaving spiders. Cladistic analysis disproved this hypothesis and showed that cobweb-weaving spiders had evolved from spiders with more orderly webs. Many other groups and many other characteristics – including parasites, distribution and pollination – have been scrutinised in this way.

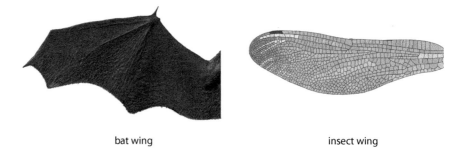

bat wing   insect wing

Figure 5.29 A bat's wing and an insect's wing are analogous structures.

Nature of science

Falsification of theories – re-classifying the figwort family

Figwort is a family of flowering plants that includes black root, Culver's root and mulleins (*Verbascum* sp.). In the past the family was named **Scrophulariaceae** and contained about 275 genera and 5000 different species. Since the early 1990s, research into the DNA sequences of three plastid genes has revealed at least five separate clades. The traditional group Scrophulariaceae was found to be an unnatural group, which was not made up of clades. There were few, clear distinguishing physical characteristics to separate members of the old group which meant that taxonomists were unable to identify clades until molecular evidence became available. The new classification now places some genera into completely different families as the molecular studies have shown that they are not closely related to the figwort family after all.

Test yourself

15 Define the term 'clade'.
16 Outline the difference between analogous and homologous characteristics.
17 Explain how evidence from cladistics can lead to new classifications.

Exam-style questions

1 Which of the following is the correct sequence for the hierarchy of taxa?

    A kingdom, domain, phylum, family, order, species
    B kingdom, domain, order, class, genus, species
    C domain, kingdom, class, phylum, order, species
    D domain, kingdom, phylum, family, genus, species. [1]

2 If two organisms are members of the same order, then they are also members of the same:

    A genus
    B class
    C family
    D species. [1]

3 Outline the role of variation in evolution. [3]

4 Why should analogous characteristics not be used when constructing a cladogram? [2]

5 Discuss how the analysis of biochemical molecules can be used to investigate the common ancestry of living organisms. [6]

6 There are a number of differences between the β chain of hemoglobin of humans and other vertebrates. These are shown in the table below.

Organism	Number of amino acid differences in β-chain of hemoglobin compared to humans
chimpanzee	0
gorilla	0
rhesus monkey	8
horse	25
chicken	45
frog	67

a Construct a cladogram to represent the relationships between the organisms in the table. [5]

b The organisms in the table all belong to the Phylum Chordata. List **three** features of the Phylum Chordata. [3]

c i Identify the organisms in the table which are mammals. [1]
   ii Outline **two** features which you used to make your choice. [2]

7 A common species of British snail occurs in two forms which vary in the pattern of bands found on the surface of their shells. Figure **5.30** shows a banded snail and an unbanded snail. Birds, especially song thrushes, feed on the snails by breaking them open to eat the soft snail body inside. The shells are broken on a stone known as an anvil. Shell fragments are found on the ground near these anvil stones.

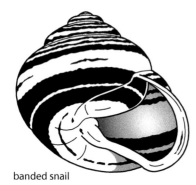
banded snail

unbanded snail

Figure 5.30 Banded snails.

In an investigation, two areas of woodland A and B were cleared of all snails. Area A was a habitat for thrushes, but no thrushes lived in area B. Equal numbers of banded and unbanded snails were placed in areas A and B. After a period of one month in autumn, the number of snails eaten in area A were estimated by counting the snail shell fragments found around the anvil stones and the numbers compared with the shells of living snails found in area B. The data are shown in the table.

Area	Numbers of each snail type found	
	Unbanded	Banded
area A (dead snails)	74	130
area B (living snails)	103	107

Using the data in the table:

a Describe the effect of predation by thrushes on the snails in this investigation. [3]

b The banding on the shells the snails is controlled by a single gene with two alleles.

   Explain how natural selection might account for different proportions of these alleles found in snail populations in different habitats in the UK. [4]

Human physiology 6

Introduction

Human physiology is the study of the organs and organ systems of the body and how they interact to keep us alive. Physiologists examine anatomy, which is the physical structure of the organs, and investigate the biochemical processes occurring inside cells and tissues to keep our organs working efficiently.

6.1 Digestion and absorption

The digestive system

The **digestive system** consists of a long, muscular tube, also called the gut or alimentary canal (Figure 6.1). Associated with it are a number of glands that secrete enzymes and other digestive juices. The gut extends from the mouth to the anus and is specialised for the movement, digestion and absorption of food.

Digestion is the biochemical breakdown of large, insoluble food molecules into small, soluble molecules and is an example of hydrolysis (Subtopic 2.1). This process is essential because only small molecules can enter cells and be used in the body. Digestion of large molecules occurs very slowly at body temperature so enzymes are essential to speed up the rate of digestion so that it is fast enough to process nutrients to supply our needs. There are many different enzymes in the human digestive system. Different enzymes are released in different sections of the digestive system and each one is specific for one food type. All digestive enzymes help to catalyse hydrolysis reactions and work most efficiently at about 37 °C. Examples of some important enzymes are shown in Table 6.1.

Enzyme type	Example	Source	Substrate	Products	Optimum pH
amylase	salivary amylase	salivary glands	starch	maltose	7
	pancreatic amylase	pancreas	maltose	glucose	7–8
protease	pepsin	gastric glands in stomach wall	protein	polypeptides	2
	endopeptidase	pancreas	polypeptides	amino acids	8
lipase	pancreatic lipase	pancreas	triglycerides (fats and oils)	fatty acids and glycerol	7

Table 6.1 Different types of enzyme digest different types of food molecule.

Learning objectives

You should understand that:
- Contractions of the circular and longitudinal muscles in the small intestine mix food with enzymes and move it along the gut.
- Enzymes are secreted from the pancreas into the lumen of the small intestine.
- It is the function of the enzymes in the small intestine to digest most macromolecules in food into monomers.
- Folds in the epithelium of the small intestine, known as villi, increase the surface area over which absorption is carried out.
- Mineral ions, vitamins and the monomers produced by digestion are absorbed into the villi.
- Different methods of transport are required to absorb different nutrients through membranes.

Digestion a series of biochemical reactions that converts large ingested molecules into small, soluble molecules

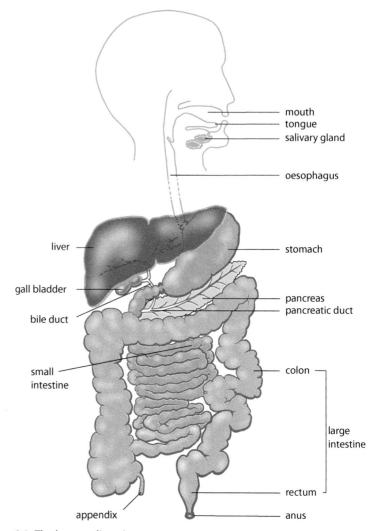

Figure 6.1 The human digestive system.

Mouth, oesophagus and stomach

In the mouth, food is broken into small pieces by the jaws and teeth, and mixed with saliva containing the enzyme salivary amylase, which begins the digestion of any starch the food contains.

Next food passes down the oesophagus to the stomach moved by a sequence of muscle contractions known as **peristalsis**. The stomach holds the food for up to four hours while digestion proceeds inside it. As muscles of the stomach contract, food and enzymes are mixed – this gives maximum contact between food and enzyme molecules, and speeds up the digestive process.

Digestion of protein begins in the stomach, catalysed by the enzyme pepsin, which is secreted in gastric juice from the gastric glands in the stomach wall. Hydrochloric acid activates the pepsin and maintains a pH of 1.5–2.0 in the stomach. This is the optimum pH for protein digestion and the acid also kills many of the bacteria present in the food we eat. Goblet cells in the stomach lining secrete mucus to protect the interior of the stomach from the acid and enzymes, which might otherwise digest it.

Food is transformed in the stomach to a semi-liquid called **chyme** and is then ready to move on to the next stage of digestion in the small intestine.

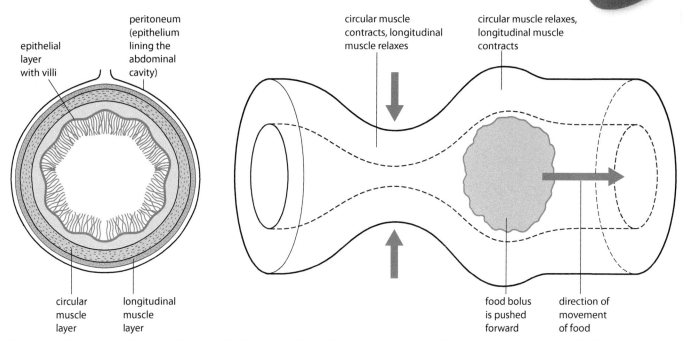

Figure 6.2 The actions of longitudinal and circular muscles in the intestine help to push food along, in a processes called peristalsis.

Roles of the small intestine

Digestion

Little by little, chyme leaves the stomach via a valve at the lower end and moves into the 5 m long small intestine. As in all parts of the alimentary canal, food is moved along the small intestine by **peristalsis**, which involves the two layers of muscles that make up the intestine wall. Longitudinal muscles run along the length of the intestine and circular muscles lie beneath them and encircle the intestine (Figure **6.2**). Contraction of the bands of circular muscle squeezes the area of the intestine behind a portion of food, while longitudinal muscles relax and extend to accommodate it. The two sets of muscles then relax and contract respectively so that food is gradually pushed along the intestine. As waves of peristalsis move along the intestine, food is mixed with enzymes that are secreted into it.

Digestion is completed in the first section of the small intestine (Figure **6.1**). Digestive juices are secreted into the lumen of the intestine from the liver, gall bladder, pancreas and the mucosa of the intestine walls. Bile is added from the liver and gall bladder, and the pancreas secretes **pancreatic juice** containing endopeptidase (a protease), lipase, amylase and bicarbonate ions. The acidity of the chyme is reduced by these ions, allowing the enzymes to work at their optimum pH (Subtopic **2.5**).

The digestion of starch, which began with the conversion of starch to maltose in the mouth, continues here as maltose is converted to glucose, which can be absorbed through the intestine walls.

Three main types of food molecule that must be digested in the small intestine are carbohydrates, polypeptides and lipids. Table **6.2** shows how these molecules are **ingested** and what is produced when they are digested.

Ingestion the act of eating

Type of molecule	Form of the molecule in ingested food	Monomers produced by digestion
carbohydrates	monosaccharides, disaccharides, polysaccharides (e.g. starch, glycogen)	monosaccharides (e.g. glucose)
proteins	proteins	amino acids
lipids	triglycerides	fatty acids and glycerol
nucleic acids	DNA, RNA	nucleotides

Table 6.2 Large food molecules (macromolecules) are broken down in digestion into small molecules (monomers), which can be absorbed.

Many enzymes are used in industrial processes (Section **2.5**). For example, amylase is of great economic importance because of its uses in brewing and in the production of sugars from starch.

Absorption

The inner surface of the small intestine is greatly folded to form thousands of tiny villi (Figure **6.3**). Each villus contains a network of capillaries and a lacteal. (A lacteal is a small vessel of the lymphatic system.) Villi greatly increase the surface area of the small intestine and improve its efficiency as an absorbing surface. Digested material must pass through the folded membranes (microvilli) of the epithelial cells in order to reach a capillary or lacteal vessel.

As monomers produced by digestion, such as glucose, amino acids, fatty acids and glycerol, as well as mineral ions and vitamins (which are already very small molecules), come into contact with a villus, they are absorbed. The molecules may pass through the wall of the intestine by diffusion, facilitated diffusion or active transport. From here, they enter the bloodstream and travel to the cells, where they are reassembled into new structures. Cells of the villi contain structures that are vital to the processes of absorption. The cells have many mitochondria, indicating that some absorption occurs using active transport, and requires energy. In addition, many vesicles are present and these structures show that some materials are taken in from the intestine by the process of pinocytosis.

Digested molecules are small enough to pass through the epithelial cells and into the bloodstream. Movement can occur by a number of means.

- **Simple diffusion** can occur if molecules are small and can pass through the hydrophobic part of the plasma membrane (Subtopic **1.3**).
- **Facilitated diffusion** occurs in the case of monomers such as fructose, which are hydrophilic. Protein channels in the epithelial cell membrane enable these molecules to move, provided they are small enough and there is a concentration gradient, which permits diffusion.
- **Active transport** is used to transport molecules that do not have a sufficiently high concentration gradient to pass by diffusion. Glucose, amino acids and mineral ions can all be absorbed by this method. Mitochondria produce the ATP needed for active transport by the membrane pumps (Subtopic **1.4**).
- **Pinocytosis** also draws in small drops of liquid from the small intestine. Each droplet is surrounded by small sections of membrane that invaginate to form a vesicle. The vesicles are taken into the cytoplasm where their contents can be released.

Amino acids and glucose enter the capillaries and are carried away in the bloodstream. Fatty acids and glycerol are taken into the lacteals and travel in the lymphatic system.

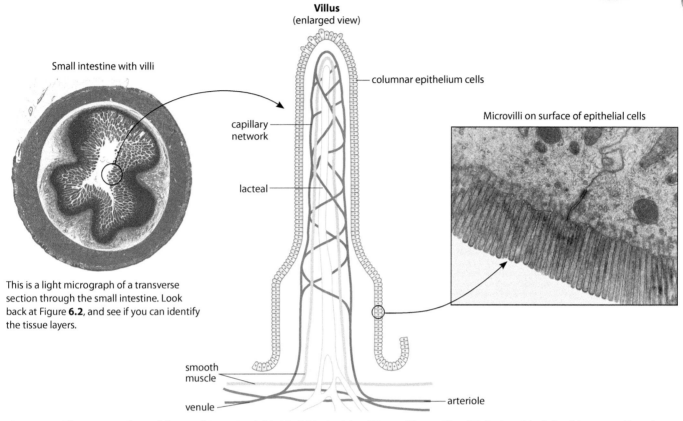

Figure 6.3 The inner surface of the small intestine is highly folded, with millions of finger-like villi. Each epithelial cell is covered in minute microvilli, so the total surface area for absorption is vast.

Blood from the capillaries in the walls of the small intestine first travels to the liver, which absorbs glucose and stores it so that it will be able to maintain a constant level of blood sugar when levels fall. Amino acids form part of the reserve of amino acids used to build new proteins in cells all over the body, and fatty acids and glycerol enter the bloodstream from lymph vessels near the heart to be used as an energy source or to build larger molecules.

Role of the large intestine

By the time food reaches the end of the small intestine, most useful substances have been removed from it. Any remaining undigested material, including fibre (cellulose), passes into the large intestine, which also contains mucus, dead cells from the intestine lining and large numbers of naturally occurring bacteria. Bacteria living here are mutualistic organisms, gaining nutrients and a suitable habitat, while synthesising vitamin K which benefits their human host.

The main role of the large intestine is reabsorbing water and mineral ions such as sodium (Na^+) and chloride (Cl^-). What remains of the original food is now referred to as feces and is egested, or eliminated from the body, via the anus. After digested food has been absorbed, it is carried around the body and enters cells to become part of the body's tissues or reserves. This is called assimilation.

Absorption the process by which small molecules are taken through the cells of the intestine and pass into the bloodstream

Assimilation the process by which products of digestion are used or stored by body cells

Egestion the process by which undigested material leaves the body at the end of the gut

Cellulose

Humans cannot digest cellulose, which makes up the walls of plant cells. Cellulose is a complex carbohydrate requiring the enzyme cellulase to catalyse its digestion. Humans and other mammals are unable to produce this enzyme so cellulose (also known as **fibre**) passes undigested out of the body in feces.

Nature of science

Using theories to explain natural phenomena – modelling the intestine using dialysis tubing

Dialysis tubing is made of a thin layer of cellulose and is used in experiments to demonstrate absorption because it is partially permeable. It will only allow small molecules to pass through the pores in its structure.

If knotted dialysis tubing is filled with starch and glucose solutions and placed in a beaker containing water, only the glucose molecules, which are small enough to pass through the tubing, will be able to leave the tubing bag (Figure 6.4). The process can be monitored by testing the solutions inside and outside the tubing for the presence of starch and glucose (using iodine solution for starch and Benedict's reagent for glucose). What control experiment would you do to ensure your findings are valid?

The experiment can also be carried out with other polymers and monomers that you might wish to investigate.

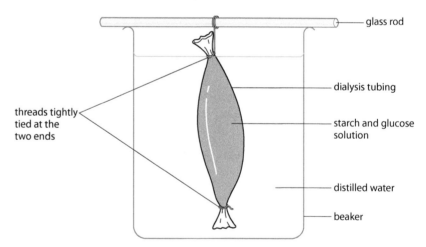

Figure 6.4 Modelling absorption in the intestine using dialysis tubing. Water molecules are also able to pass through dialysis tubing by osmosis. What effect will this have on the appearance of the dialysis tubing bag?

? Test yourself

1 Explain why circular *and* longitudinal muscles are needed to move food along the intestine.
2 Distinguish between 'absorption' and 'assimilation'.
3 State why enzymes are needed in digestion.
4 List the ways in which a villus is adapted to increase the efficiency of absorption of nutrients.

6.2 The blood system

Learning objectives

You should understand that:
- Arteries carry blood under high pressure from the ventricles of the heart to the body's tissues.
- Arteries have elastic and muscle fibres in their walls that help to maintain blood pressure between pumping cycles of the heart.
- Blood flows through tissues in capillaries. The permeable walls of the capillaries allow exchange of materials between the bloodstream and the cells in the tissue.
- Blood from capillaries in the tissues is collected at low pressure into veins, which return it to the atria of the heart.
- Valves in the veins and in the heart prevent backflow, thereby ensuring that the blood keeps circulating in one direction.
- Like all mammals, humans have a double circulation of blood, with a separate circulation for the lungs.
- The heart beat is initiated by specialised muscle cells in the right atrium called the sinoatrial node. The sinoatrial node acts as a pacemaker.
- An electrical signal from the sinoatrial node stimulates contraction of the heart muscle as it is propagated through the walls of the atria and then the ventricles.
- Heart rate can be increased or decreased by impulses to the heart from two nerves from the medulla of the brain.
- The hormone epinephrine increases heart rate.

Our blood system provides a delivery and collection service for the whole body. The heart, blood and blood vessels make up a most efficient transport system that reaches all cells, bringing the substances they need and taking away their waste. Humans and other mammals have what is known as a closed circulatory system with blood contained inside a network of arteries, veins and capillaries.

Blood vessels

Arteries are blood vessels that carry blood away from the ventricles of the heart. They branch and divide many times forming arterioles and eventually the tiny capillaries that reach all our tissues. Arteries have thick outer walls of collagen and elastic fibres (Figure 6.5), which withstand high blood pressure and prevent vessels becoming overstretched or bursting. Just beneath the outer covering is a ring of circular smooth muscle that contracts with each heart beat to maintain blood pressure and keep blood moving along. Inside an artery, the lumen is narrow to keep blood pressure high. The lumen's lining of smooth epithelial cells reduces friction and keeps blood flowing smoothly.

Capillaries are the smallest vessels – the lumen of a capillary is only about 10 μm in diameter and some are so small that red blood cells

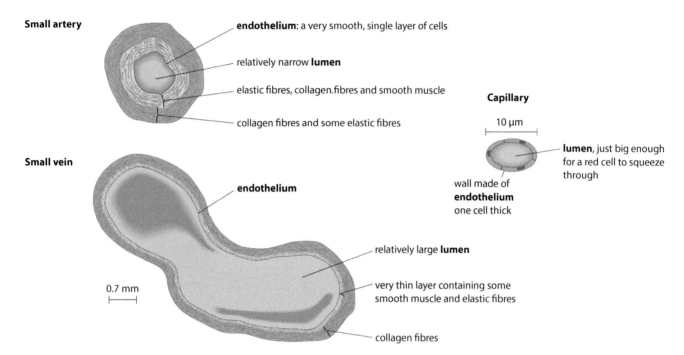

Figure 6.5 Diagrams of transverse sections of an artery, a vein and a capillary.

Plasma that leaks out of tiny capillaries bathes the nearby tissues and supplies oxygen and nutrients to the cells. Once out of the capillary, the fluid is known as **tissue fluid**.

Lymphatic system collects tissue fluid that leaks from the capillaries and returns it to the large veins close to the heart

must fold up in order to pass along. Networks of these tiny capillaries reach almost every cell in the body. Blood flow here is very slow, at less than 1 mm per second, and capillary walls are only one cell thick so the distance for diffusion of materials in and out of them is as small as possible. Some capillary walls have spaces between their cells enabling plasma and phagocytes (white blood cells) to leak out into the tissues.

Veins carry blood back towards the atria of the heart from body tissues. Small veins called venules join up to form large veins, which can be distinguished from arteries by their much thinner walls, which contain few elastic and muscle fibres. Blood inside a vein does not pulse along and the lumen is large to hold the slow-moving flow. The relatively thin walls can be compressed by adjacent muscles and this helps to squeeze blood along and keep it moving. Many veins contain valves to prevent blood flowing backwards, a problem which can arise if flow is sluggish.

Table 6.3 summarises some differences and similarities between the three types of blood vessel.

Artery	Vein	Capillary
thick walls	thin walls	walls one cell thick
no valves (except in aorta and pulmonary artery)	valves sometimes present	no valves
blood pressure high	blood pressure low	blood pressure low
carries blood from the heart	carries blood to the heart	links small arteries to small veins

Table 6.3 Comparing arteries, veins and capillaries.

The heart and circulation

In the human circulatory system, blood is kept on the move by the pumping action of the powerful heart muscle. It has been estimated that a normal human heart beats more than 2.5×10^9 times in a lifetime, sending a total of more than 1.5 million dm^3 of blood from each ventricle.

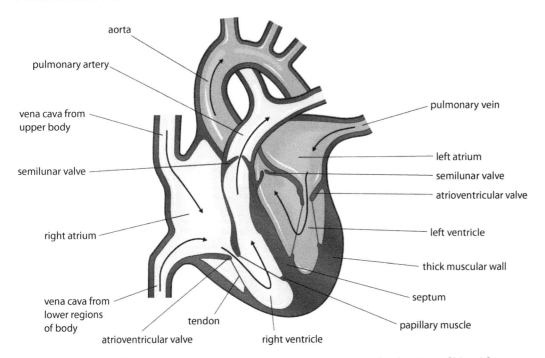

Figure 6.6 Diagram of the human heart, in longitudinal section, showing the direction of blood flow.

A human heart is about the size of a clenched fist. It is a double pump with two separate sides (Figure **6.6**). The right-hand side receives deoxygenated blood from all over the body and pumps it to the lungs via the pulmonary artery to pick up more oxygen. The left-hand side receives oxygenated blood from the lungs via the pulmonary vein and pumps it to cells all over the body where the oxygen is unloaded. So humans, like all mammals, have a **double circulation**: a pulmonary circulation between the heart and lungs and a larger circulation that carries blood from the heart to the rest of the body and back again (Figure **6.7**). On any complete journey round the body, blood passes through the heart twice.

The heart has four chambers – two smaller **atria** (singular **atrium**) at the top and two larger **ventricles** below. The right- and left-hand sides are completely separated from one another. Atria have thin walls as the blood they receive from the veins is under relatively low pressure. Ventricles are stronger and more muscular as their job is to pump blood out of the heart. Both ventricles hold the same volume of blood but the left ventricle wall

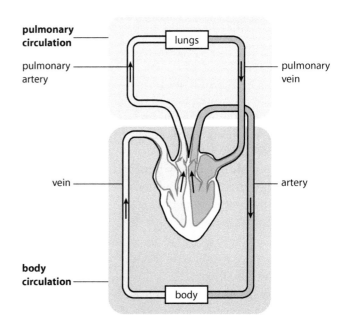

Figure 6.7 Diagram to show the double circulation of blood through the heart.

6 HUMAN PHYSIOLOGY 195

is thicker than the right as it must generate enough pressure to pump blood all round the body. The right ventricle pumps blood a much shorter distance to the lungs.

Atria are separated from ventricles by **atrioventricular valves**, which prevent the blood flowing backwards into the atria. A second set of valves in the aorta and pulmonary arteries – the **semilunar valves** – prevent backflow into the ventricles as they relax after a contraction.

Heart muscle works continuously, beating about 75 times per minute when a person is resting, and so it has a large demand for oxygen. Coronary arteries extend over the surface of the heart and penetrate deep into the muscle fibres to supply oxygen and nutrients for this unremitting activity (Figure **6.8**).

Control of the heart beat

Heart tissue is made of a special type of muscle that is different from other muscles in our bodies. **Cardiac muscle** is unique because it contracts and relaxes without stimulation from the nervous system. It is said to be **myogenic**. Natural myogenic contractions are initiated at an inbuilt pacemaker, which keeps cardiac muscle working in a coordinated, controlled sequence. The pacemaker, or **sinoatrial node** (SAN), is a special region of muscle cells in the right atrium that sets the basic pace of the heart. The rate set by the SAN is also influenced by stimulation from the nervous system and by hormones.

At the start of every heart beat, the SAN produces an impulse that stimulates both atria to contract. A second structure, the **atrioventricular node** (AVN) at the base of the right atrium, is also stimulated. It delays the impulse briefly until the atrial contraction finishes and then transmits it on down a bundle of modified muscle fibres – the bundle of His and Purkinje fibres – to the base of the ventricles. Impulses radiate up through the ventricles, which contract simultaneously about 0.1 s after the atria (Figure **6.9**).

The natural rhythm of the pacemaker is modulated by the nervous system so that the heart rate is adjusted to our activity levels. It speeds up when we are exercising and need extra oxygen and nutrients, and slows down as we sleep. Changes to our heart rate are not under our conscious control but result from impulses sent from a control centre in the part of the brain stem known as the medulla. Impulses to speed up the heart pass along the sympathetic nerve, which stimulates the pacemaker to increase its rate. Impulses sent along the parasympathetic (vagus) nerve cause the heart rate to slow down. The medulla monitors blood pressure and carbon dioxide levels using information it receives from receptors in arteries.

Emotions such as stress, as well as increases in activity level, can cause an increase in heart rate. During periods of excitement, fear or stress the adrenal glands release the hormone **epinephrine** (also called adrenalin), which travels in the blood to the pacemaker and stimulates it to increase the heart rate.

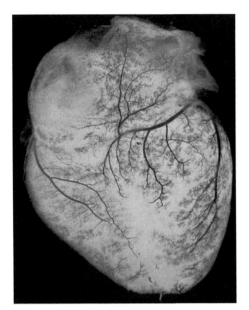

Figure 6.8 A human heart. Clearly visible are the coronary arteries, which supply oxygen to the heart muscle.

Heart transplantation

The first adult heart transplant operation was carried out by Christiaan Barnard on 3 December 1967, in Cape Town, South Africa. The recipient was Louis Washkansky, who was 54 at the time. Unfortunately, Washkansky died from pneumonia 18 days after the operation because his immune system was weakened by the immunosuppressant drugs used to avoid rejection of the 'foreign' organ. However, the transplant was considered a success, because the new heart was beating in Washkansy's body without external stimulation, and now heart transplant operations are considered routine, with about 3500 performed each year around the world.

Cardiac muscle will contract rhythmically in tissue culture if it is supplied with oxygen and glucose. Its normal resting rate in culture is about 50 beats per minute.

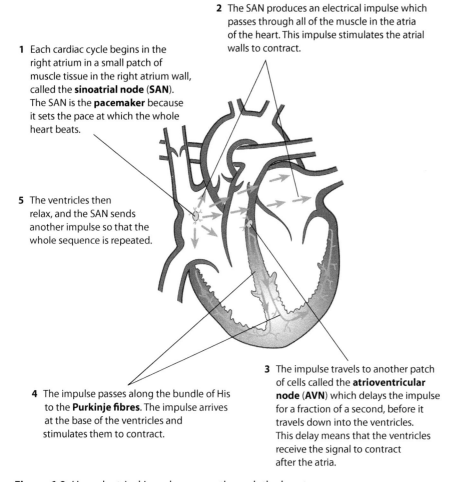

Figure 6.9 How electrical impulses move through the heart.

1. Each cardiac cycle begins in the right atrium in a small patch of muscle tissue in the right atrium wall, called the **sinoatrial node** (**SAN**). The SAN is the **pacemaker** because it sets the pace at which the whole heart beats.

2. The SAN produces an electrical impulse which passes through all of the muscle in the atria of the heart. This impulse stimulates the atrial walls to contract.

3. The impulse travels to another patch of cells called the **atrioventricular node** (**AVN**) which delays the impulse for a fraction of a second, before it travels down into the ventricles. This delay means that the ventricles receive the signal to contract after the atria.

4. The impulse passes along the bundle of His to the **Purkinje fibres**. The impulse arrives at the base of the ventricles and stimulates them to contract.

5. The ventricles then relax, and the SAN sends another impulse so that the whole sequence is repeated.

The cardiac cycle

The cardiac cycle is the sequence of events that takes place during one heart beat (Figure **6.10**). As the heart's chambers contract, blood inside them is forced on its way. Valves in the heart and arteries stop the blood flowing backwards through the heart. The pressure and volume of blood in each of the chambers of the heart change during the cardiac cycle. Figure **6.11** shows these changes for one complete cycle.

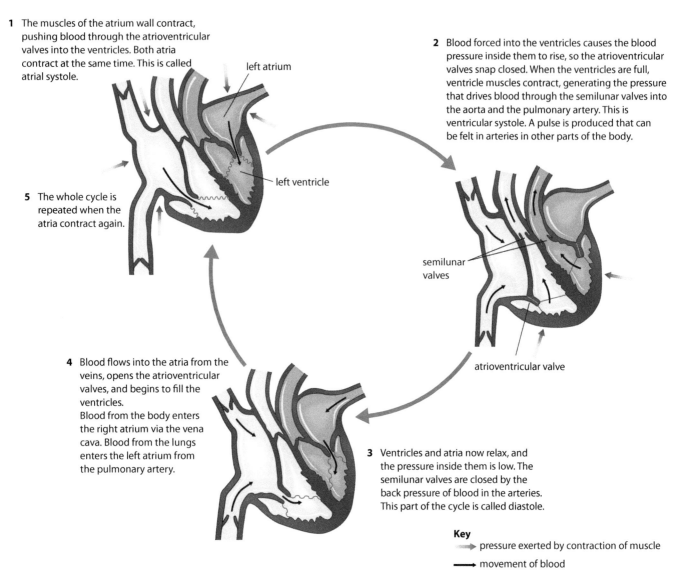

Figure 6.10 The events of a heart beat – the cardiac cycle. The heart normally beats about 75 times per minute, and a complete heart beat takes about 0.8 seconds.

Coronary heart disease

Three large coronary arteries branch from the aorta and supply heart muscle with oxygen-rich blood (Figure **6.8**). If any of these arteries is blocked, an area of the heart muscle will receive less oxygen and cells in that region may stop contracting or even die. A blockage in a coronary artery or one of its branches is known as a **coronary thrombosis** or heart attack.

One common cause of coronary heart disease (CHD) is **atherosclerosis**, a slow degeneration of the arteries caused by a build-up of material known as **plaque** inside them. Plaque becomes attached to the smooth endothelium lining where it can accumulate. Over time, the diameter of the artery becomes restricted so that blood cannot flow along it properly, and it loses elasticity (Figure **6.12**). As the rate of flow slows down, blood may clot in the artery, further restricting the movement of blood along it. Clots may also break free and travel to block another smaller artery elsewhere in the body. If this artery is in the brain, the clot may cause a stroke.

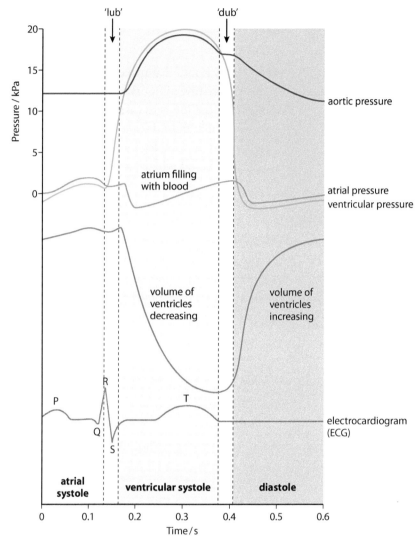

Figure 6.11 Pressure and volume changes in the heart during the cardiac cycle.

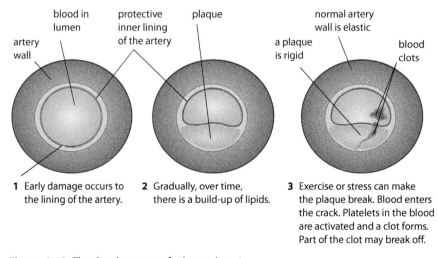

Figure 6.12 The development of atherosclerosis.

Sympathetic and parasympathetic nerves

Sympathetic and parasympathetic nerves are part of the autonomic nervous system, which controls activities, such as heart rate, that are not under our conscious control.

Eating a diet that is high in saturated fatty acids has been shown to have a positive correlation with an increased risk of CHD. Saturated fatty acids can be deposited inside the arteries, and if the deposits combine with cholesterol they may lead to plaque and atherosclerosis, which reduces the diameter of the lumen and leads to high blood pressure. Reliable evidence suggests that in countries where the diet is high in saturated fatty acids and many high-fat foods, animal products and processed foods are eaten there is likely to be a high incidence of CHD. Since all fatty acids are high in energy, an excess of these foods in the diet can also lead to obesity, which places a further strain on the heart.

The blood

Composition of blood

Blood **plasma** is a pale yellow watery liquid that makes up 50–60% of our blood volume. Suspended in plasma are three important groups of cells:
- erythrocytes (red blood cells), whose job is to carry oxygen
- leucocytes (white blood cells), which fight disease
- platelets (cell fragments), which are needed for blood clotting.

Figure **6.13** shows the appearance of blood when examined under a light microscope.

Functions of blood

Blood has two important roles: it is a vital part of the body's transport network, carrying dissolved materials to all cells, and it helps to fight infectious disease. Table **6.4** summarises the important substances the blood carries. Its role in infection control will be explored in the next section.

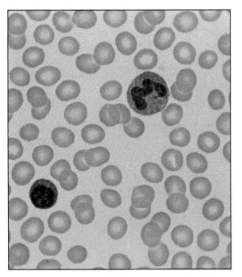

Figure 6.13 Stained light micrograph of a smear of healthy human blood. Two types of leucocyte can be seen – a lymphocyte (lower left) and a phagocyte (upper right). Lymphocytes produce antibodies, and phagocytes engulf foreign particles, including bacteria, that enter the body. Many erythrocytes can also be seen.

Substance transported	Source and destination
nutrients	glucose, amino acids, vitamins and minerals carried in plasma from the small intestine to the cells
oxygen	carried by red blood cells from the lungs to all tissues
carbon dioxide	returned to the lungs in plasma and red blood cells from all respiring tissues
urea	carried in plasma from cells to the kidneys for disposal
hormones	transported in plasma from glands to target cells
antibodies	protein molecules produced by certain lymphocytes to fight infection; distributed in plasma
heat	distributed from warm areas to cooler ones to maintain core temperature

Table 6.4 Some important substances that are transported by the blood.

Nature of science

Falsification of theories – Harvey overturned the theories of Galen

William Harvey (1578–1657), an English physician, is widely acknowledged as being the first man to describe accurately how blood flows in a continuous circulation round the body. Before Harvey, few people questioned the classical writings of authors such as Aristotle (384–322 BCE), Hippocrates (460–c.370 BCE) and Galen (c.130–c.200 CE). Galen believed that two types of blood existed. Darker, venous blood formed in the liver and red, arterial blood flowed from the heart. Blood was thought to be 'consumed' by different organs and converted to 'vital spirits' in the lungs. Harvey was unwilling to believe this doctrine and sought evidence to produce and support his own theories. He carefully recorded all his observations and carried out dissections on which to base his conclusions. He calculated the volume of blood that might pass through the heart in one day to disprove Galen's suppositions. In his famous 1628 publication, *Exercitatio Anatomica de Motu Cordis et Sanguinis in Animalibus* (*An Anatomical Study of the Motion of the Heart and of the Blood in Animals*), Harvey proposed that blood circulated in two separate, closed circuits. Illustrations in the book also showed his demonstration of the existence and function of valves in veins (Figure **6.14**).

Sir William Osler (1849–1919), a famous Canadian physician, later wrote of Harvey's book:

'It marks the break of the modern spirit with the old traditions ... here for the first time a great physiological problem was approached from the experimental side by a man with a modern scientific mind, who could weigh evidence and not go beyond it ...'

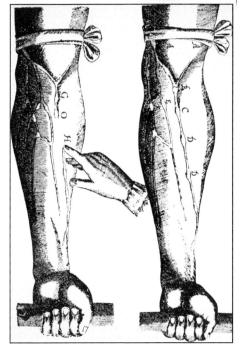

Figure 6.14 A woodcut from William Harvey's book, showing the veins in the forearm.

Questions to consider

- Why did people in the 1600s still believe writings of Greek and Roman authors from more than 1400 years earlier?
- Harvey proposed the existence of capillaries but was never able to prove his theory. Why not?
- Harvey would not accept doctrines without evidence. Can doctrines ever be accepted if the evidence is provided by authority?
- Some 21 years after the original publication of his book, Harvey produced a volume of responses to those who had made comments and criticisms of it. Why was this important?
- What did Osler mean by the phrase 'a modern scientific mind'? How could this be defined today?

Test yourself

5 List the structures on the route taken by a red blood cell on a journey from the vena cava to the aorta. Name all the chambers of the heart, valves and blood vessels it passes through.
6 Outline the role of the sinoatrial node.
7 Explain why arteries have thicker walls than veins.
8 Explain why the left-hand side of the heart has thicker walls that the right-hand side.
9 Outline the role of the atrioventricular valves.
10 State **two** reasons why the heart rate increases when we exercise.
11 State the function of epinephrine.

Learning objectives

You should understand that:
- The skin and mucous membranes form the first line of defence against pathogens that cause infectious disease.
- Cuts in the skin are sealed as the blood forms clots.
- Clotting occurs in a cascade of reactions which begin with the release of clotting factors from platelets and end in the conversion of fibrinogen to fibrin by thrombin.
- Phagocytic leucocytes ingest pathogens and give non-specific immunity to diseases.
- Lymphocytes produce antibodies in response to particular pathogens and give specific immunity.
- Antibiotics can block process that occur in bacterial cells but not in human cells and are used to treat bacterial infections.
- Antibiotics cannot be used to treat viral diseases because viruses use metabolic processes of the host cell.
- Some strains of bacteria have evolved with genes giving them resistance to one or more types of antibiotic.

6.3 Defence against infectious disease

A **pathogen** is a living organism or virus that invades the body and causes disease. Most pathogens are bacteria and viruses, but protozoa, parasitic worms and fungi can also be pathogenic.

Relatively few bacteria and fungi are pathogens, but no virus can function outside the cell of its host organism so all viruses have the potential to be pathogenic. A virus takes over the nucleic acid and protein synthesis mechanisms of its host cell and directs them to make more viruses.

The body's first line of defence

Despite the fact that we come into contact with many pathogens every day, we are seldom ill. This is due to our effective **immune system**, which both prevents pathogens entering the body and also deals with any that do.

The first line of defence against infection is our skin. Unbroken skin is a tough barrier to any potential invaders. It is waterproof and its secretions repel bacteria. Openings in the skin, such as eyes and nose, can provide entry points for pathogens but these are protected by mucous membranes, which line the respiratory, urinary, reproductive and intestinal tracts. Secretions such as tears, mucus and saliva all contain the enzyme lysozyme, which attacks the cell walls of bacteria. In addition, if pathogens are swallowed in food or water, the acidic environment of the stomach helps to kill them.

Blood clotting

If the protective layer of skin is broken or cut and blood vessels are broken, pathogens have a route into the bloodstream. To prevent blood loss and the entry of pathogens, any blood that escapes from a damaged vessel quickly forms a **clot**, which plugs the gap.

Platelets, **erythrocytes** (red blood cells) and **leucocytes** (a type of white blood cell) are all important in the clotting process. Platelets are small cell fragments, which form in the bone marrow and circulate in the bloodstream. Also important are two plasma proteins, which are present in the blood in their inactive forms until they are activated when needed (Figure **6.15**). These two inactive proteins are **prothrombin** and **fibrinogen**.

If a small blood vessel is damaged, injured cells or platelets release **clotting factors**, which cause platelets to stick to the area. These factors activate prothrombin, which is converted to its active form, **thrombin**. Thrombin, in turn activates the soluble protein fibrinogen, converting it to active **fibrin**, which is insoluble and forms long threads. This cascade of reactions ensures a speedy response to any damage. Fibrin forms a mesh of fibres that covers the damaged area and traps passing blood cells, forming a soft clot (Figure **6.16**). If a clot is exposed to air, it dries and forms a scab, which will protect the area until the tissue beneath has been repaired.

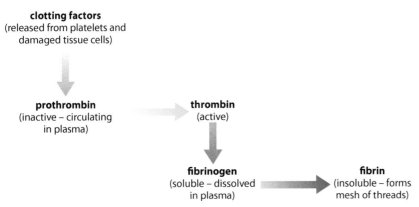

Figure 6.15 The sequence of reactions in the blood-clotting cascade.

Non-specific immunity

Pathogens that do enter the body are soon recognised by phagocytic leucocytes, which form a vital part of the body's immune system. These specialised white blood cells circulate in the blood system and, because they are easily able to change their shape, can also squeeze in and out of capillaries. Phagocytic leucocytes respond to invaders by engulfing and destroying them in a process called phagocytosis (Figure **6.17**).

This type of response provides **non-specific immunity**, which is so called because the phagocytes respond in the same way no matter what the pathogen.

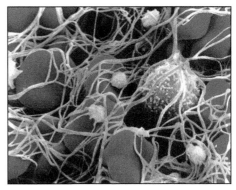

Figure 6.16 False-colour TEM showing red blood cells and threads of fibrin forming a clot (× 3600).

Specific immunity and antibody production

Antigens (**anti**body **gen**erating **s**ubstances) are proteins found embedded in the plasma membranes or cell walls of bacteria or in the protein coat of a virus. These antigens enable the body to recognise a pathogen as being 'not self' – that is, not a part of the body – and they give a clear signal to switch on the immune response, with the rapid production of antibodies.

Antibodies are protein molecules that are produced by lymphocytes in response to any antigen that enters the body. There are millions of

1. Phagocytic leucocyte detects a bacterium and moves towards it. The bacterium attaches to receptors on the cell's plasma membrane.

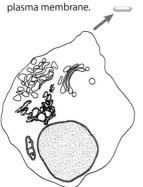

2. Bacterium is engulfed by phagocytosis into a vacuole.

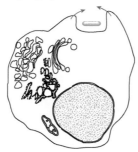

3. Lysosomes inside the cell fuse with the vacuole and release hydrolytic enzymes.

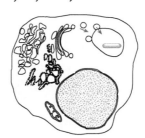

4. Bacterium is destroyed and any chemicals that are not absorbed into the cell are egested.

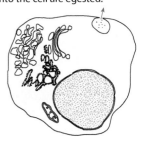

Figure 6.17 Phagocytosis of a pathogen.

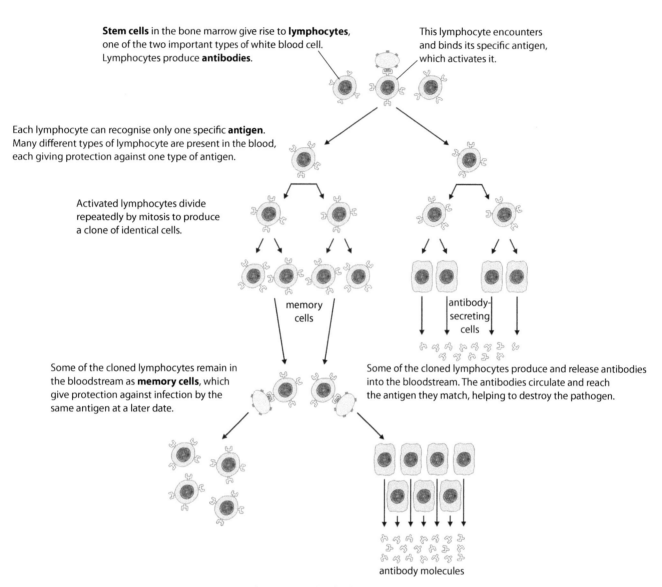

Figure 6.18 Antibody production.

A **vaccination** protects us from a specific disease by stimulating our immune system to produce antibodies against antigens carried by the disease-causing agent. Memory cells then remain in our bloodstream so that if the actual disease-causing agent is encountered later, antibodies can be released quickly and an infection avoided (Subtopic **11.1**).

different antibodies and each one is specific to an antigen. For example, the antibodies that lymphocytes produce in response to infection by an influenza virus are quite different from those produced by different lymphocytes in response to a tuberculosis bacterium. Even fragments of pathogens, or their toxins, can stimulate the release of antibodies. After an infection has passed, some of the lymphocytes giving rise to antibodies specific to the infecting antigen remain in the bloodstream as memory cells. This means that the immune system can respond quickly if the same antigen enters the body again later, by producing antibodies and preventing a widespread infection. The person is said to have acquired immunity to the antigen. Figure **6.18** explains how antibodies are made.

Each antibody molecule has a basic Y shape but the arrangement of molecules at the top of the Y form specific binding sites that give every antibody its unique properties (Figure **6.19**). These specific binding sites attach to the corresponding antigen site on the surface of the pathogen or its toxin. Once an antibody has bound to an antigen, it can destroy it in one of a number of ways. Some cause bacterial cells to clump together, making the job of phagocytes easier. Others cause cell walls to rupture, deactivate toxins, or act as recognition signals for phagocytes, giving a clear indication that action is needed (Figure **6.20**).

Antibiotics

If the body's natural defences are unable to deal with an infection, medical intervention may be needed. Most bacterial infections can be treated with antibiotics. Antibiotics are natural substances that slow the growth of bacteria. Since the discovery of penicillin in 1928, many antibiotics have been isolated and about 50 are now manufactured for medical use. These antibiotics work in different ways but are effective because prokaryotic and eukaryotic cells (Subtopic **1.2**) have different metabolic pathways. Some antibiotics block the protein synthesis mechanism in bacteria while not affecting the process in human cells. Others interfere with the formation of the bacterial cell wall and prevent bacteria growing and dividing.

Viruses are not living and have no metabolic pathways of their own. Since they use their human host's metabolism to build new viruses, antibiotics have no effect against viral infections.

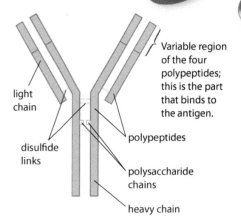

Figure 6.19 The basic structure of an antibody molecule.

Antibiotic resistance

Antibiotics kill or block the growth of bacteria but not all bacteria are susceptible to them. In any population of bacteria some individuals will have a natural resistance to the antibiotic used to kill them, which may arise spontaneously by mutations. Resistant strains multiply along with susceptible strains but if antibiotics are used, only the sensitive bacteria will be killed while the resistant ones survive. Resistant bacteria are also able to pass on their resistance to other bacteria via their plasmids (Subtopic **1.2**). Treating a disease caused by resistant strains of bacteria becomes very

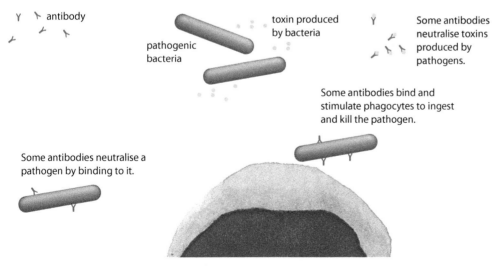

Figure 6.20 The various ways in which different antibodies can destroy bacteria or their toxins.

difficult. Doctors may have to prescribe stronger doses of antibiotic or try different antibiotics to kill the resistant bacteria (Subtopic **5.2**).

The more often antibiotics are used, and the more different types that are used, the greater the risk that resistance will develop so over-use and the improper use of antibiotics are thought to have contributed to the development of resistance. Bacteriologists are concerned that some diseases will become untreatable with currently available antibiotics. The so-called superbug MRSA now has multiple resistance to many antibiotics and recently strains of the bacteria that cause tuberculosis and the sexually transmitted disease gonorrhoea have been found to be resistant to *all* the antibiotics which have been used to treat them.

HIV and AIDS

Human immunodeficiency virus (HIV, Figure **6.21**), first identified in the early 1980s, causes the series of symptoms together known as acquired immune deficiency syndrome, or AIDS. HIV infects only the helper T-cells, a type of lymphocyte that is important in maintaining communication between cells of the immune system. After a latent period of months or years, helper T-cells are gradually destroyed and, as their numbers fall, so does the body's ability to fight infection. Helper T-cells instruct other lymphocytes to clone and generate antibodies, and without them an infected person can no longer fight off pathogens. Secondary infections result and the person is said to be suffering from AIDS.

Cause and consequences of AIDS

HIV is transmitted in blood, vaginal secretions, semen, breast milk and sometimes across the placenta. In some countries, HIV has been transmitted in blood transfusions but in most places with medical care facilities, blood for transfusion is now screened for the virus. The virus is most frequently passed from person to person in bodily fluids during sex and also when non-sterile syringe needles are used to administer either legal or illegal drugs.

HIV is a retrovirus, which means it can insert its DNA into that of a host cell using a protein called reverse transcriptase. Even if all the viruses in the body could be removed, the infected T-cells would continue to make new viruses.

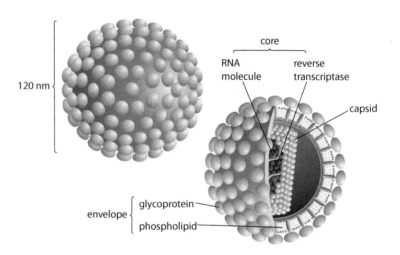

Figure 6.21 HIV viruses consist of a spherical glycoprotein and lipid coat enclosing two strands of RNA. The virus is 60 times smaller than a red blood cell.

AIDS is the end stage of an HIV infection. It is caused by a severe failure of the immune system as the HIV virus selectively infects helper T-cells. Some infected individuals have no symptoms in the early stages of the disease while others may be slightly unwell when first infected. Symptoms of AIDS develop as the number of active helper T-cells decreases. The symptoms occur as a result of secondary infections caused by bacteria, fungi and viruses that the body is unable to resist due to its compromised immune system (Figure **6.22**).

International aspects of disease

AIDS is a worldwide pandemic but some regions are more seriously affected than others. In 2008, the number of people living with HIV and AIDS was estimated at 25 million in sub-Saharan Africa and almost 8 million in South and South East Asia. AIDS is the main cause of death for men and women aged between 16 and 50 years in these countries, Latin America and the Caribbean.

The spread of the HIV virus, and other pathogens such as the bird flu virus, are problems for the whole world. International travel means that pathogens can travel further and faster than ever before and governments

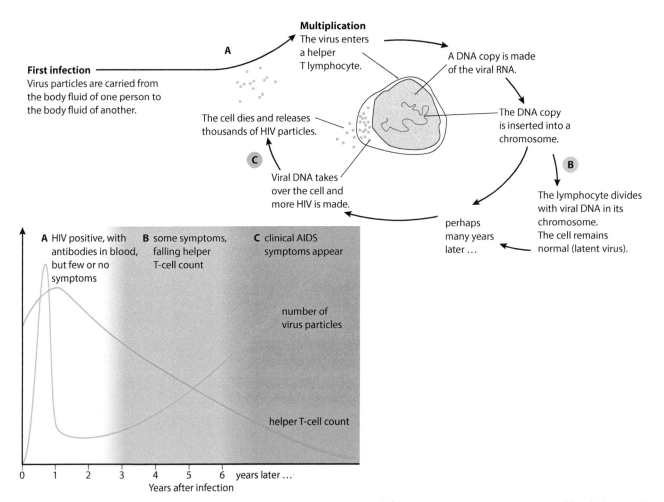

Figure 6.22 HIV infection proceeds through three stages – HIV positive with few symptoms, some symptoms and low helper T-cell count, and clinical AIDS with associated symptoms.

and health authorities must work together to co-ordinate their responses.

As new viruses and antibiotic-resistant bacterial strains evolve, scientists have to develop new vaccines and medication to try and combat them. Research and development may take a long time and there is a constant battle to keep up. The importance of this research is crucial to international health, which in turn can affect not only the life of citizens but also the economy of countries. Family incomes and national productivity decline if people of working age fall ill. Food production is affected and causes more problems. Caring for the sick and providing medical care form an expensive burden for individual families, companies and governments.

Nature of science

Assessing risk in science – collaboration, safety and new medicines

The discovery of antibiotics began by accident. On 3 September 1928, Professor Alexander Fleming was examining a batch of culture plates on which he had grown *Staphylococcus* bacteria. He noticed that one of the plates had a green mould growing on it. The mould was *Penicillium notatum*. The mould was circular in shape, and the area around it seemed to be free of *Staphylococcus*. On other areas of the plate, the bacteria were continuing to grow well. Fleming deduced that the bacteria around the circular mould had been killed off by a substance produced by the mould.

Fleming discovered that the mould could kill other bacteria and that it could be given to small animals without any harmful effects. However, he then moved on to other research and it was not until 10 years later that Howard Florey and Ernst Chain, working at Oxford University, isolated the bacteria-killing substance, penicillin, produced by the mould. Chain was a German chemist and Florey an Australian pathologist. It was Chain who isolated and purified penicillin and Florey who tested its safety to use on animals and humans. The first tests the team carried out on mice in 1940 would not have met the stringent standards for testing on animals today. Eight mice were given lethal doses of streptococci bacteria. Half the mice were then given injections of penicillin. The following day all the untreated mice were found to be dead but those that had been given penicillin survived.

One of the first uses of penicillin was in 1941, when Dr Charles Fletcher gave it to patient at a hospital in Oxford who was near to death as a result of bacterial infection in a wound. Fletcher used some penicillin on the patient and the wound made a spectacular recovery. Unfortunately, Fletcher did not have sufficient penicillin to clear the patient's body of bacteria and he died a few weeks later as the pathogen regained a hold.

An American brewing company began mass production of penicillin and soon sufficient was available to treat all the bacterial infections among the troops fighting in World War II. Penicillin was nicknamed 'the wonder drug' and in 1945 Fleming, Chain and Florey shared the Nobel Prize for Medicine, for its discovery and development.

Deductive reasoning occurs when a scientist works from general information to the more specific. It is known as a 'top-down' approach because it begins with a very broad spectrum of information and leads to a specific conclusion.

Inductive reasoning works the opposite way, moving from specific observations to broader theories. It is known as a 'bottom up' approach. A scientist begins with specific observations, starts to look for patterns and then formulates a hypothesis to investigate which may lead to more general conclusions or theories. Inductive reasoning is a method of thinking that involves using a set of specific facts to come to a general conclusion.

Questions to consider

- Was the discovery of penicillin an example of inductive reasoning? If not, why not?
- How useful are the two types of reasoning in science?

Questions to consider
- Why did the discovery of penicillin have such a profound effect on people at the time?
- Why was the collaboration of three scientists vital to the discovery of penicillin?
- What are the ethical issues involved in using a new drug for the first time? Was Fletcher right to use penicillin on his patient?
- The test on the safety of penicillin used by Florey would not be accepted today. What are the risks associated with the development of new medicines?

Test yourself
12 Define the term 'pathogen'.
13 Describe what is meant by the term 'antigen'.
14 State why antibiotics are not effective in treating viral diseases.
15 Distinguish between the roles of a 'phagocyte' and a 'lymphocyte'.

6.4 Gas exchange

All living cells need energy for their activities. Energy is released from the breakdown of glucose and other substances during the process of cell respiration. **Respiration** is a chemical reaction that occurs in mitochondria and the cytoplasm and releases energy as ATP, a form that can be used inside cells (Subtopics **1.2** and **8.2**).

Our cells use oxygen to carry out **aerobic respiration** and produce carbon dioxide as a waste product. Oxygen is taken in from the air and carbon dioxide is returned to it in a passive process known as **gas exchange**. Gas exchange occurs in the alveoli of the lungs where oxygen from the air diffuses into blood capillaries, and carbon dioxide passes in the opposite direction. Gases are also exchanged in the tissues where oxygen diffuses into respiring cells and is exchanged for carbon dioxide.

Whenever diffusion occurs, there must always be a **concentration gradient** with a higher level of the diffusing substance in one area than in another. Air inside the alveoli contains a higher concentration of oxygen than the blood, so oxygen diffuses into the blood. Blood contains a higher level of carbon dioxide than inhaled air, so carbon dioxide diffuses into the alveoli.

For gas exchange to continue, these concentration gradients must be maintained. As oxygen diffuses out of the alveoli, the level of oxygen inside them gradually falls and the level of carbon dioxide rises. Stale air with high levels of carbon dioxide and low levels of oxygen must be expelled regularly and replaced with a fresh supply to restore the concentration gradients of the two gases. This is achieved by breathing in and out, a process known as **ventilation**.

Learning objectives

You should understand that:
- Ventilation of the lungs maintains a concentration gradient of both oxygen and carbon dioxide between air in the alveoli and blood in adjacent capillaries.
- Thin alveolar cells (type I pneumocytes) are adapted to carry out gas exchange.
- Type II pneumocytes create a moist surface inside the alveoli by producing a solution containing surfactant, which reduces surface tension and ensures that the sides of alveoli do not adhere to each other.
- Air is carried to the lungs and alveoli via the trachea, bronchi and bronchioles.
- Muscle contractions allow ventilation of the lungs by causing pressure changes in the thorax that force air in and out.
- Different muscles are used for inspiration and expiration because muscles only do work as they contract.

The human ventilation system

Our lungs are protected inside the thorax in an air-tight cavity formed by the ribs and diaphragm. The inside of the ribcage is lined with membranes that secrete fluid to lubricate the lungs, making them slippery and reducing friction during breathing. Air is drawn in through the nose and passes down the **trachea** to the two **bronchi** (singular **bronchus**), one of which passes to each lung (Figure **6.23**). Bronchi divide into smaller and smaller tubes, called **bronchioles**, which end in tiny air sacs or **alveoli** (singular **alveolus**). The alveoli are covered with a network of capillaries and together provide us with a very large surface area for the exchange of oxygen and carbon dioxide.

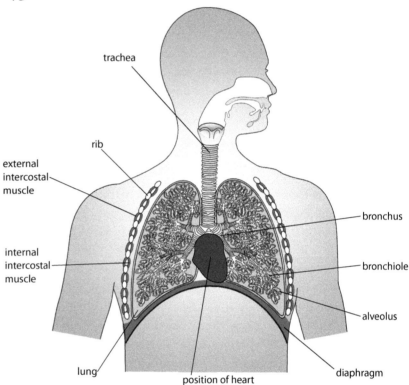

Figure 6.23 The ventilation system.

Tidal volume volume of air breathed in and out in a single breath

Inspiratory reserve volume breathed in by a maximum inhalation at the end of a normal inhalation

Expiratory reserve volume exhaled by a maximum effort after a normal exhalation

Residual volume volume of air remaining in the lungs at the end of a maximum exhalation

Measuring changes in lung volume

A spirometer (Figure **6.24**) is used to measure the amount of air that is exchanged during breathing. It can also measure the rate of breathing – for example, when a person is at rest or during or after exercise. A simple spirometer has a chamber filled with oxygen or air, which is inverted over a container of water. The subject is connected to the chamber via a tube and mouthpiece (the nostrils are closed with a nose clip). As the subject inhales and exhales a trace is produced on a rotating drum or computer monitor. Inhalation causes the chamber to fall, producing a falling line on the trace. Exhalation causes the chamber to rise and produces a rising line. The trace is called a spirogram and various volume measurements can be made from it (Figure **6.24**). Usually soda lime is used to absorb carbon dioxide that is exhaled.

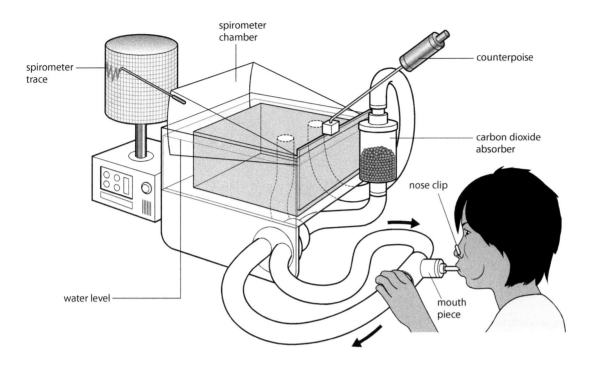

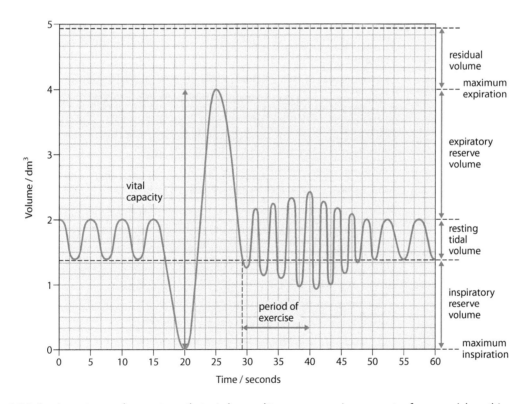

Figure 6.24 A spirometer produces a trace that can be used to measure various aspects of a person's breathing.

6 HUMAN PHYSIOLOGY

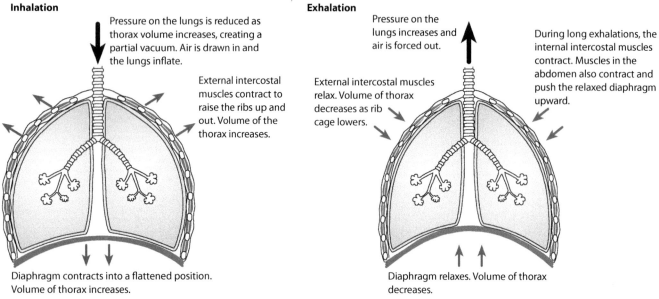

Figure 6.25 The mechanism of ventilation.

Blowing a trumpet

External and internal intercostal muscles are antagonistic muscles because they have opposite effects when they contract. During relaxed breathing, internal intercostal muscles do not contract, but they are used to force air out of the lungs when we sing or play a wind instrument.

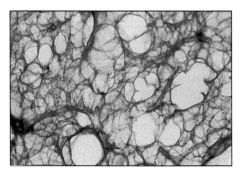

Figure 6.26 Healthy human lung tissue, showing many alveoli (×34).

Mechanism of ventilation

Lungs have no muscles and cannot move by themselves. Breathing is brought about by two sets of **intercostal muscles** between the ribs, and by the **diaphragm**, the sheet of muscle separating the thorax from the abdomen (Figure **6.25**).

During **inhalation**, contraction of the external intercostal muscles raises the ribs and contraction of the diaphragm lowers the floor of the thorax. These movements increase the volume of the chest cavity and lower the pressure on the lungs to below that of the air outside. As a result, air is drawn down the trachea to fill the lungs.

Gentle **exhalation** occurs as the intercostal and diaphragm muscles relax, reducing the volume of the chest cavity. Elastic fibres around the alveoli return to their original length and pressure forces air out of the lungs.

Long or forced exhalations involve the internal intercostal muscles, which contract to lower the ribs. Muscles in the abdominal wall also contract and push the relaxed diaphragm upward. Pressure inside the chest cavity increases and air is forced out of the lungs.

External and internal intercostal muscles are an example of antagonistic muscles. The work of external intercostals produces an opposite effect to the contraction of the internal intercostals. The muscles of the diaphragm are also antagonistic to the abdominal muscles. Remember that muscles can only produce an effect and do work when they contract.

Importance of alveoli

Alveoli are the body's gas exchange surfaces. Formed in clusters at the ends of the smallest bronchioles, more than 300 million alveoli in each lung together provide a surface area of about 75 m². Alveoli are roughly spherical in shape and made of pneumocyte cells less than 5 μm thick (Figure **6.26**). The capillaries that wrap around them also have thin walls

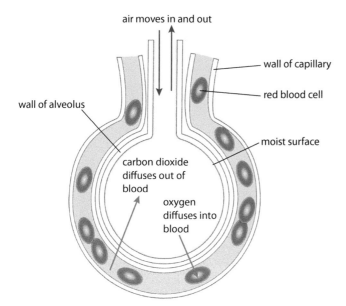

Figure 6.27 Gas exchange in the alveolus.

of single epithelial cells. These two thin layers make the distance for diffusion of gases as small as possible. Oxygen diffuses through the alveolus and capillary into the blood and carbon dioxide diffuses in the opposite direction (Figure 6.27). So long as the diffusion gradient is maintained by regular breathing, diffusion continues.

The walls of the alveoli contain two types of cells called **pneumocytes**. **Type I pneumocytes** cover most of the surface (97%) of the air sac and are responsible for gas exchange. Type I cells are about twice as numerous as **type II pneumocytes**, which are larger, rounder cells that produce and secrete a liquid containing a surfactant. The surfactant reduces surface tension and prevents the sides of the alveolus sticking to one another. Type I pneumocytes cannot divide but if they are damaged type II cells can divide to replace them (see Option D, Figure D.27).

Table **6.5** summarises ways in which the alveoli are well adapted for their role in gas exchange.

Feature of alveoli	Importance
many small, spherical alveoli	provide a large area for gas exchange
thin walls of flattened single cells	short diffusion distance
rich blood supply from capillaries	maintains concentration gradient and carries absorbed gases away rapidly

Table 6.5 Adaptations of alveoli for gas exchange.

Emphysema

Emphysema is a condition in which the surface area of the alveoli available for gas exchange is reduced. It can be caused by environmental factors including cigarette smoking. Chemicals in tobacco smoke cause white blood cells, known as neutrophils, to secrete an enzyme called neutrophil elastase, which breaks down elastic fibres in the alveoli. The walls of the alveoli may break down so that millions of tiny alveoli are replaced with larger air spaces. The area for gas exchange decreases and a person suffering from emphysema has difficulty getting sufficient oxygen into their blood. Alveoli also become less elastic and so it becomes even more difficult to ventilate the lungs. The damage is irreversible and patients with emphysema may need to rely on a supply of oxygen from a cylinder to help them walk and carry out everyday tasks without becoming breathless.

Premature babies have immature type II pneumocytes, which only begin to develop fully after 24 weeks. These infants are likely to suffer with Infant Respiratory Distress syndrome due to the lack of surfactant in their alveoli. After about 35 weeks' gestation the production of surfactant is sufficient. A gene MUC1 which is associated with type II pneumocytes has also been identified as a marker in lung cancer.

Nature of science

Obtaining evidence – understanding the causes of lung cancer

Epidemiology is defined as the study of the state of health of a population and the use of the information that is collected to address health-related issues. Our understanding of lung cancer has grown as a result of pioneering epidemiological studies, which began in the 1950s and continued until the early part of this century.

Lung cancer was uncommon before cigarette smoking became popular and in 1912 only 374 deaths from lung cancer were recorded worldwide. But additional data from post mortems later revealed that the incidence of lung cancer increased from 0.3% in 1852 to approximately 6% in 1952. Fritz Lickint, a German doctor, was one of the first to consider a link between smoking and lung cancer in 1929, but his evidence was largely circumstantial.

In 1951 the Medical Research Council in the UK set up a study, known as The British Doctors' survey, to try to understand why rates of lung cancer had risen to such an extent. The study was run by Dr Richard Doll and Austin Bradford Hill and involved the collection of data and medical information from 34 439 male doctors. The doctors responded to questionnaires and the study was one of the first to demonstrate the importance of epidemiology and statistics. The doctors who replied were divided into categories based on the decade of their birth, as well as general physical health and smoking habits and their cause-specific mortality. The study continued, using the same subjects, for 50 years and additional data was collected using further questionnaires in 1957, 1966, 1971, 1978, 1991 and finally in 2001.

The preliminary results of the study were published in 1956 at a time when smoking was not considered a serious health problem, but the true understanding of the link between smoking and lung cancer and other respiratory diseases only followed later as additional data was added.

A major conclusion of the study is that smoking decreases life expectancy up to 10 years. Men born between 1900 and 1930, who smoked, died on average 10 years younger than non-smokers. Their deaths were significantly more likely to be caused by diseases such as lung cancer and diseases of the cardiovascular system.

You can read Richard Doll's summary of the study 'Mortality in relation to smoking: 50 years' observations on male British doctors' in the *British Medical Journal* – use a search engine to find the article using key words such as Richard Doll smoking study *BMJ*.

When a person has lung cancer, the processes that usually control cell division by mitosis (Subtopic **1.6**) break down, so that cells in the lung tissue grow and divide uncontrollably. This causes a tumour to develop, which occupies space and gradually inhibits the ability of the lungs to exchange gases, putting a strain on the heart and eventually causing premature death.

? Test yourself

16 List the structures that a molecule of oxygen would pass on its way into an alveolus from the atmosphere.
17 List the **three** key characteristics of a gas exchange surface.
18 List the muscles involved in inhalation.
19 Distinguish between 'gas exchange' and 'ventilation'.

6.5 Neurons and synapses

The human nervous system

The nervous system consists of **neurons**, or nerve cells, which transmit information in the form of nerve impulses. Figure **6.28** shows the basic layout of the human nervous system.

The **central nervous system**, or CNS, is made up of the neurons of the brain and the spinal cord. The CNS receives information from sensory receptors all over the body. Information is processed and interpreted before the CNS initiates suitable responses.

The **peripheral nerves** are the network of neurons that carry information to and from the CNS. Peripheral nerves include sensory neurons, which carry information to the CNS, and motor neurons, which transmit impulses from the CNS to muscles and glands that then cause a response.

Three types of neuron are found in the nervous system. **Sensory** and **motor neurons** transmit information to and from the CNS, while **relay neurons** within the CNS form connections between them.

The structure of a motor neuron is shown in Figure **6.29**. Many small **dendrites** receive information from relay neurons and transmit the impulses to the cell body. One long **axon** then carries impulses away. The cell body contains the nucleus and most of the cytoplasm of the cell. The axon is covered by a **myelin sheath** formed from Schwann cells, which wrap themselves around it. Myelin has a high lipid content and forms an electrical insulation layer that speeds the transmission of impulses along the axon.

Transmission of nerve impulses

Neurons transmit information in the form of **impulses**, which are short-lived changes in electrical potential across the membrane of a neuron. All neurons contain sodium (Na^+) and potassium (K^+) ions. Impulses occur as these important ions move in and out through the plasma membrane.

Learning objectives

You should understand that:
- Cells called neurons transmit electrical impulses.
- Nerve fibres that are covered with myelin sheaths allow salutatory conduction.
- Sodium and potassium ions are pumped across the membranes of neurons to generate a resting potential.
- An action potential consists of the rapid depolarisation and repolarisation of the neuron membrane.
- Nerve impulses are action potentials that are propagated along the axons of neurons.
- A nerve impulse is transmitted as a result of local currents that cause each successive part of the axon to reach the threshold potential.
- A synapse is the junction between two neurons or between neurons and receptor or effector cells.
- Presynaptic neurons release a neurotransmitter into the synapse when they are depolarised.
- A nerve impulse is only initiated if the threshold potential is reached.

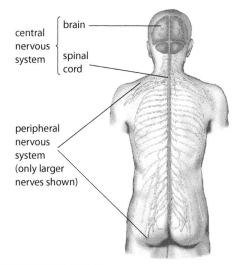

Figure 6.28 The human nervous system.

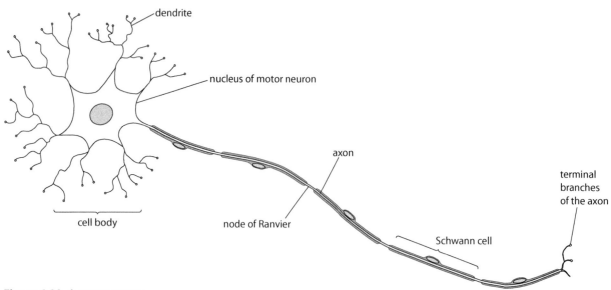

Figure 6.29 A motor neuron.

When a neuron is not transmitting an impulse, it is said to be at its **resting potential**. The resting potential is the potential difference across the plasma membrane when it is not being stimulated – for most neurons, this potential is −70 mV. The inside of the axon is negatively charged with respect to the outside (Figure 6.30).

As a nerve impulse occurs, the distribution of charge across the membrane is reversed. For a millisecond, the membrane is said to be **depolarised**. As charge is reversed in one area of the axon, local currents depolarise the next region so that the impulse spreads along the axon (Figure 6.31). An impulse that travels in this way is known as an **action potential**.

Figure 6.32 explains what is happening at the plasma membrane of the neuron as an action potential is generated.

1 When a neuron is stimulated, gated sodium channels in the membrane open and sodium ions (Na⁺) from the outside flow in. They follow both the electrical gradient and the concentration gradient, together known as the electrochemical gradient, to move into the cell. The neuron is now said to be depolarised.

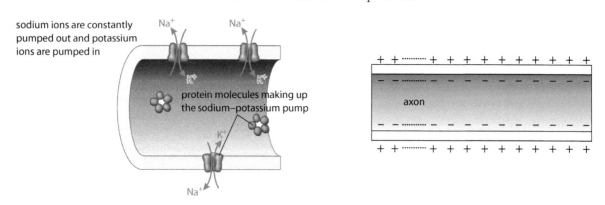

Figure 6.30 At rest, sodium ions are pumped out of the neuron and potassium ions are pumped in, to establish the resting potential. Inside the neuron is negatively charged because of the presence of chloride and other negative ions.

2 For a very brief period of time, the inside of the axon becomes positively charged with respect to the outside as sodium ions enter. At this point, the sodium channels close.
3 Now, gated potassium channels open and potassium ions (K⁺) begin to leave the axon, moving down their electrochemical gradient to re-establish the resting potential, a process known as repolarisation.
4 Because so many potassium ions start to move, the potential difference falls below the resting potential. At this point, both sodium and potassium channels close. The resting potential is re-established by the action of sodium–potassium pumps, which move ions back across the membrane.

An action potential in one part of an axon causes the depolarisation of the adjacent section of the axon. This occurs because local currents are set up between adjacent regions and these cause ion channels to open, allowing sodium ions in and potassium ions out of the axon and cause each successive part of the axon to reach its threshold potential and become depolarised.

The action potential travels along the neuron rather like a 'Mexican wave'. The impulse can only pass in one direction because the region behind it is still in the recovery phase of the action potential and is temporarily unable to generate a new action potential. The recovery phase is known as the refractory period.

The speed of conduction along an axon is affected by the diameter of the axon. A larger diameter means faster conduction. Larger axons are myelinated while smaller ones are not. At intervals along myelinated axons are gaps between the myelin covering known as nodes of Ranvier (Figure 6.29). The sheath prevents the flow of ions across the membrane so the current must jump from node to node – a process known as saltatory conduction, which speeds up the transmission of the nerve impulse.

The synapse

A synapse is the place where two neurons meet. Two neurons do not touch one another and the tiny gap of about 20 nm between them is known as the **synaptic cleft**. Action potentials must be transmitted across this gap for the impulse to pass on its way and this is achieved by the presence of chemicals known as neurotransmitters. Neurotransmitters are held in vesicles in the pre-synaptic cell until an action potential arrives. They are then released into the synaptic cleft, and diffuse across to the post-synaptic membrane. There they can cause another action potential to be produced and a nerve impulse to be initiated, provided the threshold potential is reached.

The synapse shown in Figure 6.33 uses the neurotransmitter acetylcholine (ACh) and is a cholinergic synapse. ACh binds to receptors and causes depolarisation of the post-synaptic membrane and the initiation of an action potential. Once an action potential is generated in the post-synaptic membrane, ACh in the synaptic cleft is deactivated by acetylcholinesterase enzymes and the products are reabsorbed by the pre-

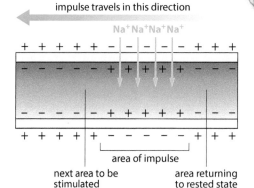

Figure 6.31 When an impulse passes along the neuron, sodium ions diffuse via ion channels and the potential is reversed. This process is called an action potential (only a small part of a long neuron is shown).

Resting potential the electrical potential across the plasma membrane of a neuron that is not conducting an impulse

Action potential the reversal and restoration of the resting potential across the plasma membrane of a neuron as an electrical impulse passes along it

Threshold potential the electrical potential across the plasma membrane of a neuron that is required in order to trigger an action potential

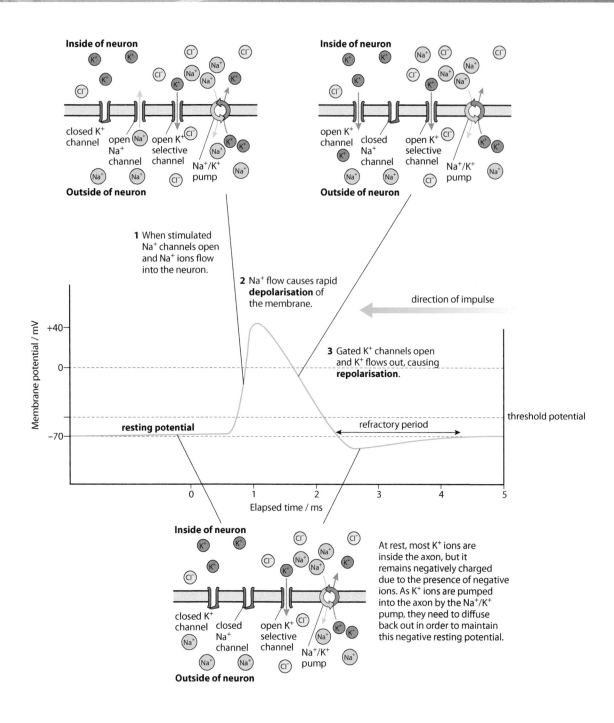

Figure 6.32 The action potential.

synaptic membrane to be remade and repackaged in vesicles.

There are more than 40 different neurotransmitters in the body. Acetylcholine and noradrenalin are found throughout the nervous system, others (for example, dopamine) are found only in the brain.

Many drugs and toxins affect synapses and influence the way nerve impulses are transmitted. Nicotine is an excitatory drug. It has a similar molecular shape to acetylcholine and affects the post-synaptic membrane so that it transmits an action potential. Cocaine and amphetamines are also

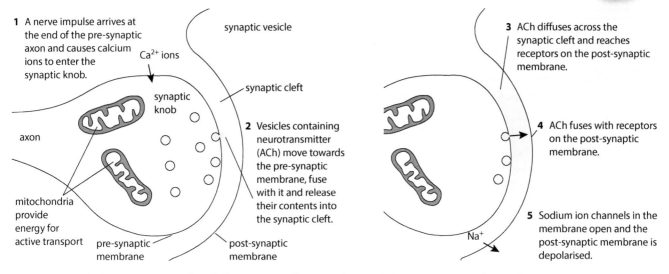

Figure 6.33 A cholinergic synapse. The whole sequence of events of transmission at a synapse takes 5–10 ms.

excitatory drugs, which stimulate synaptic transmission. Benzodiazepenes, cannabis and alcohol suppress the activity of the nervous system. Cannabis binds to receptors in the brain and blocks synaptic transmission. These drugs affect a person's behaviour and some are highly addictive. Used incorrectly they can cause social problems and may result in long-term damage to the body.

Neonicotinoids are chemical pesticides used in insecticides. They are similar in structure to nicotine and block transmission at the synapses of insects by binding to acetylcholine receptors. Neonicotinoid pesticides have been linked to the decline of bee populations throughout the world. In 2013, the European Food Safety Authority declared that these insecticides pose an unacceptable high risk to bees and the European Commission imposed restrictions on their use.

Tetanus toxin is a neurotoxin produced by the tetanus bacterium *Clostridium tetani*. The disease tetanus affects the muscles causing them to go into spasm as the toxin binds irreversibly to acetylcholine receptors in post-synaptic membranes and prevents the transmission of nerve impulses at synapses at neuromuscular junctions. Another neurotoxin produced by the bacterium *Clostridium botulinum* is one of the most toxic substances known. Nevertheless it has been used in recent years for cosmetic and medical procedures under the trade name Botox. In the early 1980s, it was used to treat muscle spasm, strabismus (squint) and uncontrollable blinking. More recently it has been used to induce temporary muscle paralysis and conceal the appearance of wrinkles.

Test yourself

20 Explain how sodium and potassium ions establish a resting potential.
21 Define the term 'action potential'.
22 List the key events of synaptic transmission.
23 State what is meant by the term 'saltatory conduction'.
24 Name the substance released by a pre-synaptic membrane.

Learning objectives

You should understand that:
- α and β cells in the pancreas secrete insulin and glucagon to control blood glucose.
- Thyroxin, secreted by the thyroid gland, regulates the metabolic rate and helps control body temperature.
- Leptin is a hormone secreted by adipose cells that acts on the hypothalamus in the brain to control appetite.
- Melatonin, secreted by the pineal gland, controls circadian rhythms.
- A gene on the Y chromosome causes the development of gonads in a male embryo so that they form testes and secrete testosterone.
- Testosterone causes the development of male genitalia before birth and also sperm production and development of male secondary sexual characteristics at puberty.
- Estrogen and progesterone cause the development of female genitalia before birth and development of female secondary sexual characteristics at puberty.
- The menstrual cycle is controlled by positive and negative feedback involving hormones from the ovary and pituitary gland.

One decimetre cubed is the same volume as one litre.

Negative feedback occurs when a deviation to the normal level is detected and corrective mechanisms act to return the system to normal. Many homeostatic mechanisms work by negative feedback.

6.6 Hormones, homeostasis and reproduction

Homeostasis

The internal environment of the body remains constant, within certain limits, despite changes that occur in the external environment. The control process that maintains conditions within these limits is known as **homeostasis**.

The factors that are controlled include water balance, blood glucose concentration, blood pH, carbon dioxide concentration and body temperature.

Each of these has a 'normal' or set point although they may vary slightly above or below it. For example, the normal body temperature for humans is about 37 °C, which is optimum for the efficient functioning of the body's enzymes and cell processes.

Both the nervous system and the endocrine system are involved in homeostasis. The **endocrine system** consists of ductless **endocrine glands**, which release different hormones (Figure **6.34**). **Hormones** circulate in the bloodstream but each one is a chemical messenger that only affects the metabolism of specific target cells.

Insulin and glucagon, and control of blood glucose

Blood glucose level is the concentration of glucose dissolved in blood plasma. It is expressed as millimoles per decimetre cubed ($mmol\,dm^{-3}$). Normally blood glucose level stays within narrow limits, between $4\,mmol\,dm^{-3}$ and $8\,mmol\,dm^{-3}$, so that the osmotic balance (Subtopic **1.4**) of the blood remains constant and body cells receive sufficient glucose for respiration. Levels are higher after meals as glucose is absorbed into the blood from the intestine and usually lowest in the morning as food has not been consumed overnight.

Glucose levels are monitored by cells in the pancreas. If the level is too high or too low, α and β cells in regions of the pancreas known as the islets of Langerhans produce hormones that turn on control mechanisms to correct it. This is an example of **negative feedback**. Table **6.6** summarises these responses.

	Responses to a rise in blood glucose above normal	Responses to a fall in blood glucose below normal
Pancreas	β cells in the pancreas produce the hormone **insulin**	α cells in the pancreas produce the hormone **glucagon**
Glucose uptake or release	insulin stimulates cells in the liver and muscles to take in glucose and convert it to **glycogen** and fat, which can be stored inside the cells – blood glucose levels fall	glucagon stimulates the hydrolysis of glycogen to glucose in liver cells – glucose is released into the blood

Table 6.6 The body's responses to changes in blood glucose.

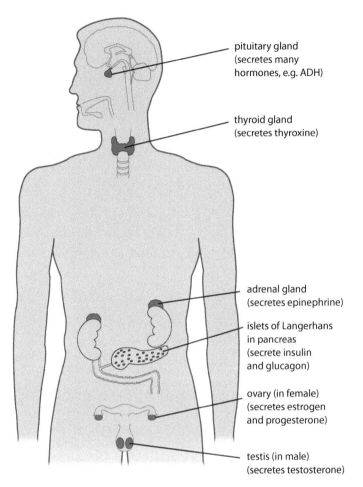

Figure 6.34 The positions of some endocrine glands in the human body. Endocrine glands have no ducts and secrete hormones directly into the bloodstream, which carries them to target cells.

Diabetes

Diabetes is the inability of the body to control blood glucose level. A diabetic person will experience wide fluctuations in their blood glucose above and below the normal limits (Figure **6.35**).

In **Type I diabetes**, the β cells in the pancreas do not produce insulin. Without insulin, glucose is not taken up by body cells so blood levels remain high, a condition known as hyperglycemia (Figure **6.36**). Excess glucose is excreted in urine and its presence is used to diagnose diabetes. About 10% of diabetics have type I diabetes. Symptoms usually begin in childhood, which is why type I diabetes is sometimes known as 'early onset' diabetes.

Type II diabetes is the most common form of diabetes, accounting for nine out of ten cases worldwide. The pancreas does produce insulin although levels may fall as the disease progresses. Type II diabetes occurs when body cells fail to respond to the insulin that is produced. Again, the result is that blood glucose levels remain too high. This form of diabetes is also known as late-onset diabetes or non-insulin-dependent diabetes mellitus. Individuals who have the condition develop insulin resistance, which means that the receptor cells that normally respond to insulin fail

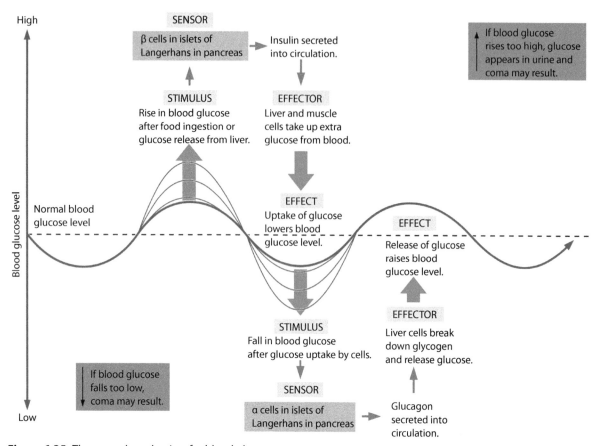

Figure 6.35 The control mechanism for blood glucose.

to be stimulated by it, even though the β cells in the pancreas still produce insulin. This type of diabetes is often associated with obesity, age, lack of exercise and genetic factors. Diabetics must monitor their blood glucose level carefully so that they can control it, since the body's internal control mechanism is not working properly.

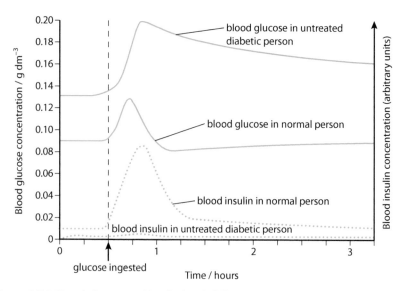

Figure 6.36 Blood glucose and insulin levels following intake of glucose in a normal person and a person with untreated type I diabetes.

Causes and symptoms of diabetes

Type I diabetes is caused when the β cells in the pancreas do not produce insulin. This can be a result of autoimmune disease in which the body's immune system destroys its own β cells.

The causes of type II diabetes are not fully understood but there is a strong correlation of risk with weight and diet. High levels of fatty acids in the blood may be a factor causing the condition and people whose diets are high in fat but low in fibre seem to be most at risk. Obesity, associated with a lack of exercise or a genetic makeup that influences fat metabolism, is a key risk factor. The condition is more common in older people but there are an increasing number of cases in overweight children.

Studies of ethnic groups worldwide have shown that some are more likely to develop type II diabetes, and this provides evidence for a genetic predisposition to the condition. Aboriginal Australians, people of Asian and Afro-Caribbean origin, Native Americans and Polynesian Maori peoples all have a higher incidence of diabetes than would occur by chance.

The symptoms of diabetes include:
- high glucose levels in the blood
- glucose in the urine
- frequent need to urinate, which leads to dehydration and increased thirst
- tiredness and fatigue
- some loss of weight.

In type II diabetes, these symptoms tend to develop slowly.

In the last decade, there has been a large increase in the number of people in industrialised countries affected by type II diabetes. Can you suggest a reason for this?

Treatment of diabetes

Type I diabetes must be controlled by regular insulin injections, but many people who have type II diabetes are advised to control their blood sugar levels by following a healthy diet, taking exercise and losing weight. They are advised to eat foods that are low in saturated fat and salt but high in fibre and complex (slowly absorbed) carbohydrates, such as wholegrain cereals, pulses, beans and lentils. These foods, especially if they are taken at regular intervals during the day, help to keep blood sugar levels steady. Foods that should be avoided include sugary snack foods and drinks, and food with a high level of saturated fat. These foods cause a rapid rise in blood sugar level that the diabetic person is unable to deal with.

If left untreated, type II diabetes can lead to long-term health problems such as kidney disease, retinal damage, high blood pressure, stroke and heart attack.

Glycaemic index

Some health professionals recommend that people who have or are at risk from type II diabetes follow a low GI (glycaemic index) diet. Low GI foods allow the body to absorb carbohydrates more slowly, which helps to stabilise glucose levels through the day. Low GI foods include apples, oranges, pasta, sweet potato, sweetcorn and noodles.

Thyroxin and control of metabolic rate

Thyroxin is a hormone produced by the thyroid gland situated in the neck (Figure **6.34**). Thyroxin contains iodine, which is a dietary requirement for a healthy thyroid gland. Thyroxin controls the body's basal metabolic rate (the rate of the body's metabolism at rest) and so the hormone is also important in controlling growth. If a person has insufficient thyroxin present during childhood, their physical and mental development does not proceed properly and a condition known as cretinism may result. A person with this condition will be short in stature and have severe mental retardation. In adults, a shortage of thyroxin caused by an underactive thyroid (hypothyroidism) leads to a decreased metabolic rate and an accumulation in fat, as well as general decrease in physical and mental activity. In contrast, an overactive thyroid gland (hyperthyroidism) can cause greatly increased metabolic activity, loss of body mass, an increased heart rate and a general increase in physical and mental activity. Both hypothyroidism and hyperthyroidism can lead to a condition known as goitre, which is a swelling of the thyroid gland in the neck.

The majority of cases of goitre used to result from thyroxin deficiency caused by a lack of iodine in the diet, but a UNICEF programme promoting the use of iodised salt has reduced the incidence of the condition worldwide over recent years.

Normal levels of thyroxin are maintained by homeostasis, in a negative feedback system (Figure **6.37**). In normal conditions, thyroxin secretion is triggered by a stimulating hormone produced by the pituitary gland. If too much thyroxin is present, it inhibits the pituitary gland and production is decreased. The stimulating hormone from the pituitary gland is in turn influenced by the hypothalamus. The hypothalamus responds to other influences such as temperature so that, in some animals, metabolism can be aligned to environmental factors such as the seasons. Thyroxin produced as a result of stimulation by the hypothalamus can speed up metabolism of protein, fat and carbohydrates and increase body temperature by generating metabolic heat.

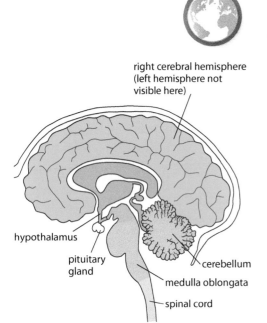

Figure 6.37 The hypothalamus and pituitary gland. The hypothalamus links the hormonal and nervous systems by producing releasing hormones that regulate pituitary gland secretions.

Control of body temperature

Body temperature is monitored and controlled by the hypothalamus in the brain. The 'set point' for body temperature is 36.7 °C. The hypothalamus responds to nerve impulses from receptors in the skin and also to changes in the body's core temperature. If body temperature fluctuates above or below the set point, the hypothalamus coordinates responses to bring it back to normal. This is another example of negative feedback. Nerve messages are carried from the hypothalamus to organs that bring about warming or cooling of the body. Table **6.7** lists some of the body's responses to changes in temperature.

	Responses to a rise in body temperature	**Responses to a fall in body temperature**
Arterioles in the skin	dilate (widen) so that more blood flows to skin capillaries and excess heat is lost from the skin	narrow to restrict flow of warm blood to the skin capillaries – heat is retained in the body
Sweat glands	produce more sweat, which evaporates from the skin surface to cool it	cease production of sweat
Muscles	remain relaxed	muscular activity such as shivering generates heat
Metabolic rate	may decrease to minimise heat production	thyroxin increases metabolic rate

Table 6.7 The body's responses to changes in core temperature.

Leptin and control of appetite

Leptin is a protein hormone that is important in regulating energy intake and expenditure, including appetite and metabolism. Human leptin is a protein of 167 amino acids. It is manufactured mainly by cells in white adipose (fat) tissue, and the level of circulating leptin is directly proportional to the total amount of fat in the body. It can also be produced by brown adipose tissue, cells in the placenta, ovaries, skeletal muscle and gastric cells in the stomach.

Leptin acts on receptor cells in the hypothalamus and it inhibits appetite by:
- counteracting the effects of chemicals secreted by the gut and hypothalamus that stimulate feeding
- promoting the synthesis of α-MSH, a long-term appetite suppressant.

Medical use of leptin

Leptin signals to the brain that the body has had enough to eat. The absence of leptin, or the receptor for it, leads to uncontrolled food intake. A very small group of humans possess homozygous mutations (Subtopic **3.4**) for the leptin gene and this leads to a constant desire for food, resulting in severe obesity. Attempts have been made to control the condition by giving patients doses of human leptin. However, clinical trials have been inconclusive. It was found that large and frequent doses were needed and only provided small improvements because leptin is quickly broken down in the blood stream and has low potency and poor solubility.

The effects of leptin were observed by studying mutant obese mice that appeared within a laboratory mouse colony in 1950. These mice were massively obese (Figure **6.38**) and excessively voracious. Since that time several strains of laboratory mice have been found to be homozygous for single-gene mutations that cause them to become grossly obese, and they fall into two classes: those having mutations in the gene for the protein hormone leptin, and those having mutations in the gene that encodes the receptor for leptin. When mice from the first group are treated with injections of leptin, they lose their excess fat and return to normal body weight. Leptin itself was discovered by Jeffrey M. Friedman in 1994 through the study of such mice.

Figure 6.38 Mice with a mutation in the gene for leptin production are hugely obese compared to a normal mouse.

Melatonin and control of circadian rhythms

Melatonin is a hormone produced by the pineal gland, a pea-sized gland located just above the middle of the brain. One of the key influences of melatonin is to maintain the body's **circadian rhythms** and especially sleep–wake cycles. A vital factor in human sleep regulation is the exposure to light or darkness. Exposure to light stimulates a nerve pathway from the retina in the eye to the hypothalamus. Cells in the hypothalamus (the supra-chiasmatic nucleus, SCN) send signals to parts of the brain that control hormones, body temperature and other functions that have a role in our feelings of sleep or wakefulness. The SCN produces a signal that can keep the body on an approximately 24-hour cycle of activity. But the 'internal clock' is not exactly set to 24 hours and environmental clues, the most important of which is light, are needed to reset the clock each morning and keep a person in step with their external environment.

A number of hormones are used in medical treatments to correct imbalances that may occur.

Hormone replacement therapy (HRT) – given to some women to alleviate the symptoms caused by the sudden change in hormone levels at menopause

Thyroxin – given to patients suffering from hypothyroidism and a shortage of the hormone.

Human Growth Hormone – given to children to treat growth disorders.

Circadian rhythms biological processes that vary naturally in a cycle lasting about 24 hours; these rhythms indicate the right times for eating and sleeping, and also regulate hormone production

During the day, the pineal gland is inactive, but during the hours of darkness, it is 'turned on' by the SCN and begins to produce melatonin, which is released into the blood. Rising levels of melatonin cause our feelings of sleepiness. The level of melatonin remains high for about 12 hours until the following morning when light causes it fall to a minimal level in the blood.

Even if the pineal gland is stimulated, it will not produce melatonin unless a person is in a dimly lit environment – even artificial indoor lighting can be bright enough to prevent the release of melatonin. The amount of melatonin released at night also varies between individuals, but on average children secrete more melatonin than adults.

Jet lag

Jet lag is caused by the disruption of the body's day–night rhythms caused by long-distance travel and arrival in a new time zone. After a long journey the day–night patterns may no longer correspond to the new environment and some people need many days to adjust.

Jet lag upsets the body clock because the expected patterns of light and darkness are out of alignment. Light is the strongest stimulus for the sleep–wake pattern so jet lag can be controlled by avoiding bright light so that the body clock is reset. Melatonin tablets are sometimes used to adjust a person's body clock but the effectiveness of these treatments is not proved. The effect of melatonin may be very short term and the correct doses and times to take the hormone are not easy to determine. Melatonin is not approved for sale in some countries. In the USA is can be bought in pharmacies but in other countries it is only available with a doctor's prescription.

Hormones and reproduction

Reproduction is one of the important characteristics of living things. Human male and female reproductive systems produce the gametes (the sperm cell and egg cell) that must come together to begin a new life. The two reproductive systems enable the gametes to meet and the female reproductive system provides a suitable place for fertilisation to occur and an embryo to develop (Figures **6.39** and **6.40**). The ovaries and testes also produce hormones that regulate sexual development and reproduction.

In the first weeks after conception, the sex of a fetus cannot be identified from its appearance. Only its karyotype (Subtopic **3.2**) will distinguish a male from a female fetus. The formation of the testes from undifferentiated gonadal tissue is dependent on **SRY gene** on the **Y chromosome**. From about 8 weeks after fertilisation, the fetal testes form and begin to secrete testosterone which causes the formation of the male genitalia. If no Y chromosome (and so no SRY gene) is present, female reproductive organs and ovaries form during the period between the second and sixth month of fetal development and ovarian follicles develop.

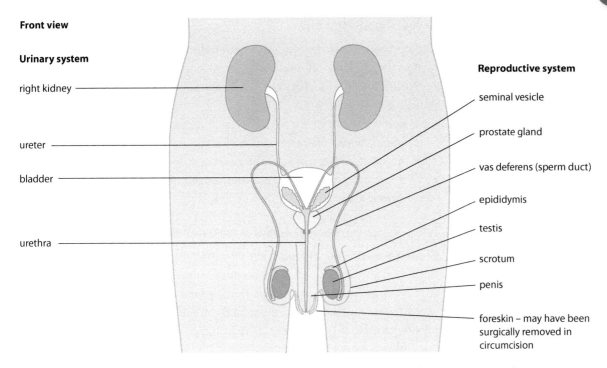

Figure 6.39 The male reproductive system. (The diagram also shows the organs of the urinary system.)

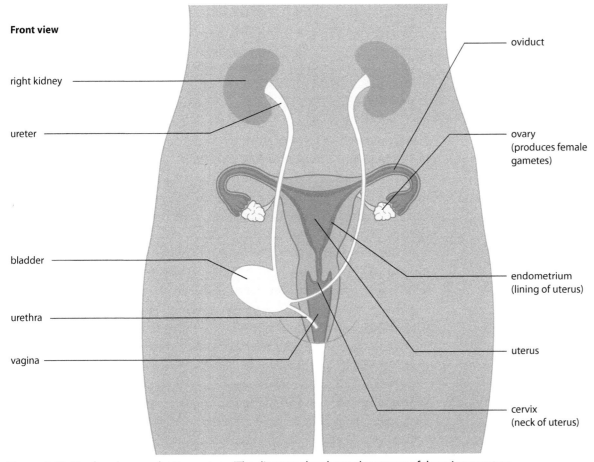

Figure 6.40 The female reproductive system. (The diagram also shows the organs of the urinary system – the bladder has been drawn to one side, to reveal the uterus.)

Roles of testosterone

The hormone **testosterone**, which is produced by the testes, has important roles in the sexual development and reproductive behaviour in males.
- During fetal development, testosterone causes the development of the male genitalia.
- At puberty, levels of testosterone rise and cause the development of male secondary sexual characteristics including growth of muscle, deepening of the voice, enlargement of the penis and growth of body hair.
- Testosterone stimulates the continuous production of sperm and behaviour associated with the sex drive.

Female sex hormones and the menstrual cycle

Ovaries produce two hormones, **estrogen** and **progesterone**. These hormones stimulate the development of female genitalia before birth and at puberty are responsible for the development of female secondary sexual characteristics including the onset of the menstrual cycle, development of breasts, growth of body hair and widening of the hips. The two hormones also influence the changes in the uterus lining during the menstrual cycle and pregnancy. The pituitary gland in the brain produces two further hormones, **luteinising hormone (LH)** and **follicle-stimulating hormone (FSH)**. FSH stimulates the development of immature follicles in the ovary, one of which will come to contain a mature egg cell. LH stimulates the follicle to release the egg and subsequently to form the corpus luteum.

Production of female gametes is a cyclical process, which lasts approximately 28 days. During the first half of this **menstrual cycle** the egg cell is produced and in the second half the uterus lining thickens to prepare for implantation of a fertilised egg. The cycle involves hormones that are released by the ovaries and the pituitary gland.

The sequence of events begins at the start of **menstruation**, which is often called a period (Figure **6.41**). During the first four or five days of the cycle, the endometrium (lining) of the uterus is shed and leaves the body through the vagina. This indicates that fertilisation has not occurred during the previous month.

In this early part of the cycle, the pituitary gland secretes FSH, which stimulates the development of an immature follicle in the ovary. The follicle then secretes estrogen, which enhances the follicle's response to FSH. At this stage of the cycle, estrogen has a positive feedback relationship with FSH (Figure **6.42**). Increasing levels of estrogen cause an increase in the level of FSH released by the pituitary gland. As the level of estrogen rises, it also stimulates the repair of the uterus lining.

As the follicle grows, estrogen levels rise to a peak at around day 12, when they stimulate the release of LH from the pituitary gland. As LH levels reach their highest point, **ovulation** – the release of the egg cell from the follicle – takes place. Ovulation usually occurs at around the day 14 of the cycle. Immediately afterwards, LH stimulates the empty follicle to form the **corpus luteum**. Levels of estrogen fall and as a result FSH and LH levels fall.

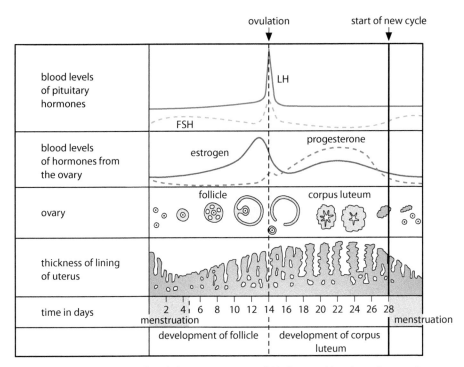

Figure 6.41 The menstrual cycle lasts an average of 28 days and involves changes in hormone levels that influence the follicles and lining of the uterus.

The corpus luteum secretes progesterone, which stimulates the thickening of the endometrium and prepares the uterus to receive an embryo. It also inhibits the production of FSH and LH.

If the egg cell is not fertilised, the corpus luteum degenerates and progesterone and estrogen levels fall. The fall in progesterone stimulates the breakdown of the uterus lining. FSH is no longer inhibited, so a new follicle is stimulated and the cycle begins again.

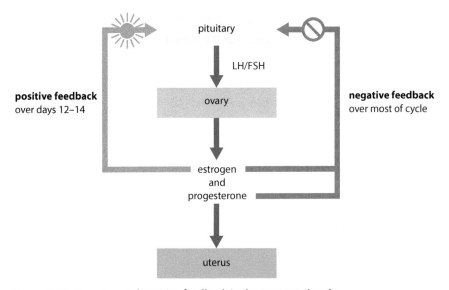

Figure 6.42 Negative and positive feedback in the menstrual cycle.

6 HUMAN PHYSIOLOGY

In vitro fertilisation

In vitro fertilisation (IVF) is a technique used to help couples who have been unable to conceive naturally. There are many reasons for infertility. Males may have a low sperm count, blocked or damaged sperm ducts or be unable to achieve an erection. Females may fail to ovulate or have blocked or damaged oviducts, or produce antibodies in cervical mucus that destroy sperm.

The first step in IVF treatment is an assessment of whether the couple are suitable for treatment. If so, the woman may first be injected with hormones to suppress her natural hormones before being given injections of FSH for about 10 days. In some treatments FSH may be given alone. This hormone causes a number of egg cells to mature at the same time in her ovaries (superovulation). Just before the egg cells are released from the follicles, they are collected using a laparoscope (a thin tubular instrument that is inserted through an incision in the abdominal wall). The egg cells are 'matured' in culture medium for up to 24 hours before sperm cells are added to fertilise them. Fertilised egg cells are incubated for about three days until they have divided to form a ball of cells. These embryos are checked to make sure they are healthy and developing normally. Usually two will be selected and placed into the woman's uterus for implantation. The pregnancy is then allowed to continue in the normal way. Any remaining embryos can be frozen and stored for use later. Figure **6.43** summarises the stages in IVF treatment.

The first successful IVF baby was Louise Brown, born in 1978 following her mother's IVF treatment. Professor Robert Edwards, who pioneered the technique, was awarded the Nobel Prize in Physiology or Medicine in 2010.

Ethical issues associated with IVF treatment

IVF has enabled men and women who would naturally be infertile to have children but it has also produced some serious ethical issues. Some of these are outlined in Table **6.8**. Each society needs to think about these issues and decide what should be done about them.

Arguments in favour of IVF	Arguments against IVF
• Enables infertile couples to have a family. • Couples willing to undergo IVF treatment must have determination to become parents. • Embryos used in IVF treatment can be screened to ensure they are healthy and do not have certain genetic conditions that would be inherited. • IVF techniques have led to further understanding of human reproductive biology.	• Unused embryos produced by IVF are frozen for a limited period and then destroyed. • Multiple births often result from IVF and this increases the risks to mother and babies. • Infertility is natural whereas IVF is not, some religions object to it on this basis. • Some causes of infertility are due to genetic conditions, which may be passed on to children born as a result of IVF. • There may be risks to the health of women who are treated with hormones during IVF.

Table 6.8 Arguments for and against IVF treatment.

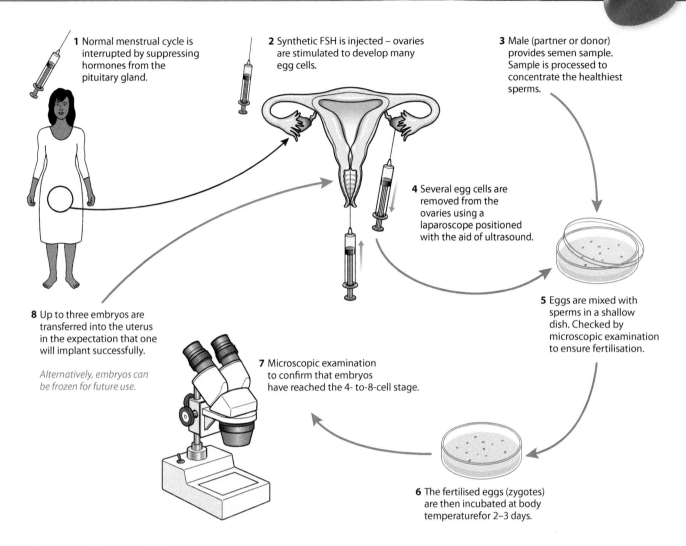

Figure 6.43 The stages in IVF treatment.

Born to give

My Sister's Keeper is a novel by Jodi Picoult, published in 2004. The book, which was made into a film in 2009, follows the life of 13-year-old Anna, who enlists the help of an attorney to sue her parents for rights to her own body. Anna's older sister Kate suffers from leukemia and Anna was conceived through IVF to be a genetic match and donor for her sister. Anna donates bone marrow and blood for her sister throughout her life. As Kate's condition worsens, their parents want Anna to donate a kidney to Kate after she goes into renal failure, but Anna files a lawsuit against her parents for medical emancipation from them despite the consequences for her sister's health.

This novel is a fictional story but already there have been real examples of children being conceived by IVF and born to provide stem cells for siblings who would otherwise die.

Questions to consider

1 Should IVF conceptions such as Anna's be permitted?
2 Who should have rights over a child's genetic material in cases like this?
3 Is it right that a child should be created simply to keep another person alive?

Nature of science

Scientific advance follows technical innovation – why was William Harvey unsuccessful in his research into reproduction?

William Harvey (1578–1657) was the leading anatomist of his day. As well as his discoveries on the circulation of blood (Section **6.2**), he also studied reproduction. He believed that humans and other mammals must reproduce through the joining of an egg and sperm in the same way as the chickens that he was able to observe. Before Harvey's research, scientists believed in a theory known as 'pre-formation', which assumed an embryo was a fully formed tiny version of a mature individual that simply grew in size in the mother's womb.

Harvey studied the development of hens' eggs and recorded his observations of the fetus, which he saw began as what looked like a single drop of blood and then further differentiated into an embryo which later became the chick. He rejected the idea that an exact replica of the organism could be found in reproductive material of either the male or the female.

William Harvey carried out meticulous experiments. He used hen's eggs, because they were inexpensive and abundant, but he also carried out detailed examinations of deer. Harvey was physician to King Charles I, who was fascinated by his research and supplied deer carcasses from his hunting trips for Harvey to dissect. This support enabled Harvey to develop his concepts of reproduction in mammals. He was never able to prove his theories and observe eggs and sperm because sufficiently powerful microscopes were not available at the time. But he was the first to suggest that humans and other mammals reproduced via the fertilisation of an egg by sperm. It took a further 200 years and advances in the production of microscopes before a mammalian egg was finally observed, but nonetheless Harvey's stature and experimental evidence meant that his theory was generally accepted during his lifetime.

Test yourself

25 State **two** hormones that help regulate blood glucose level.
26 State the site of action of the hormone leptin.
27 State the location of the gene that controls the development of sex in the fetus.
28 Outline the functions of the hormone progesterone in the menstrual cycle.

Exam-style questions

1 Which of the following statements is correct?

    A Large food molecules need to be digested so that they can be absorbed through the gut wall.
    B The liver and pancreas both release their digestive enzymes into the small intestine.
    C The approximate pH of the stomach is 8.
    D Assimilation is the transfer of the end products of digestion from the gut lumen to the blood. [1]

2 Which of the following statements is correct?

   A Blood passes from the atria to the ventricles through the semilunar valves.
   B The coronary and pulmonary arteries supply oxygenated blood to the heart muscle.
   C The pacemaker, brain and adrenal gland are all involved in the regulation of the heart beat.
   D A red blood cell travels from the lungs to the stomach via the right atrium, right ventricle and aorta. [1]

3 Which of the following statements is correct?

   A Antibiotics are effective against viruses but not bacteria.
   B Antigens are made by lymphocytes.
   C Phagocytes are white blood cells that can be found in the tissue fluid around cells.
   D HIV is a bacterium that reduces the number of active lymphocytes in the blood. [1]

4 Which of the following statements is correct?

   A During breathing in, the rib cage moves up and out and the pressure inside the thorax increases.
   B Alveoli are small sac-like structures that have walls only two cells thick.
   C To assist with breathing out, the abdominal and external intercostal muscles contract.
   D The purpose of ventilation is to increase the rate of gas exchange in the alveoli. [1]

5 Which of the following statements is correct?

   A In a motor neuron, myelin always covers the axon.
   B A nerve impulse travels from a receptor to the CNS via a sensory neuron and from the CNS to an effector via a motor neuron.
   C The endocrine system produces hormones, which pass from the endocrine glands to the blood through ducts.
   D The resting potential is divided into two parts, depolarisation and repolarisation. [1]

6 Which of the following statements is correct?

   A The sex of a fetus is determined by the mother's hormones.
   B During the process of IVF, a woman is injected with FSH to stimulate egg cells to mature.
   C FSH and estrogen are hormones released from the ovaries whereas LH and progesterone are hormones released from the pituitary gland.
   D In the menstrual cycle, the levels of progesterone begin to rise around day 5. [1]

7 Outline the functions of the hormones shown here and complete the table below. [4]

Hormone	Where produced	Main effects
estrogen	ovaries	
	pancreas	
melatonin		
	adipose cells	

8   Explain the relationship between the structure and function of arteries, veins and capillaries. [9]

9   Explain antibody production. [3]

10 Explain how the skin and mucous membranes prevent entry of pathogens into the body. [3]

11 Explain how blood glucose concentration is controlled in humans. [8]

12 Summarise the stages needed for IVF in humans. [6]

13 The diagram A below shows a cross-section through the small intestine and diagram B shows a longitudinal section through a single villus.

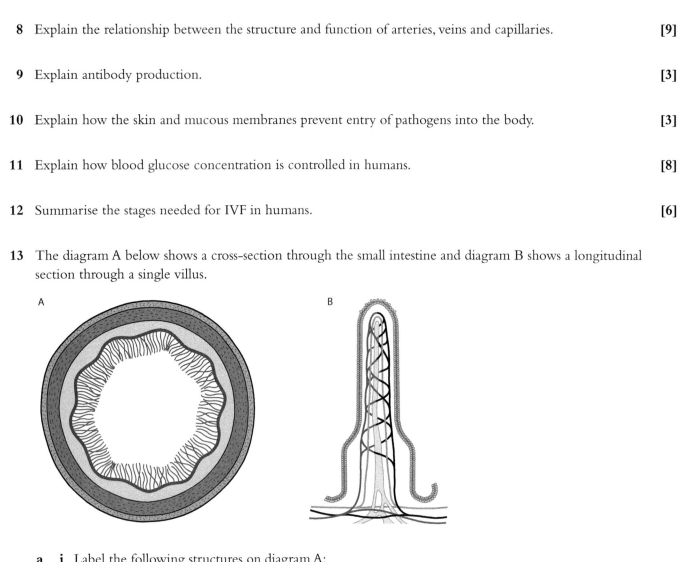

    a   i Label the following structures on diagram A:
           longitudinal muscle, circular muscle [1]
        ii Outline how these muscles move food along the gut. [2]

    b   i Label the following structures on diagram B:
           epithelium, capillaries [1]
        ii Outline the importance of these two structures to the absorption of digested food. [2]

14 a   Which of the following is most likely to cause the heart sounds heard through a stethoscope?

            A   contraction of the atria and ventricles
            B   contraction of ventricles
            C   heart valves closing
            D   ventricles filling [1]

    b   Suggest a reason for the answer you have chosen. [2]

    c   Explain how the rhythmical action of the human heart is converted into a steady flow of blood to the muscles. [3]

Nucleic acids (HL) 7

Introduction

Nucleic acids are very large macromolecules composed of a backbone of sugar and phosphate molecules each with a nitrogenous base attached to the sugar. In Subtopics **2.6** and **2.7**, the role of DNA and RNA is outlined. In this topic, the detailed structure of different nucleic acids is considered, as well as the vital role of nucleic acids in converting the genetic code contained in the chromosomes into the protein molecules that are needed to make all cells function.

7.1 DNA structure and replication

DNA structure

The 3'–5' linkage

A DNA nucleotide consists of the sugar deoxyribose to which are attached a phosphate group and a nitrogenous base. The carbons in the sugar are numbered from 1 to 5 in a clockwise direction starting after the oxygen at the apex (Figure **7.1**).
- The base is attached to carbon 1.
- Carbon 2 has just a hydrogen attached instead of an OH group – this is the reason the sugar is called **deoxy**ribose.
- Carbon 3 is where the next nucleotide attaches in one direction.
- Carbon 5 has a phosphate group attached to it, which is where the next nucleotide attaches in the other direction.

This means that each nucleotide is linked to those on either side of it through carbons 3 and 5. The linkages are called **3'–5' linkages**.

Antiparallel strands

Look back at Figure **2.21**. This shows part of a DNA molecule, in which two polynucleotide strands, running in opposite directions, are held together by hydrogen bonds between pairs of bases. Notice that the deoxyribose molecules are orientated in opposite directions. Figure **7.2** also shows this – at one end of each DNA strand there is a free 3' carbon and at the other there is a free 5' carbon. (Ignore the fact that there is a phosphate group attached to this 5' carbon.) One strand runs in a 5' → 3' direction whereas the other runs in a 3' → 5' direction. The strands are described as being **antiparallel**.

The bases and hydrogen bonding

The four DNA bases are cytosine, thymine, adenine and guanine, and they fall into two chemical groups called **pyrimidines** and **purines**. Cytosine and thymine are pyrimidines, and adenine and guanine are purines.

Learning objectives

You should understand that:
- Nucleosomes are structures that help to supercoil DNA and regulate transcription.
- Understanding the structure of DNA suggested a mechanism for DNA replication.
- During replication, DNA polymerase enzymes can only add nucleotides to the 3' end of a primer.
- DNA replication is continuous on the leading strand but discontinuous on the lagging strand.
- DNA replication is achieved by means of a complex system of enzymes.
- Some regions of DNA do not code for proteins but have other vital functions.

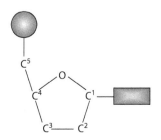

Figure 7.1 The structure of a nucleotide.

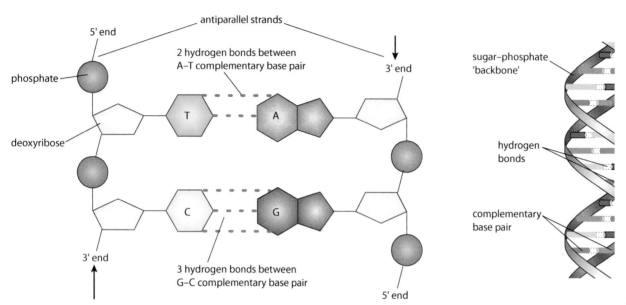

Figure 7.2 Hydrogen bonding between antiparallel strands of DNA.

Exam tip
An easy way to remember the two groups of DNA bases is to look at the words. Thymine and cytosine, which both contain a letter y, are in the group pyrimidines, which also contains a letter y.

Cytosine pairs with guanine and thymine pairs with adenine – that is, a pyrimidine always pairs with a purine. This is because they are different sizes: pyrimidines are smaller than purines. The pairing of a pyrimidine with a purine ensures that the strands are always the same distance apart (Figure 7.2). Understanding that DNA strands are antiparallel and that hydrogen bonds between the bases can be broken were crucial steps in working out how DNA is replicated.

Nucleosomes

A eukaryotic chromosome is composed of a double strand of DNA combined with proteins. Some of these proteins, called **histones**, combine together in groups of eight to form a bead-like structure (Figure 7.3). The strand of DNA takes two turns around this bead before continuing on to the next bead. It is held in place on the bead by a ninth histone. The group of nine histones with the DNA is called a **nucleosome**. The function of nucleosomes is to help supercoil the chromosomes during mitosis and meiosis and also to help regulate transcription.

Nature of science

Scientific advance follows technical innovation – the Hershey and Chase experiments

Although DNA has been well known to science since the 19th century, it is surprising to think that it was not until the middle of the 20th century that scientists discovered its role as the genetic material. Until that time, most people believed that proteins were the molecules responsible for inheritance. Then, in 1952, Alfred Hershey (1908–1997) and Martha Chase (1927–2003) conducted a series of experiments with T2 phage (a virus that infects bacteria), which confirmed that DNA was the genetic

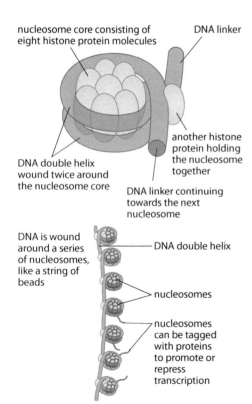

Figure 7.3 The structure of a nucleosome.

material. They were able to carry out these investigations thanks to two relatively new techniques – electron microscopy and radioactive labelling.

The structure of the T2 phage had recently been revealed using the electron microscope. The virus injects its DNA into the cell it infects, but leaves behind its protein coat. Hershey and Chase labelled the viral DNA with radioactive ^{32}P. Phosphorus is present in DNA but not in amino acids, so as they followed the transfer of the labelled material into the cytoplasm of the bacterium, Hershey and Chase knew that it was only DNA that was being transferred.

They then labelled viruses with ^{35}S (sulfur is present in amino acids but not in DNA). After these viruses had infected the bacterial cells, Hershey and Chase examined the discarded protein coats and found that they contained radioactive sulfur, but the bacterial cytoplasm did not. These results supported their hypothesis that DNA is the genetic material that infects the bacteria, and protein (found in the protein coats) was not (Figure **7.4**).

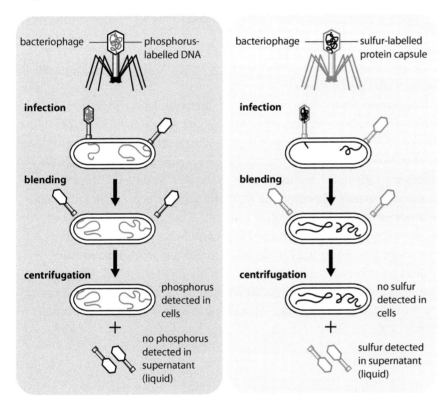

Figure 7.4 The Hershey and Chase experiment showed that the genetic material transferred to bacterial cells by infecting T2 phages is DNA, and not protein as previously believed.

The process of DNA replication

DNA **replication** ensures that exact copies of existing molecules are produced before a cell divides. The process is said to be semi-conservative and each strand of an existing DNA molecule acts as a template for the production of a new strand.

When Watson and Crick proposed their double-helix model for the structure of DNA in 1953, one of the most striking things they realised was that it immediately suggested a mechanism for replication – if the two strands were unwound, each one could provide a template for the synthesis of a new strand.

Most prokaryotes continuously copy their DNA, while in eukaryotes, replication is controlled through interactions between proteins, including cyclins and CDKs (Subtopic **1.6**). Unlike prokaryotes, which can duplicate their DNA in as little as 20 minutes, the process in a eukaryotic cell takes up to 24 hours to complete. There are a number of similarities between prokaryotic and eukaryotic replication:

- both are bi-directional processes
- both require primers to start the process
- DNA polymerase enzymes work from the direction of the 5' end of the strand towards the 3' end in both cases, so that new nucleotides are added to the 3' end of a primer
- both have leading and lagging strands.

The process of replication is **semi-conservative** – that is, each original strand acts as a template to build up a new strand (Figure **7.5**). The DNA double helix is unwound to expose the two strands for replication by the enzyme **helicase**, at a region known as a **replication fork**. The action of helicase creates single-stranded regions, which are less stable than the double-stranded molecule. To stabilise these single strands, **single-stranded binding proteins** (SSBs) are needed. SSBs protect the single-stranded DNA and allow other enzymes involved in replication to function effectively upon it.

Replication must occur in the 5' → 3' direction (and also in transcription and translation, described in Subtopics **7.2** and **7.3**), because the enzymes involved only work in a 5' → 3' direction (adding new nucleotides to the 3' end of the newly forming molecule). As the two strands are antiparallel, replication has to proceed in opposite directions on the two strands. However, the replication fork where the double helix unwinds moves along in one direction only. This means that on one of the strands replication can proceed in a continuous way, following the replication fork along, but on the other strand the process has to happen in short sections, each moving away from the replication fork (Figure **7.6**). The strand undergoing continuous synthesis is called the **leading strand**. The other strand, in which the new DNA is built up in short sections, is known as the **lagging strand**.

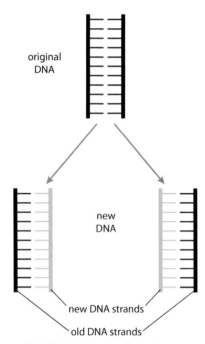

Figure 7.5 DNA replication is semi-conservative. As it is copied one original strand becomes paired with one new strand. One out of the two strands in each new DNA molecule is conserved, hence 'semi-conservative'.

SSB proteins bind to and modulate the function of numerous proteins involved in replication, recombination and repair. They protect single-stranded DNA from being digested by nucleases and they also remove secondary structure from the DNA – that is, they prevent hydrogen bonds forming between the bases along the single strands, which would cause them to coil back on themselves.

Leading strand

Replication to produce the leading strand begins at a point on the molecule known as the 'origin of replication' site. First **RNA primase** adds a short length of RNA, attached by complementary base pairing, to the template DNA strand. This acts as a **primer**, allowing the enzyme **DNA polymerase III** to bind. DNA polymerase III adds free building units called **deoxynucleoside triphosphates** (dNTPs) to the 3' end of the primer and then to the forming strand of DNA. In this way the new molecule grows in a 5' → 3' direction, following the progress of helicase as it moves the replication fork along the DNA double helix. The RNA primer is later removed by DNA polymerase I.

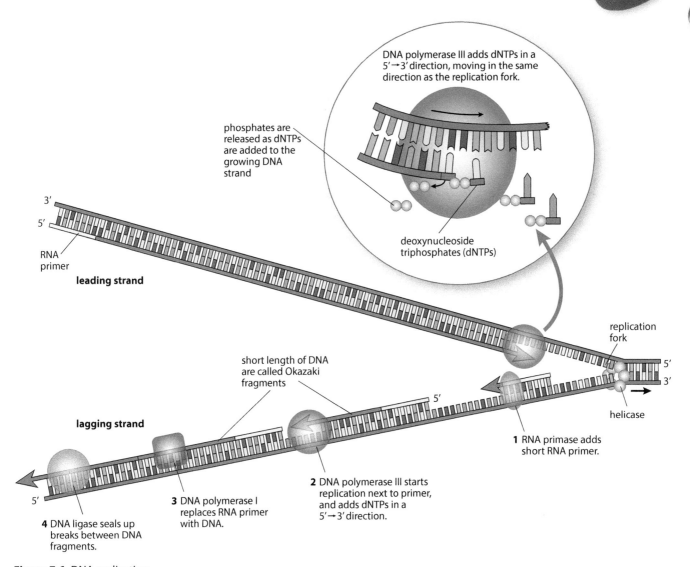

Figure 7.6 DNA replication.

The dNTPs have two extra phosphate groups attached, and are said to be 'activated'. They pair up with their complementary bases on the exposed DNA strand and DNA polymerase III then links together the sugar and the innermost phosphate groups of adjacent nucleotides. The two extra phosphate groups are broken off and released. In this way, a continuous new DNA strand is built up on the leading strand.

Lagging strand

Synthesis of the lagging strand is a little more complicated, as it has to occur in discontinuous sections, which are then joined together.
1 As for the leading strand, **RNA primase** first synthesises a short RNA primer, complementary to the exposed DNA. This happens close to the replication fork.
2 **DNA polymerase III** starts replication by attaching at the 3' end of the RNA primer and adding dNTPs in a 5' → 3' direction. As it does so, it moves away from the replication fork on this strand.

Replication fork the point where the DNA double helix is being separated to expose the two strands as templates for replication
Leading strand the new strand that is synthesised continuously and follows the replication fork
Lagging strand the new strand that is synthesised in short fragments in the opposite direction to the movement of the replication fork
Okazaki fragments short fragments of a DNA strand formed on the lagging strand

3. **DNA polymerase I** now removes the RNA primer and replaces it with DNA using dNTPs. Short lengths of new DNA called **Okazaki fragments** are formed from each primer. The new fragment grows away from the replication fork until it reaches the next fragment.
4. Finally, **DNA ligase** seals up each break between the Okazaki fragments by making sugar–phosphate bonds so that a continuous strand of new DNA is created.

As DNA replication proceeds, supercoiling in the region of the molecule that is ahead of the replication fork is affected by the uncoiling process controlled by helicase. **DNA gyrase** is an enzyme that relieves the tension put upon a DNA molecule as it is being unwound. DNA gyrase is able to create negative supercoils in bacterial DNA. This is important during the replication process because DNA is a 'right handed' helix and without gyrase positive supercoils would accumulate in advance of DNA polymerase and prevent replication continuing. Gyrase is also important during transcription (Subtopic **7.2**) because without it, portions of the circular bacterial chromosome could be prevented from re-attaching to one another.

Bacterial DNA gyrase is the target for a number of antibiotics. Without gyrase, the replication of bacterial DNA cannot occur, so an antibiotic that destroys gyrase can slow down bacterial growth that is causing an infection.

Sanger's method of stopping replication

DNA samples are prepared for profiling (Subtopic **3.5**) using the polymerase chain reaction (PCR), which replicates tiny amounts of DNA until there is sufficient to test. In this process, and in the preparation of DNA samples for sequencing, special chain-terminating nucleotides that inhibit DNA polymerase are used to stop the replication process when required. These **dideoxynucleotides** (ddNTPs) have no 3' hydroxyl group (Figure **7.7**) so that after they have been added by DNA polymerase to a nucleotide chain no further nucleotides can be added and no further chain elongation can occur. ddNTPs were first used by Frederick Sanger in 1977 and his method, used in sequencing, is known as the dideoxy chain-termination method.

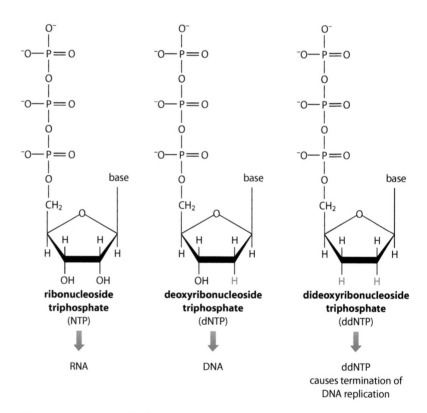

Figure 7.7 Structure of ddNTP molecule.

Non-coding regions of DNA

DNA molecules are very long but every strand has regions that do not code for proteins. For many years these regions were poorly understood but many of them have now been found to have important functions in the regulation of gene expression and other cell activities.

Regulator genes

Regulator genes are present in DNA to control the expression of one or more other genes. Some may code for protein but many work at the level of RNA. Prokaryotic regulator genes often code for repressor proteins that bind to sections of DNA known as promoters and operators. Here they block the transcription of RNA by inhibiting RNA polymerase. Some regulator genes code for activator proteins that bind to DNA and cause a gene to be transcribed. One well-studied activator protein (known as an inducer) is involved in switching on the LAC operon (see box). Activators and repressors are both important in controlling gene expression. You can read more about them in Subtopic **7.2**.

Introns

Introns (or intragenic regions) are sequences of nucleotides *within* genes (they '**in**tervene' in genes). The corresponding sequences in the mRNA transcribed from these genes are removed in the nucleus before the mRNA moves to the cytoplasm for translation (Subtopic **7.2**). Introns were discovered in 1977 by Phillip Allen and Richard Sharp who shared a Nobel prize for their work. Introns occur in many genes in all organisms and the number of introns per gene varies considerably between species. The human genome has been found to have about 8 introns per gene but simpler organisms such as fungi may have fewer than 20 in their entire genome. It seems that there are more introns in larger species with smaller populations, and evolutionary and biological factors are thought to influence this.

Once introns have been removed by **RNA splicing**, the remaining RNA is known as mature RNA. Mature RNA contains **exons**, which are the sequences that will be translated (or '**ex**pressed'). Introns may be spliced out in different ways so that a number of different, but similar, protein sequences can be produced from a single gene. Splicing is controlled by molecules that respond to signals from both inside and outside the cell.

Telomeres

Telomeres are regions of repeated nucleotide sequences that are found at each end of the chromatids of chromosomes. Prokaryotes, which do not have linear chromosomes, do not have telomeres. Telomeres protect the ends of chromosomes from damage during cell division. Without them chromosomes would gradually become shorter and important genes might be lost. Telomeres are needed because, as Okazaki fragments are synthesised, RNA primers must attach in advance of the replication site on the lagging strand. But enzymes involved in this process cannot

Prokaryotic gene control – the LAC operon in *Escherichia coli*

The operon model explains how prokaryotic cells can regulate gene activity and respond to changes in their environment. An operon is a group of genes that act together to code for the enzymes needed for a particular metabolic pathway. The structural genes coding for the enzyme (called a transcription unit) are controlled by a promoter sequence (which is the site of attachment for RNA polymerase and the starting point for mRNA formation) and a region just before the structural genes, which is called the operator. In the case of the LAC operon, a regulator gene elsewhere on the DNA molecule produces a repressor molecule that interacts with the operator region to switch off the genes for lactose metabolism, because lactose is not usually an energy source for *E. coli*. If enzymes to metabolise lactose should be needed, the repressor is removed from the operator region by the action of an activator molecule, known as an inducer, which binds to it and alters its shape. This means that RNA polymerase can bind and the transcription of enzymes for lactose metabolism can begin. (Eukaryotes do not have operons.)

As a person ages, their telomeres become shorter and cell division is less efficient.

continue all the way to the end of the chromosome. With every cell division, the telomere becomes shorter, but the DNA replicates without causing harm to genes. Telomeres can be reconstructed by the enzyme telomerase reverse transcriptase.

Genes for transfer RNA (tRNA)

In humans, the genes that code for tRNA molecules (Subtopic **7.3**) are found on all chromosomes except 22 and Y. These genes do not code for protein but code either for cytoplasmic tRNA or for mitochondrial tRNA. The number of genes that code for tRNA is related to evolutionary history so that organisms in the Domains Archaea and Eubacteria have fewer than those in the Domain Eukarya. This seems to be due to the duplication of the genes over time.

Tandem repeats

DNA profiling is the technique used by forensic scientists to identify a person from their DNA. It is important in criminal investigations, paternity testing and in establishing family relationships. Samples of DNA for profiling are replicated using the polymerase chain reaction (PCR), fragmented using restriction enzymes and then electrophoresis is used to separate the fragments and produce banded patterns (Subtopic **3.5**). The DNA sequences that are important in creating an individual's unique profile, or in determining the degree of relatedness between two individuals, are repeated sequences of non-coding DNA called variable number tandem repeats (VNTRs), which are very similar in close relatives but very different in unrelated individuals.

Nature of science

Careful observation – Rosalind Franklin's meticulous work

Rosalind Franklin and Maurice Wilkins worked on the structure of DNA at King's College London in the 1950s (Subtopic **2.6**, Figure **2.23**). They used the technique of **X-ray crystallography**, which was developed in the early 1900s by the physicist Sir Lawrence Bragg. The process involves directing a beam of X-rays at the atoms in a molecule. The beam is deflected so that it produces an X-ray diffraction pattern, which is unique to the molecule. The molecule is rotated several times and X-rays directed towards it from different angles so that its three dimensional structure can be studied. Large molecules containing many thousands of atoms produce very complex patterns and considerable skill is required to interpret them. Today this is done by computers, which can locate the positions of atoms within the molecules with precision and compile 3D images, but in the 1950s, Franklin and Wilkins did not have the help of a computer. This makes their achievements all the more impressive. It was Franklin who showed in 1952 that phosphate groups in a DNA molecule must be positioned on the outside of its structure. Without this knowledge, Watson and Crick might not have been able to establish the structure of DNA. Although Watson, Crick and Wilkins were awarded a Nobel Prize for their work, Franklin – who died in 1958 – did not receive the award because Nobel Prizes are not awarded posthumously.

X-ray crystallography has established the structure of many complex biological molecules including the triple helix of collagen, the structure of antibodies and the respiratory pigments, myoglobin and hemoglobin. Our understanding of the structure of these molecules has enabled biochemists to work out not only their properties but also how they function.

Exam tip
There are several chemical names involved in replication, so it is helpful to list them.
Nucleoside triphosphate (NTP) a building unit for RNA – a ribose nucleotide with two additional phosphates, which are chopped off during the RNA-synthesis process
Deoxynucleoside triphosphate (dNTP) the same as NTP but with deoxyribose instead of ribose, so it is used to build DNA rather than RNA through base-pairing to the parent DNA strand
Helicase an enzyme found at the replication fork, with two functions – to unwind the two DNA strands, and to separate them by breaking the hydrogen bonds
Single-stranded binding proteins (SSBs) bind to single-stranded regions of DNA to stabilise and protect their structure
DNA gyrase relieves the tension put upon a DNA molecule as it is being unwound
RNA primase this enzyme adds NTPs to the single-stranded DNA that has been unzipped by helicase, in a 5' → 3' direction, to make a short length of RNA (a primer) base-paired to the parent DNA strand
DNA polymerase III this enzyme adds dNTPs in a 5' → 3' direction where RNA primase has added a short length of complementary RNA as a primer; it is unable to add dNTPs directly to the single-stranded parent DNA that has been unzipped by helicase
DNA polymerase I this enzyme removes the RNA nucleotides of the primers on the lagging strand in a 5' → 3' direction and replaces them with DNA nucleotides using dNTPs
DNA ligase this enzyme joins adjacent Okazaki fragments by forming a covalent bond between the deoxyribose and the phosphate of adjacent nucleotides

Test yourself

1. Outline what is meant by the term 'antiparallel'.
2. Outline the structure of a nucleosome.
3. State the direction in which DNA replication occurs.
4. Explain why Okazaki fragments must be produced on one of the DNA strands.
5. Outline the role of telomeres.

What is junk?

The nucleus of a human cell contains nearly 2 m of DNA, but genes make up only a small proportion of this. The DNA in a eukaryotic cell can be divided up into two types:
- **unique** or **single-copy genes**, which make up 55–95% of the total
- highly repetitive sequences, or satellite DNA, which account for 5–45%.

The repetitive sequences are typically between 5 and 300 base pairs long and may be duplicated as many as 100 000 times in a genome.

Many genes include sections called **introns** that are transcribed but not translated. Only sections of genes known as **exons** are both transcribed and translated.

When the highly repetitive sequences of DNA were first discovered, they appeared to have no function. Scientists at the time called them 'junk DNA' and thought they were simply 'excess baggage'. Before scientists began mapping several animal genomes, they had a rather restricted view about which parts of the genome were important. According to the traditional viewpoint, the really crucial things were genes and a few other sections that regulate gene function were also considered useful.

But new findings suggest that this interpretation was not correct. In 2004, David Haussler's team at the University of California, Santa Cruz, USA, compared human, mouse and rat genome sequences. They were astonished to find that several long sequences of repeated DNA were identical across the three species. As David Haussler exclaimed:

'It absolutely knocked me of my chair.'

When the Human Genome Project was planned, there were calls from some people to map only the bits of genome that coded for protein – mapping the rest was thought to be a waste of time. Luckily, entire genomes were mapped and have proved vital to the study of so-called 'junk DNA'. Most geneticists now accept that 'junk DNA' regulates and controls the activity of vital genes and possibly embryo development.

Questions to consider

- Do you think it is appropriate to label something as 'junk' simply because it is thought to have no function?
- Does such labelling hinder scientific progress?
- Should experiments be carried out to answer fundamental questions even if they have no obvious application?
- Who should decide which research is most likely to be valuable?

7.2 Transcription and gene expression

The process by which the DNA code is used to build polypeptides occurs in two stages. The first is transcription, which transfers sections of the genetic code from DNA to an mRNA molecule. Transcription happens in the nucleus. The second stage, known as translation, occurs in the cytoplasm and uses the mRNA, together with ribosomes, to construct the polypeptide. Translation is discussed in Subtopic **7.3**.

Transcribing DNA into mRNA

At the start of transcription, the DNA molecule is separated into two strands by the enzyme RNA polymerase, which binds to the DNA near the beginning of a gene. Hydrogen bonds between the bases are broken and the double helix unwinds (Figure **7.8**).

Transcription begins at a specific point on the DNA molecule called the promoter region, which is a short sequence of non-coding DNA just before the start of the gene. Only one of the two strands is used as a template for transcription and this is called the antisense strand. The other DNA strand is called the sense strand. RNA polymerase uses free nucleoside triphosphates (NTPs) to build the RNA molecule, using complementary base pairing to the DNA and condensation reactions between the nucleotides. This produces a primary mRNA molecule that is complementary to the antisense strand being transcribed, and has the same base sequence as the sense strand (except that it contains the base U in place of T). RNA polymerase moves along the antisense DNA strand in a 3' → 5' direction. As it does so, the 5' end of a nucleotide is added to the 3' end of the mRNA molecule so that the construction of the mRNA proceeds in a 5' → 3' direction.

RNA polymerase checks the mRNA molecule as it forms to ensure that bases are paired correctly. As the mRNA molecule is extended, the DNA is rewound into a helix once a section has been transcribed. Eventually RNA polymerase reaches another specific sequence on the DNA called the **terminator region**, which indicates the end of the gene. The RNA polymerase releases the completed RNA strand and finishes rewinding the DNA before breaking free.

Regulation of transcription (gene expression) in eukaryotes

Gene expression in eukaryotes is controlled by several regulatory processes including:
- **transcriptional** regulation – mechanisms that prevent transcription
- **post-transcriptional** regulation – mechanisms that control or regulate mRNA after it has been produced
- **translational** regulation – mechanisms that prevent translation (Subtopic **7.3**).

Learning objectives

You should understand that:
- Transcription is carried out in a 5' → 3' direction.
- In eukaryotes, nucleosomes help to regulate transcription.
- In eukaryotes, mRNA is modified after transcription.
- Splicing of mRNA increases the number of different proteins that a cell can produce.
- Gene expression is related by proteins that attach to specific base sequences in DNA.
- Gene expression is affected by the environment of a cell and of the organism.

A single chromosome contains DNA that codes for many proteins. During transcription, genes (short lengths of DNA that code for single polypeptides) are used to produce mRNA. Most genes are about 1000 nucleotides long, a few are longer and a very small number are less than 100 nucleotides. The size of the gene corresponds to the size of the polypeptide it codes for.

Nucleoside triphosphates (NTPs) are the molecules used by RNA polymerase to build mRNA molecules during transcription (Figure **7.7**). As they move into place, two phosphates are removed from them so they become converted into nucleotides.

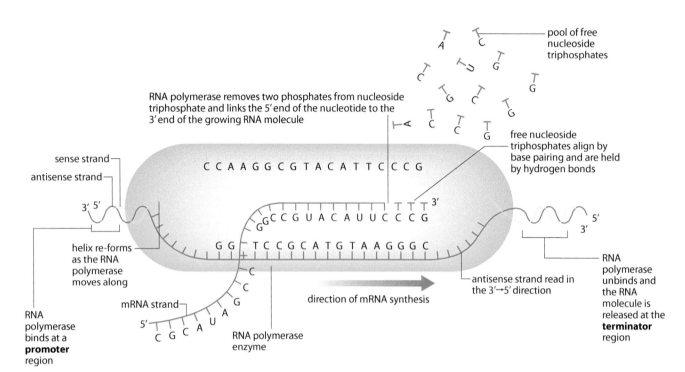

Figure 7.8 Transcription.

Transcriptional regulation

Regulation by nucleosomes

DNA in eukaryotes is incorporated into nucleosomes so that the genetic material can be stored in a compact form (Figure **7.3**). But in order to transcribe genes, **activators** and enzymes involved in transcription must be able to gain access to DNA. It has been found that in all eukaryotic species, the regions of DNA that contain promoters and regulators have fewer nucleosomes than other areas, allowing greater access for binding proteins, while regions that are transcribed have a higher density of nucleosomes. This suggests that nucleosomes have an important role in determining which genes are transcribed and this in turn can influence other factors such as cell variation and development.

Researchers have shown that DNA in nucleosomes may be either 'wrapped' or 'unwrapped' and suggest that DNA does not need to be completely released from a nucleosome to be transcribed, but that there is a significant period of time during which DNA is accessible. Although nucleosomes are very stable protein–DNA complexes, they are not static and can undergo different structural rearrangements including so-called 'nucleosome sliding' and DNA site exposure. Nucleosomes are important because they can either inhibit or allow transcription by controlling whether the necessary molecules can bind to DNA.

Regulation by binding proteins

RNA polymerase requires the presence of a class of proteins known as 'general transcription factors' before transcription can begin. Interactions between the transcription factors, RNA polymerase, and the promoter region of the DNA molecule allow the polymerase to move along the gene so that transcription can occur. There is greater transcription of certain genes when specific transcription factors are present. Many different transcription factors have been found and each one is able to recognise and bind to a specific nucleotide sequence in DNA. A specific combination of transcription factors is necessary to activate a particular gene.

The role of activators is shown in Figure 7.9. These proteins bind to a region of the DNA called the **enhancer**, which may be some distance from the gene. 'Bending proteins' may then assist in bending the DNA so that the enhancer region is brought close to the promoter. Activators, transcription factors and other proteins attach, so that an 'initiation complex' is formed and transcription can begin.

Transcription factors are regulated by signals produced from other molecules. For example, hormones are able to activate transcription factors and thus control transcription of certain genes. Many other molecules in the environment of a cell or an organism can also have an impact on gene expression and protein production. Understanding these molecules and the way they work will add to our understanding of problems arising when proteins are produced incorrectly.

General transcription factors (GTFs)

GTFs are proteins that bind to particular sites on DNA and activate transcription. Together with RNA polymerase and other proteins (activators), GTFs form the transcription apparatus and have a key role in regulating genes. Bacteria need RNA polymerase and just one GTF for their transcription apparatus, while in eukaryotes (and archaeans) RNA polymerase and several GTFs are needed.

One important feature of transcription factors is that they have the ability to bind to specific enhancer or promoter sequences of DNA. Some will bind to a DNA promoter sequence near the transcription start site and form part of the transcription initiation complex. Others bind to regulatory sequences, such as enhancer sequences, and can either stimulate or repress transcription of the related gene.

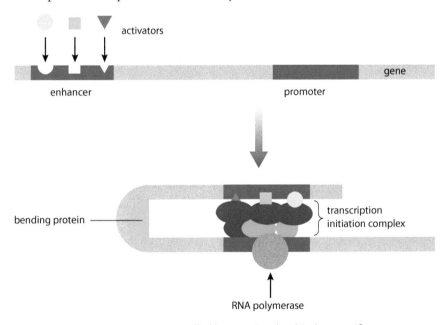

Figure 7.9 Gene expression is controlled by proteins that bind to specific sequences in DNA.

Post-transcriptional modification

In eukaryotes, many genes contain sequences of DNA that are transcribed but not translated. These sequences, which appear in mRNA, are known as introns. After transcription of a gene, the introns are removed in a process known as post-transcriptional modification. The sequences of bases that remain are known as **exons** and these are spliced together to form the **mature mRNA** that is then translated (Figure 7.10). Mature mRNA leaves the nucleus via the nuclear pores and moves to the cytoplasm.

RNA splicing can result in variations in the mature mRNA that is produced, depending on which introns are removed – so the removal of

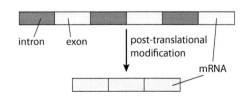

Figure 7.10 Introns and exons in mRNA.

introns enables a gene to code for more than one polypeptide. An average human gene is thought to code for three different proteins.

Environment and gene expression

The expression of genes can be influenced by the environment – not only the organism's external environment, but also its internal environment, which is affected by chemicals such as hormones and various products of metabolism. Temperature, light, chemicals including drugs and the organism's sex are just some of the environmental factors that can cause some genes to be turned on or off and influence how an organism functions or develops.

'Sex-influenced' characteristics are those which are expressed differently in the two sexes. They are genes found on autosomes, not on the sex chromosomes, and one example is male baldness. The baldness allele is influenced by high levels of two hormones, testosterone and dihydrotestosterone. Males usually have much higher levels of these hormones than females, so the baldness allele is more active in men, who are more likely to go bald than women. But sometimes, in very stressful conditions, a woman's adrenal glands may produce testosterone, which can be converted to dihydrotestosterone. In this case the genes for baldness may be expressed so that women may also suffer from hair loss.

The presence of drugs or chemicals in the external or internal environment can also influence gene expression. One example is the drug thalidomide, which was used from the mid 1950s to treat morning sickness during pregnancy. It has no effect on gene expression in adults but has severe effects on limb development in fetuses.

Temperature is an environmental factor that has been known to influence gene expression for many years. A well-studied example is the Himalayan rabbit, which has a gene for the development of pigment in its fur. The gene remains inactive above 35 °C, but becomes very active between 15 °C and 25 °C. In the warm parts of the rabbit's body, the gene is inactive so that the animal's fur is white, but in parts of the body where the temperature is lower, such as ears, feet and nose, the gene becomes active and pigment is produced making the fur in these areas black.

Exam tip
Remember: **in**trons 'intervene' in genes, but only **ex**ons are '**ex**pressed'.

Nature versus nurture

Francis Galton (1822–1911) was one of the first scientists to consider the question of 'nature versus nurture' – that is, the relative importance of an individual's genetic makeup versus the influence of the experiences and environment they are exposed to during their life. The philosopher John Locke (1632–1704) had previously proposed that humans develop as a result of their economic, educational and social status. Today, studies of identical twins raised separately counter Locke's simple viewpoint. But the relative importance of a person's innate characteristics and their personal experiences is still widely discussed today.

Question to consider

- Is it important that science considers the relative importance of innate genetic characteristics and personal experiences in the development of an individual?

Nature of science

Looking for patterns and trends

Epigenetics is usually defined as heritable changes in gene expression function that occur without a change in the nucleotide sequence of the DNA molecule. These changes may remain through cell divisions for the remainder of the cell's life. Since there is no change in the DNA sequence, non-genetic factors must influence how genes come to be expressed differently.

We know that an organism's development is influenced by genes being switched on or off at specific times and there has been much debate about how and whether environmental factors can lead to such epigenetic modifications. If so, the environment could have an important influence on gene expression. Pollutants, diet, temperature and stress all have effects on development. It is also known that these factors can lead to changes in histones or the methylation of DNA (Figure **7.11**). So perhaps these changes could cause epigenetic modifications to gene expression, which affect development.

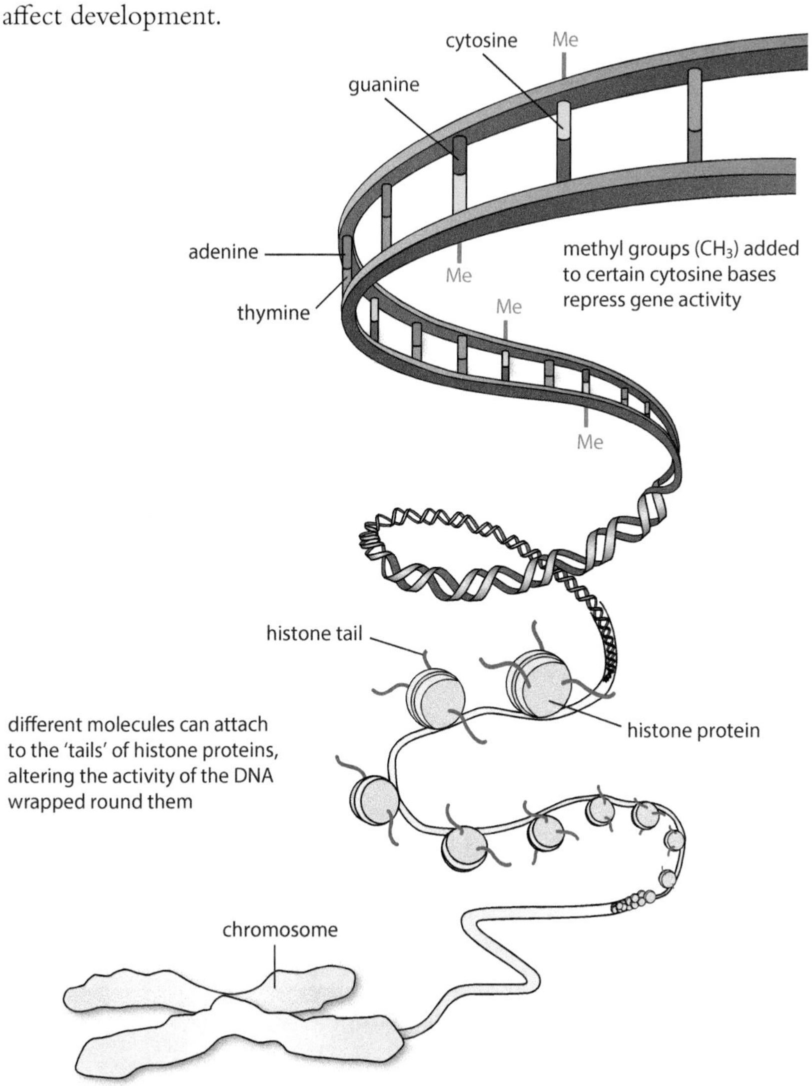

DNA methylation is a process that involves the addition of a methyl group to cytosine. It is an example of an epigenetic marker. The presence of the methyl group changes the way that the DNA sequence is read. In most cases methylation of a section of DNA turns off a gene, leading to reduced gene expression (Figure **7.11**).

Figure 7.11 Methylation of DNA is one epigenetic factor affecting gene expression; histone modification is another.

Modifications can be passed on during mitosis or meiosis, and in plants it is well known that they can pass from one generation to the next. There is now some evidence to suggest this may also happen in animals. One example involves the *Avy* (agouti variable yellow) gene in mice, which influences the animals' coat colour. If there is a high degree of methylation of the gene, it is inactive and the mouse coat colour is dark. Without methylation, the gene is active and the coat is yellow. An active gene is also linked to an increased likelihood of obesity and diabetes. Research into the *Avy* gene conducted by Cooney and co-workers in 1998 established that coat colour was related to the degree of methylation. But, in 2003, Waterland and Jirtle found that increasing the level of methylated molecules such as folic acid and zinc in a pregnant mouse's diet altered the coat colour distribution of offspring by modifying DNA methylation at the agouti locus. Baby mice were born with darker coats and leaner bodies (Figure **7.12**). It was discovered that if pregnant mice were fed in this way, the changes affected not only the offspring but also the next generation. That is, a mother's diet affected the condition of not only her own offspring, but also of her daughters offspring'. Scientists have previously believed that methylation markers were removed from DNA as sperm and egg cells were produced but the experiments suggest that for at least one gene, some markers must remain.

It is an enzyme, DNA methyl transferase, that adds methylation markers to the base cytosine. It usually operates if cytosine is followed by guanine in the base sequence, an area known as CpG. Recent work has revealed that promoter regions of genes often lie within areas known as 'CpG islands' and if CpG is methylated, the gene will not be expressed. A special binding protein can 'read' these epigenetic markers and without it, genes which should not be expressed are transcribed. Mutations in the genes for the binding protein have also been linked to Rett syndrome, a severe form of autism.

The evidence suggests that epigenetic effects caused by the environment may be important in the development of diseases such as obesity and type II diabetes. Epigenetic factors can also help explain how cells with identical

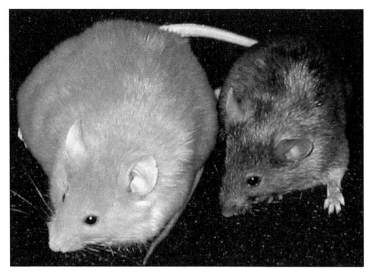

Figure 7.12 These mice are genetically identical and the same age. The mother of the left-hand mouse received a normal 'mouse diet' during pregnancy, while the mother of the mouse on the right was fed supplements including folic acid.

DNA can differentiate into different cell types with different appearances. Recent data also suggests that epigenetic patterns may change during the course of life, so that key genes in vital processes may be affected with age.

Test yourself

6 State what is meant by an 'exon'.
7 Distinguish between the 'sense' and 'antisense' strands of DNA.
8 Outline the role of the promoter region in transcription.
9 Outline the importance of the environment to the expression of any named gene.

7.3 Translation

Translation is the process by which the information carried by mRNA is decoded and used to build the sequence of amino acids that eventually forms a protein molecule. During translation, amino acids are joined together in the order dictated by the sequence of codons on the mRNA to form a polypeptide. This polypeptide eventually becomes the protein coded for by the original gene.

Transfer RNA (tRNA)

The process of translation requires another type of nucleic acid known as transfer RNA or tRNA. tRNA is made of a single strand of nucleotides that is folded and held in place by base pairing and hydrogen bonds (Figure **7.13**). There are many different tRNA molecules but they all have a characteristic 'clover leaf' appearance with some small differences between them.

At one position on the molecule is a triplet of bases called the anticodon, which pairs by complementary base pairing with a codon on the mRNA strand. At the 3' end of the tRNA molecule is a base sequence CCA, which is the attachment site for an amino acid.

An amino acid is attached to the specific tRNA molecule that has its corresponding anticodon, by an **activating enzyme**. As there are 20 different amino acids, there are also 20 different activating enzymes in the cytoplasm. The tRNA-activating enzymes are substrate-specific and recognise the correct tRNA molecules by their shapes. Energy for the attachment of an amino acid comes from ATP. A tRNA molecule with its attached amino acid is called a **charged tRNA**.

Ribosomes

Ribosomes are the site of protein synthesis. Some ribosomes occur free in the cytoplasm and these synthesise proteins that will be used within the cell. Others are bound to the endoplasmic reticulum, forming rough endoplasmic reticulum, and these synthesise proteins for secretion from the cell or for use within lysosomes.

Learning objectives

You should understand that:
- Initiation of translation involves the assembly of components that carry out the process.
- Polypeptide synthesis involves a repeated cycle of events.
- Components used in the process are disassembled when translation is terminated.
- Free ribosomes synthesise proteins for use within the cell, while bound ribosomes synthesise proteins primarily for secretion or inclusion in lysosomes.
- In prokaryotes, translation can occur immediately after transcription because there is no nuclear membrane for mRNA to move through.
- The number and sequence of amino acids in a protein make up its primary structure.
- Secondary structure consists of α helices and β pleated sheets, stabilised by hydrogen bonding.
- Tertiary structure is folding of the secondary structure stabilised by interactions between R groups of amino acids.
- Quaternary structure occurs in proteins with more than one polypeptide chain.

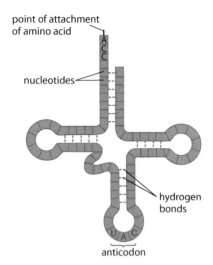

Figure 7.13 Transfer RNA – a 'clover leaf' shape.

Ribosomes are composed of two subunits, one large and one small. The subunits are built of protein and ribosomal RNA (rRNA). On the surface of the ribosome are three tRNA binding sites (site 1, site 2 and the exit site), and one mRNA binding site (Figure 7.14). Two charged tRNA molecules can bind to a ribosome at one time. Polypeptide chains are built up in the groove between the two subunits.

Building a polypeptide

Translation is the process that decodes the information of mRNA into the sequence of amino acids that eventually form a protein. In eukaryotes, mRNA must move out of the nucleus and into the cytoplasm of the cell before translation can occur, but in prokaryotes mRNA can be translated immediately after transcription because it does not have to cross the nuclear membrane. In both eukaryotes and prokaryotes, translation consists of four stages:

1 initiation
2 elongation
3 translocation
4 termination.

Initiation

Translation begins at a start codon (AUG) near the 5' end of the mRNA strand. This codon codes for the amino acid methionine and is a signal to begin the process of translation (Figure 7.15). This is called **initiation**. The mRNA binds to the small subunit of a ribosome. Then an activated tRNA molecule, carrying the amino acid methionine, moves into position at site 1 of the ribosome. Its anticodon binds with the AUG codon using complementary base pairing. Hydrogen bonds form between the complementary bases of the mRNA and tRNA and, once this has happened, a large ribosomal subunit moves into place and combines with the small subunit.

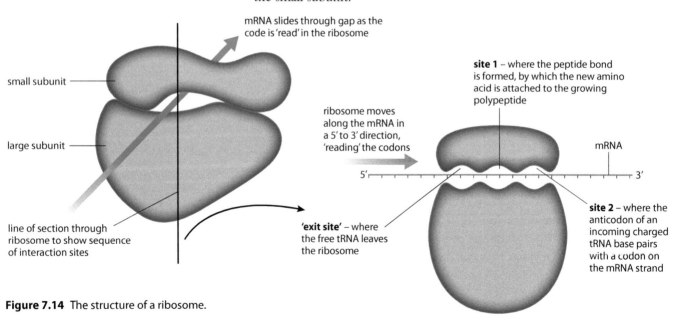

Figure 7.14 The structure of a ribosome.

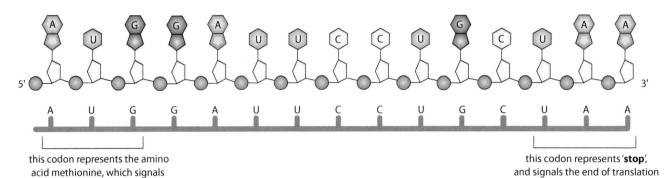

this codon represents the amino acid methionine, which signals the **'start'** of translation

this codon represents **'stop'**, and signals the end of translation

Figure 7.15 mRNA 'start' and 'stop' codons.

Elongation

Initiation is followed by **elongation** and the formation of peptide bonds (Figure **7.16**). tRNA molecules bring amino acids to the mRNA strand in the order specified by the codons. To add the second amino acid, a second charged tRNA with the anticodon corresponding to the next codon enters site 2 of the ribosome and binds to its codon by complementary base pairing. The ribosome catalyses the formation of a **peptide bond** between the two adjacent amino acids (Figure **7.16**). The ribosome and tRNA molecules now hold two amino acids. The methionine becomes detached from its tRNA. Now the ribosome moves along the mRNA and the first tRNA is released to collect another methionine molecule.

Translocation

Translocation is the movement of the ribosome along the mRNA strand one codon at a time. As the ribosome moves, the unattached tRNA moves into the exit site and is then released into the cytoplasm, where it will pick up

Look back at Table **3.1** in Subtopic **3.1** to see all the mRNA codons and what they represent.

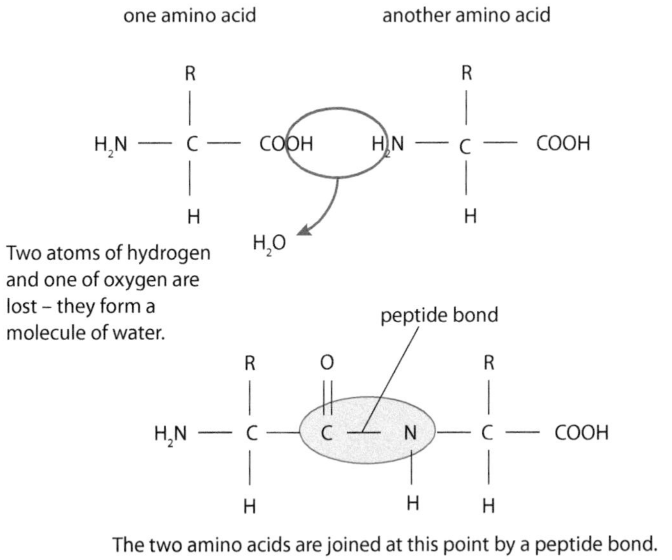

one amino acid   another amino acid

Two atoms of hydrogen and one of oxygen are lost – they form a molecule of water.

peptide bond

The two amino acids are joined at this point by a peptide bond.

Figure 7.16 The formation of a peptide bond by a condensation reaction.

7 NUCLEIC ACIDS (HL)

another amino acid molecule. The growing peptide chain is now positioned in site 1, leaving site 2 empty and ready to receive another charged tRNA molecule to enter and continue the elongation process. Figure **7.17** shows how initiation, elongation and translocation occur as mRNA is translated.

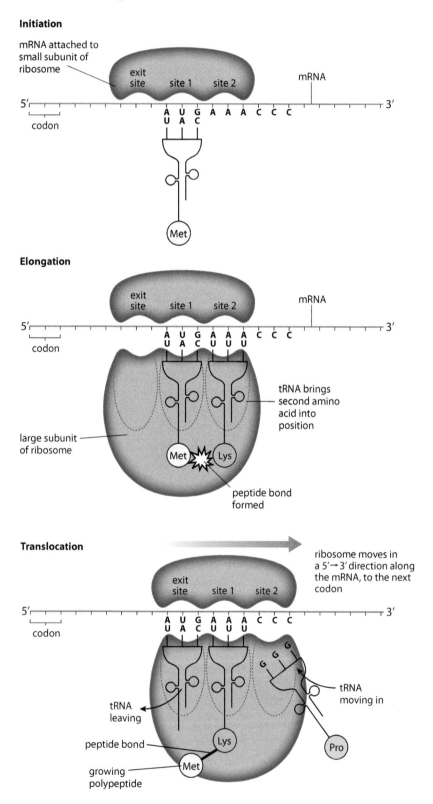

Figure 7.17 The stages of translation.

Termination

Translocation and elongation are repeated until one of the three 'stop' codons aligns with site 2, which acts as a signal to end translocation. There are no tRNA molecules with anticodons corresponding to these stop codons. The polypeptide chain and the mRNA are released from the ribosome and the ribosome separates into its two subunits. This final stage of translation is called **termination**.

Polysomes

Translation occurs at many places along an mRNA molecule at the same time. The electron micrograph in Figure 7.18 shows transcription and translation occurring simultaneously in a bacterium. A **polysome** is a group of ribosomes along one mRNA strand (Figure 7.19). Part of the bacterial chromosome can be seen as the fine pink line running horizontally along the bottom of the micrograph and two growing polypeptide chains are shown forming above it. DNA is being transcribed by RNA polymerase and the newly formed mRNA is being immediately translated by the ribosomes. In eukaryotes, the two processes occur in the nucleus and cytoplasm, respectively, and so are separated not only in time but also in location.

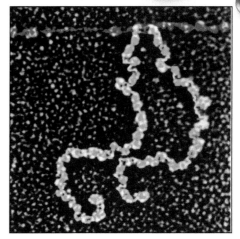

Figure 7.18 Electronmicrograph of polysomes in a bacterium (×150 000).

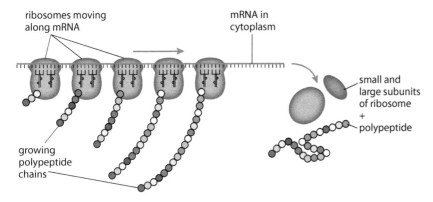

Figure 7.19 Diagram of a polysome.

Protein structure

Proteins are large, complex molecules, usually made up of hundreds of amino acid subunits. The way these subunits fit together is highly specific to each type of protein, and is vital to its function. Figure 7.20 illustrates the structure of the protein hemoglobin.

1 The first stage of protein production is the assembly of a sequence of amino acid molecules that are linked by peptide bonds formed by condensation reactions. This sequence forms the **primary structure** of a protein. There are many different proteins and each one has different numbers and types of amino acids arranged in a different order, coded for by a cell's DNA.

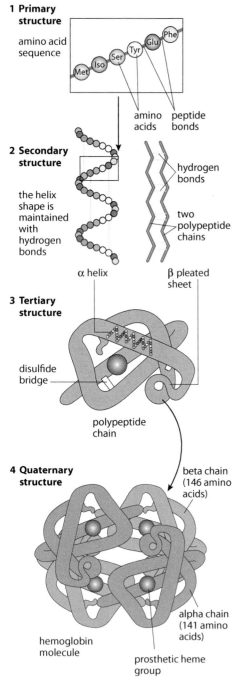

1 **Primary structure**
amino acid sequence

2 **Secondary structure**
the helix shape is maintained with hydrogen bonds
α helix
β pleated sheet

3 **Tertiary structure**
disulfide bridge
polypeptide chain

4 **Quaternary structure**
hemoglobin molecule
prosthetic heme group
beta chain (146 amino acids)
alpha chain (141 amino acids)

Figure 7.20 The structure of hemoglobin.

2 The **secondary structure** of a protein is formed when the polypeptide chain takes up a permanent folded or twisted shape. Some polypeptides coil to produce an α helix, others fold to form β pleated sheets. The shapes are held in place by many weak hydrogen bonds. Depending on the sequence of amino acids, one section of a polypeptide may become an α helix while another takes up a β pleated form. This is the case in hemoglobin.

3 The **tertiary structure** of a protein forms as the molecule folds still further due to interactions between the R groups of the amino acids (Figure 7.21) and within the polypeptide chain. The protein takes up a three-dimensional shape, which is held together by ionic bonds between particular R groups, disulfide bridges (covalent bonds) between sulfur atoms of some R groups, and by weaker interactions between hydrophilic and hydrophobic side chains. Figure 7.22 shows the different types of bond involved in maintaining the tertiary structure of proteins. Tertiary structure is very important in enzymes because the shape of an enzyme molecule gives it its unique properties and determines which substrates can fit into its active site.

4 The final level of protein structure is **quaternary structure**, which links two or more polypeptide chains to form a single, large, complex protein. The structure is held together by all the bonds that are important in the previous levels of structure. Examples of proteins that have quaternary structure are collagen (which has three polypeptide chains), hemoglobin (which has four), antibodies (which also have four) and myosin (which has six).

In addition, many proteins contain prosthetic groups and are called conjugated proteins. Prosthetic groups are not polypeptides but they are able to bind to different proteins or parts of them. For example, hemoglobin is a conjugated protein, with four polypeptide chains, each containing a prosthetic heme group.

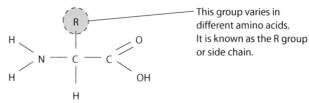

General structure of an amino acid

This group varies in different amino acids. It is known as the R group or side chain.

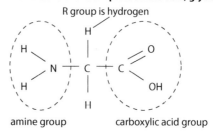

Structure of the simplest amino acid, glycine
R group is hydrogen
amine group   carboxylic acid group

Figure 7.21 The general structure of an amino acid and the structure of glycine.

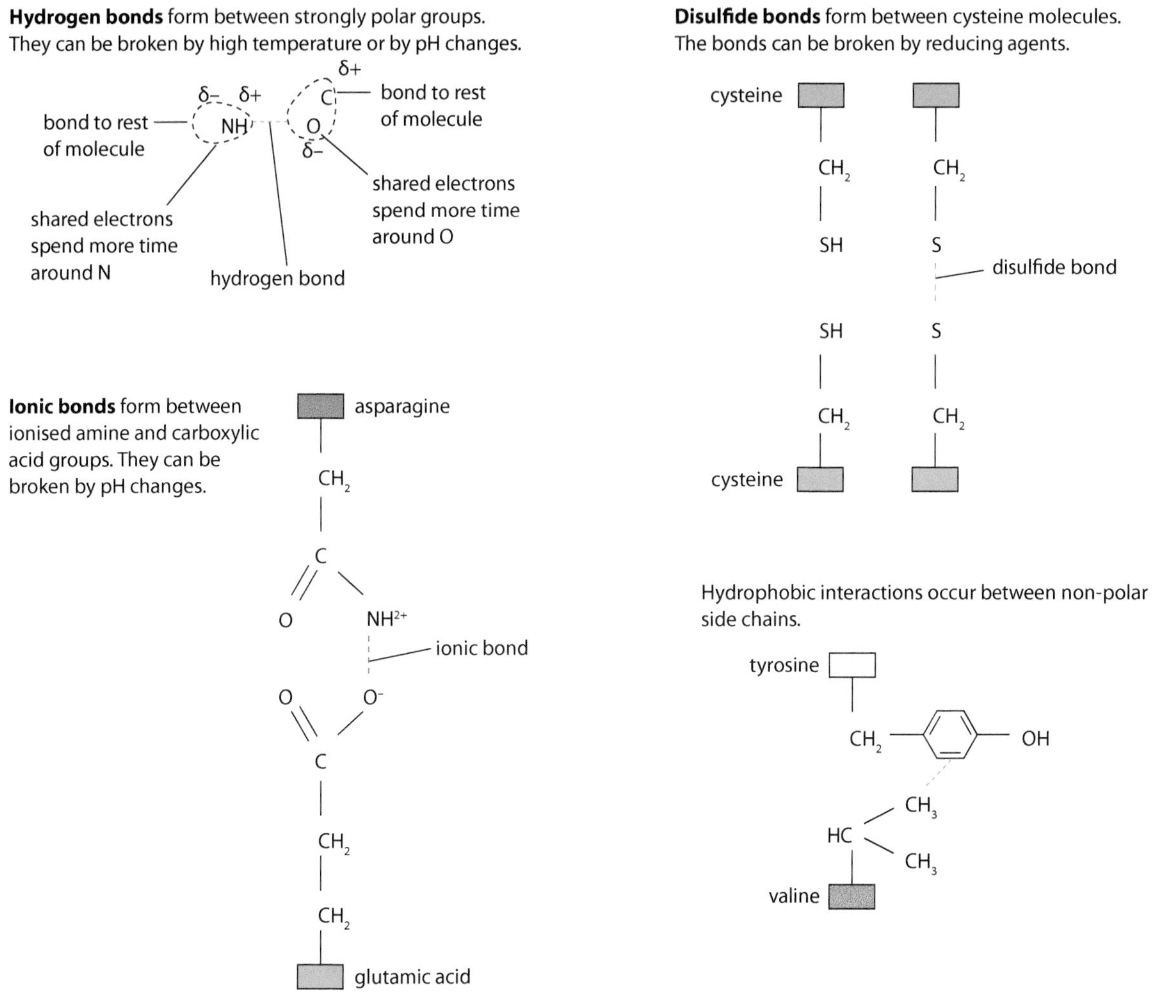
Hydrogen bonds form between strongly polar groups.
They can be broken by high temperature or by pH changes.
δ+
δ– δ+
C
NH
O
δ–
bond to rest of molecule
bond to rest of molecule
shared electrons spend more time around N
hydrogen bond
shared electrons spend more time around O
Disulfide bonds form between cysteine molecules.
The bonds can be broken by reducing agents.
cysteine
CH2
CH2
SH
S
disulfide bond
SH
S
CH2
CH2
cysteine
Ionic bonds form between ionised amine and carboxylic acid groups. They can be broken by pH changes.
asparagine
CH2
C
O
NH2+
ionic bond
O
O-
C
CH2
CH2
glutamic acid
Hydrophobic interactions occur between non-polar side chains.
tyrosine
CH2
OH
CH3
HC
CH3
valine

Figure 7.22 Types of bond that are important in protein structure.

Fibrous and globular proteins

Protein molecules are categorised into two major types by their shape. **Fibrous proteins** are long and narrow and include collagen (found in skin, tendons and cartilage), keratin (in hair and nails) and silk. Fibrous proteins are usually insoluble in water and in general have secondary structure. **Globular proteins** have a more rounded, 3D shape and have either tertiary or quaternary structure. Most globular proteins are soluble in water. Globular proteins include enzymes, such as pepsin, and antibodies. Myoglobin and hemoglobin are also globular proteins.

Polar and non-polar amino acids

Amino acids are divided into two groups according to the chemical properties of their side chains or R groups (Figure **7.21**). Polar and non-polar amino acids have different properties and their positions in a molecule affect the behaviour and function of the whole protein.

Amino acids with non-polar side chains are **hydrophobic**. Those with polar side chains are **hydrophilic**. Non-polar amino acids are found in parts of proteins that are in hydrophobic areas, while polar amino acids are in areas that are exposed to an aqueous environment such as cytoplasm or blood plasma.

For membrane proteins, the polar hydrophilic amino acids are found on the outer and inner surfaces in contact with the aqueous environment, while the non-polar hydrophobic amino acids are embedded in the core of the membrane in contact with the hydrophobic tails of the phospholipid bilayer (Figure **7.23**). This helps to hold the protein in place in the membrane. Some integral proteins act as channels, and the pore is lined with hydrophilic amino acids to enable polar substances to pass through.

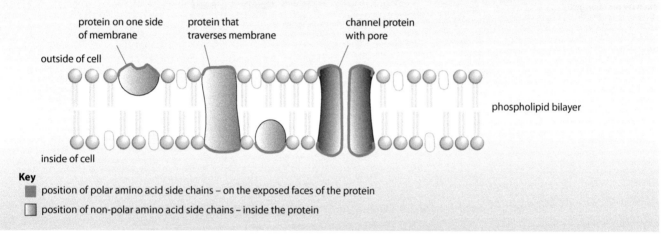

Figure 7.23 In membrane proteins, polar (hydrophilic) amino acids are found on the surfaces in contact with the aqueous environment, while non-polar (hydrophobic) amino acids are embedded inside the phospholipid bilayer.

Polar and non-polar amino acids are also important in enzymes, where they assist in the binding of substrates. An enzyme that acts on a polar substance (for example, amylase) has polar amino acids in its active site, whereas lipases have non-polar amino acids in the active site.

Polar amino acids on the surface of a protein increase its solubility while non-polar amino acids help a protein maintain its structure. Lipases are proteins that have polar groups on the outside so they are soluble in the gut contents, but non-polar groups in their active site so that lipids can bind to them.

Nature of science

Scientific advance follows technical innovation – computers help locate genes and identify conserved sequences

When the Human Genome Project was first begun, scientists could only deal with and analyse small sequences of DNA at any one time. But as the data began to accumulate to hundreds of thousands of sequences it became necessary to record how many genes there were, whether there was redundancy in the code and whether there were patterns in the sequencing. Mathematicians and computer scientists built algorithms to do this. One of the best known is the Tiger Assembler, which put together the sequences to try to work out how many genes there are in a human genome.

The DNA sequences are stored in databases such as GenBank at the US National Centre for Biotechnology Information and other organisations in Europe and Asia. The data are analysed by computer programs that determine the boundaries between genes, a process known as gene annotation, and help to identify sequences. Bioinformatics speeds up the process and uses statistical models to annotate and predict. The most recent methods for doing this use 'formal grammars' which derived from computer science. Without these tools, DNA research could not have proceeded as rapidly as it has.

Test yourself

10 Distinguish between 'transcription' and 'translation'.
11 State where the protein that is synthesised by free ribosomes is used.

Exam-style questions

1 The components of a nucleosome are:

    A ribosomal RNA and DNA
    B eight histone proteins and DNA
    C eight histones proteins in a ball + one further histone
    D nine histone proteins and DNA [1]

2 Which of the following statements is correct about the structure of DNA?

    A The purine base cytosine is linked to the pyrimidine base guanine through three hydrogen bonds.
    B The sugar–phosphate strands are antiparallel and linked by complementary base pairing.
    C The bases are linked to each other through a 3'–5' linkage.
    D Complementary base pairing of guanine with cytosine and adenine with uracil means that the two sugar–phosphate strands lie parallel. [1]

3   Which of the following statements is correct about DNA replication?

   A   The enzymes DNA ligase and RNA primase can be found on the lagging strand.
   B   Okazaki fragments are produced by DNA polymerase I and DNA polymerase III on the leading strand.
   C   On the lagging strand, the RNA primer is synthesised by RNA primase and then converted into a DNA strand with the enzyme DNA polymerase III.
   D   The enzyme DNA polymerase III uses deoxynucleoside triphosphates to build a new DNA strand only on the leading strand. [1]

4   Which of the following statements is correct about transcription?

   A   The enzyme RNA polymerase moves along the antisense strand in a 3' → 5' direction.
   B   The sequence of bases in the strand of RNA being synthesised is the same as the sequence of bases in the sense strand of DNA.
   C   In eukaryotic cells, exons are removed from the primary RNA in the nucleus to make mature RNA.
   D   Messenger RNA is synthesised by RNA polymerase in a 3' → 5' direction. [1]

5   Which of the following statements is correct about translation?

   A   Ribosomes that are free in the cytoplasm synthesise proteins that are primarily for lysosome manufacture and exocytosis.
   B   Ribosomes are made of two subunits and mRNA binds to the larger one.
   C   On the larger ribosome subunit there are three binding sites that can be occupied by tRNA molecules.
   D   During polypeptide synthesis, the ribosome moves along the mRNA strand in a 3' → 5' direction until it reaches a stop codon. [1]

6   Which of the following statements about protein structure is correct?

   A   There are four levels of protein structure. The primary level is held together by covalent and hydrogen bonding.
   B   Enzymes have an active site that is a 3D structure produced by secondary level folding of the protein.
   C   The α helix and β pleated sheet are both types of tertiary level folding.
   D   Both tertiary and quaternary level proteins can form conjugated proteins. [1]

7   Explain the process of translation. [9]

8 In a study of a gene containing 5 exons, mature mRNA produced by the gene was isolated. The two strands of the original DNA were then separated to produce a single strand. In the next step of the experiment the single-stranded DNA and mRNA were mixed. Some of the single-stranded DNA hybridised (paired) with the complementary mRNA.

Draw a simple diagram to show the appearance of the DNA–RNA hybrids as seen under an electron microscope. No details of structure are required, use single lines to represent DNA and RNA and indicate complementary base pairing. [4]

9 The diagram below shows a short stretch of replicating DNA from the bacterium *E. coli* and shows 4 Okazaki fragments (labelled 1, 2, 3 and 4).

The upper strand represents the DNA template. The grey boxes indicate RNA primers.

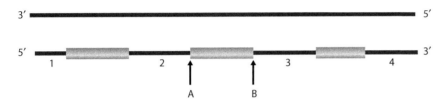

a Which of the fragments was first to be made? [1]
b Make a copy of the diagram and indicate on it the primer that will be the first to be removed. [1]

When the primer in the middle is removed and replaced by DNA, the fragments must be joined together.

c Name the enzyme that removes the RNA primer. [1]
d Name the enzyme that joins the fragments. [1]
e State where the final connection is made, at point A or point B. [1]

8 Metabolism, cell respiration and photosynthesis (HL)

Introduction

Metabolic reactions are chemical processes that occur in all cells to keep them alive. Respiration and photosynthesis are two key metabolic pathways in ecosystems. Light energy from the Sun is trapped as chemical energy in photosynthesis and then the energy is transferred through food chains and released back to the environment as heat energy from respiration. The two pathways can be simply written as:

$$6CO_2 + 6H_2O + \text{energy} \underset{\text{respiration}}{\overset{\text{photosynthesis}}{\rightleftarrows}} C_6H_{12}O_6 + 6O_2$$

Learning objectives

You should understand that:
- Metabolic pathways are made up of chains and cycles of enzyme-catalysed reactions.
- Enzymes lower the activation energy of chemical reactions.
- Enzyme inhibitors can be competitive or non-competitive.
- End-product inhibition can control metabolic pathways.

8.1 Metabolism

Metabolic pathways

Metabolic pathways consist of chains or cycles of reactions that are catalysed by enzymes. Metabolism includes all the chemical activities that keep organisms alive. Metabolic pathways may be very complex, but most consist of a series of steps, each controlled by an enzyme. Simple pathways involve the conversion of substrates to a final product:

$$\text{substrate X} \xrightarrow{\text{enzyme 1}} \text{substrate Y} \xrightarrow{\text{enzyme 2}} \text{substrate Z} \xrightarrow{\text{enzyme 3}} \text{end product}$$

Each arrow represents the specific enzyme needed to catalyse the conversion of one substrate to the next.

Other metabolic pathways, such as photosynthesis or respiration, involve chains of reactions and cycles of reactions.

Activation energy

Enzymes work by lowering the activation energy of the substrate or substrates. In order for a metabolic reaction to occur, the substrate has to reach an unstable, high-energy 'transition state' where the chemical bonds are destabilised, and this requires an input of energy, which is called the activation energy. When the substrate reaches this transition stage, it can then immediately form the product. Enzymes can make reactions occur more quickly because they reduce the activation energy of reactions they catalyse to bring about a chemical change (Figure 8.1). Most biological reactions release more energy than they take as activation energy and are said to be exothermic. This is shown as energy change on the graph.

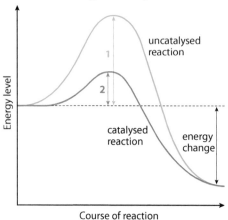

Key
1 = activation energy without catalyst
2 = activation energy with catalyst

Figure 8.1 Graph to show activation energy for an exothermic reaction with and without a catalyst.

Metabolic reactions that occur in living organisms have to occur at the body temperature of the organism, which is never high enough to bring substrates to their transition state. The active site of an enzyme is very important because it can lower the amount of energy needed to reach a transition state, so the reaction can occur at the temperature of the organism.

Induced-fit model of enzyme action

The lock-and-key hypothesis discussed in Subtopic **2.5** explains enzyme action by suggesting that there is a perfect match between the shape of the active site of an enzyme and the shape of its substrate. This theory was proposed in 1890 by Emil Fischer.

In the last century, research published by Daniel Koshland in 1958 suggested that the process is not quite this straightforward. The lock-and-key hypothesis cannot account for the binding and simultaneous change that is seen in many enzyme reactions, nor the fact that some enzymes can bind to more than one similarly shaped substrate.

A more likely explanation of enzyme action is that the shape of an enzyme is changed slightly as a substrate binds to its active site (Figure **8.2**). The substrate causes or induces a slight change in the shape of the active site so it can fit perfectly. As the enzyme changes shape, the substrate molecule is activated so that it can react and the resulting product or products are released. The enzyme is left to return to its normal shape, ready to receive another substrate molecule.

Enzymes do not change the *quantity* of product that is formed, only the *rate* at which the product is formed.

This hypothesis is known as the **induced-fit model** of enzyme action.

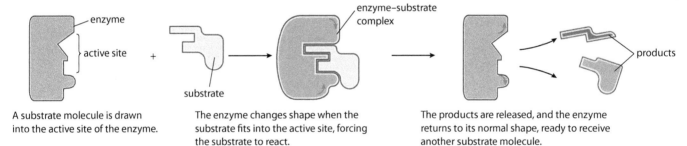

Figure 8.2 The induced-fit model of enzyme action.

Competitive and non-competitive inhibition

Enzyme inhibitors are substances that reduce or prevent an enzyme's activity. Some inhibitors are competitive and others non-competitive.

Competitive inhibitors have molecules whose structure is similar to that of the substrate molecule that normally binds to the active site. They **compete** with the substrate to occupy the active site of the enzyme, and prevent the substrate molecules from binding (Figure **8.3**). The inhibitors are not affected by the enzyme and do not form products, so they tend to remain in the active site. This means that the rate of reaction is lower because substrate molecules cannot enter the active sites of enzyme molecules that are blocked by an inhibitor. At low concentrations of substrate, competitive inhibitors have a more significant effect than at higher concentrations, when the substrate can out-compete the inhibitor (Figure **8.4**).

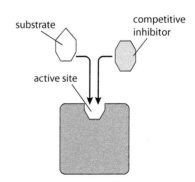

Figure 8.3 Competitive inhibition.

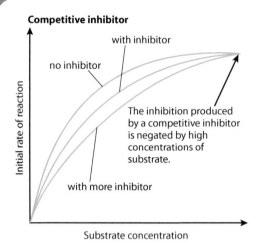

Non-competitive inhibitors also combine with enzymes but not at the active site. They bind at another part of the enzyme molecule where they either partly block access of the substrate to the active site or cause a change in the shape of the enzyme so that the substrate cannot enter the active site (Figure **8.5**). Increasing the concentration of substrate in the presence of a non-competitive inhibitor does not overcome inhibition (Figure **8.4**).

Table **8.1** compares the nature and effects of competitive and non-competitive inhibitors.

Using enzyme inhibition to treat poisoning

Many enzyme inhibitors are used in medicine. One example is fomepizal, also known as fomepizole. Fomepizal is a competitive inhibitor of alcohol dehydrogenase, an enzyme which normally catalyses the oxidation of ethanol to acetaldehyde, which is oxidised to harmless products in the liver. But alcohol dehydrogenase also catalyses steps in the metabolism of ethylene glycol (antifreeze) to toxic metabolites that cause severe damage to the kidneys. Fomepizal inhibits alcohol dehydrogenase enzyme activity in the human liver. An injection of fomepizal given after accidental ingestion of ethylene glycol prevents kidney damage by blocking alcohol dehydrogenase so that toxic metabolites are not produced.

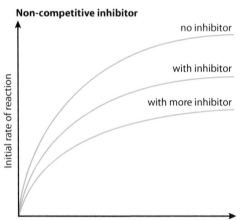

Figure 8.4 Graphs to show the effects of competitive and non-competitive inhibitors on reaction rate, as substrate concentration increases.

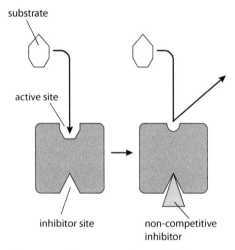

Figure 8.5 Non-competitive inhibition.

Exam tip
If you are asked to compare different types of inhibition a table is a good way to present your answers.

Competitive inhibitors	Non-competitive inhibitors
structurally similar to the substrate molecule	structurally unlike the substrate molecule
occupy and block the active site	bind at a site away from the active site, reducing access to it
if concentration of inhibitor is low, increasing the concentration of substrate will reduce the inhibition	if concentration of substrate is low, increasing the concentration of substrate has no effect on binding of the inhibitor so inhibition stays high
examples include: • oxygen, which competes with carbon dioxide for the active site of ribulose bisphosphate carboxylase in photosynthesis • disulfiram, which competes with acetaldehyde for the active site of aldehyde dehydrogenase • ethanol, which can be used in preventing antifreeze poisoning because it is a competitive inhibitor of the enzyme alcohol dehydrogenase	examples include: • cyanide and carbon monoxide, which block cytochrome oxidase in aerobic respiration, leading to death

Table 8.1 Comparing competitive and non-competitive inhibitors.

Controlling metabolic pathways by end-product inhibition

End-product inhibition means that an enzyme in a pathway is inhibited by the product of that pathway. This prevents a cell over-producing a substance it does not need at the time. Many products may be needed by the cell at a specific time or in specific amounts and over-production not only wastes energy but may also become toxic if the product accumulates.

In an assembly-line reaction, such as those described in Figure **8.6**, each step is controlled by a different enzyme. If the end-product begins to accumulate because it is not being used, it inhibits an enzyme earlier in the pathway to switch off the assembly line. In most cases, the inhibiting effect is on the first enzyme in a process, but in other cases it can act at a branch point to divert the reaction along another pathway.

When the end-product starts to be used up, its inhibiting effect reduces, the inhibited enzyme is reactivated and production begins again. This is an example of **negative feedback** (Subtopic **6.6**).

End-product inhibition may be competitive or non-competitive. Competitive inhibition will only work if the product is a similar shape to the normal substrate and there can be an induced fit of the product or inhibitor onto the enzyme. In most cases, the product will be a different shape and therefore this has to be non-competitive inhibition. In this case, the enzyme is known as an **allosteric** enzyme, the product is called an **allosteric inhibitor** and the place where it binds to the enzyme is called the **allosteric site** (Figure **8.7**).

Disulfiram is a competitive inhibitor of alcohol dehydrogenase which is used as a deterrent to drinking for people with alcohol dependency. It causes headaches and feelings of nausea if alcohol is drunk by a person who has taken it.

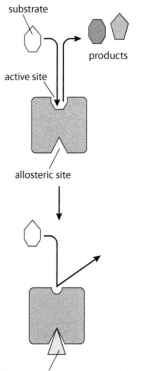

Figure 8.7 Allosteric control. Allosteric inhibitors prevent the active site functioning.

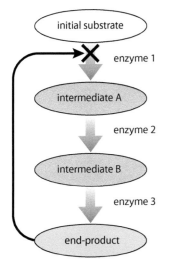

The end-product inhibits the enzyme catalysing the first reaction in the series, so all the subsequent reactions stop.

Figure 8.6 End-product inhibition.

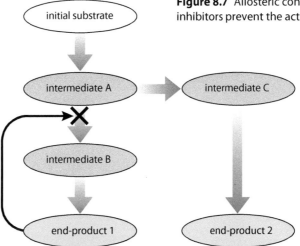

The end-product inhibits an enzyme in the pathway, which causes a different enzyme to come into play and the pathway is diverted down a different route.

An example of end-product inhibition

Threonine is converted to isoleucine in a series of five enzyme-controlled stages. Isoleucine, as the end product of threonine metabolism, can inhibit threonine deaminase, the first of the five enzymes in the process (Figure 8.8). Isoleucine inhibits the enzyme by binding to the molecule at a site away from the active site. When it is attached, the active site of the enzyme is changed so that no further substrate can bind to it. As isoleucine concentration increases, more and more isoleucine molecules attach to this inhibition site on enzyme molecules and therefore inhibit further production of isoleucine. As their concentration falls, isoleucine molecules detach from the threonine deaminase enzyme molecules and are used in the cell. Once the inhibitor has been removed, the active site can bind new substrate and the pathway is reactivated. This mechanism makes the metabolic pathway self-regulating so that there is always sufficient isoleucine present in the cell.

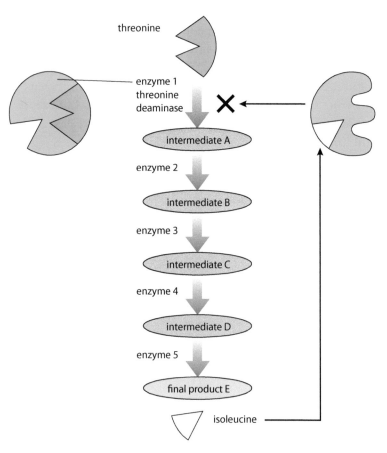

Figure 8.8 The pathway that converts threonine to isoleucine — a specific example of end-product inhibition.

Studying metabolic pathways

Metabolic pathways have been studied for centuries but one of the most significant advances were made by Eduard Buchner (1860–1917) who discovered enzymes at the start of the 20th century. At first, studies of whole animals were made, but more recently it has been possible to analyse metabolic pathways and their component reactions using modern techniques such as chromatography, X-ray diffraction, spectroscopy and radioactive isotopes. For example, in the mid-20th century, the citric acid cycle – often called the Krebs cycle after its discoverer (Subtopic **8.2**) – and the glyoxylate cycle were discovered by Hans Krebs (1900–1981) and Hans Kornberg (1928–). But metabolic pathways are very elaborate. There are many pathways that are interrelated and together make up a complex metabolic network in a cell. These pathways are vital to homeostasis and cell function. Pathways may be connected by intermediate products, and products of one pathway may be substrates for another.

Question to consider
- Most biochemical studies are made using carefully controlled experiments that look at one part of a pathway. To what extent can looking at component parts of a complex system give us knowledge of the whole?

Nature of science

Scientific advance follows technical innovation – biochemical databases facilitate research

Now that large amounts of data can be stored and curated in databases, known and predicted biochemical pathways can be catalogued and updated by scientists who are researching specific organisms. Two important databases, which have enhanced the knowledge of a vital staple food and a killer disease, are RiceCyc (which stores information on the biochemical pathways in rice) and PlasmoDB (which stores information on the genetics and biochemistry of the malarial parasite *Plasmodium falciparum*). Databases like these are regularly updated so that researchers can find new discoveries in biochemical pathways in a matter of minutes, as well as using the pathways that are known, in their research programmes – in the development of disease-resistant strains of rice, for example, or new antimalarial drugs. In this way, the data stored in a database can be turned into biologically meaningful information.

Test yourself

1 Outline what is meant by 'activation energy'.
2 Explain how an enzyme pathway can be switched off by an accumulation of the end product of the pathways.

Learning objectives

You should understand that:
- In cell respiration, electron carriers are oxidised and reduced.
- Molecules that have been phosphorylated are less stable.
- Glucose is converted to pyruvate during glycolysis, which takes place in the cytoplasm.
- Glycolysis does not use oxygen and produces a small net gain of ATP.
- In aerobic respiration, pyruvate is decarboxylated and oxidised. In the link reaction, it is converted to an acetyl compound, then attached to coenzyme A to form acetyl coenzyme A.
- During the Krebs cycle, the oxidation of acetyl groups is coupled with the reduction of hydrogen carriers and carbon dioxide is released.
- Energy released during oxidation reactions is carried by reduced NAD and FAD to the cristae of mitochondria.
- The transfer of electrons between carrier molecules in the electron transport chain (ETC) in the membrane of the cristae is coupled to proton pumps.
- During chemiosmosis protons diffuse across the membrane down a concentration gradient via ATP synthase to produce ATP.
- Oxygen binds with free protons to form water, thus maintaining the hydrogen (proton) gradient.
- The structure of mitochondria is closely linked to their function.

8.2 Cell respiration

Oxidation and reduction

Cell respiration involves several **oxidation** and **reduction** reactions. Such reactions are common in biochemical pathways. When two molecules react, one of them starts in the oxidised state and becomes reduced, and the other starts in the reduced state and becomes oxidised, as shown in Figure **8.9**.

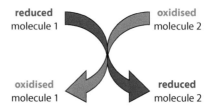

Figure 8.9 Oxidation and reduction are linked processes — as one molecule is reduced another is oxidised in a redox reaction.

There are three different ways in which a molecule can be oxidised or reduced, as outlined in Table **8.2**. In biological oxidation reactions, addition of oxygen atoms is an alternative to removal of hydrogen atoms. Since a hydrogen atom consists of an electron and a proton, losing hydrogen atoms (oxidation) involves losing one or more electrons.

Oxidation	Reduction
loss of electrons	gain of electrons
loss of hydrogen	gain of hydrogen
gain of oxygen	loss of oxygen

Table 8.2 Changes involved in oxidation and reduction.

Oxidation and reduction occur together in biochemical reactions. As one compound loses electrons, for example. In the simple equation for respiration, glucose is oxidised as hydrogen atoms, and therefore electrons, are gradually removed from it and added to hydrogen acceptors (the oxygen atoms on the left side of the equation), which become reduced.

$$C_6H_{12}O_6 + 6O_2 \rightarrow 6CO_2 + 6H_2O + \text{energy}$$

Chemical reactions like this are referred to as **redox reactions**. In redox reactions, the reduced molecule always has more potential energy than the oxidised form of the molecule. Electrons passing from one molecule to another carry energy with them.

Respiration

Cell respiration is the controlled breakdown of food molecules such as glucose or fat to release energy, which can be stored for later use. The energy is most commonly stored in the molecule adenosine triphosphate, or ATP. The respiration pathway can be divided into four parts:
- glycolysis
- link reaction
- Krebs cycle
- electron transfer chain and chemiosmosis.

Glycolysis

Glycolysis is the first stage in the series of reactions that make up respiration. It literally means 'breaking apart glucose'. The glycolysis pathway occurs in the cytoplasm of the cell. It is anaerobic (that is, it can proceed in the absence of oxygen) and produces pyruvate and a small amount of ATP. One molecule of the hexose sugar glucose is converted to two molecules of the three-carbon molecule called pyruvate with the net gain of two molecules of ATP and two molecules of NADH + H$^+$. The process is shown in detail in Figure **8.10**.

1. The first steps are to add two phosphate groups from ATP, in a process called **phosphorylation**. A hexose bisphosphate molecule is produced. (This appears contrary to the purpose of respiration, which is to *make* ATP, but the two lost ATPs are recovered later.)
2. The hexose bisphosphate is now split into two triose phosphates in a reaction called **lysis**.
3. Now, another phosphorylation takes place but this time an inorganic phosphate ion, P$_i$, is used and not ATP. Two triose bisphosphates are formed. The energy to add the P$_i$ comes from an **oxidation** reaction. The triose bisphosphate is oxidised and at the same time NAD$^+$ is reduced to NADH + H$^+$.
4. There now follows a series of reactions in which the two phosphate groups from each triose bisphosphate are transferred onto two molecules of ADP, to form two molecules of ATP – this is **ATP formation**. A pyruvate molecule is also produced.

Four molecules of ATP are formed by converting one molecule of glucose to two molecules of pyruvate. However, two molecules of ATP were required to start the pathway and so there is a net gain of two molecules of ATP per glucose. In addition two NADH + H$^+$ are formed.

To summarise, the net products of glycolysis per glucose molecule are:
- 2 ATP
- 2 NADH + H$^+$
- 2 molecules of pyruvate.

Exam tip
An easy way to remember oxidation and reduction is to think of the words OIL RIG:
Oxidation
Is
Loss of electrons
Reduction
Is
Gain of electrons

NAD$^+$ is a hydrogen carrier that accepts hydrogen atoms removed during the reactions of respiration. During glycolysis, two hydrogen atoms are removed and NAD$^+$ accepts the protons from one of them and the electrons from both of them.

NAD$^+$ + 2H → NADH + H$^+$

Note that NADH + H$^+$ must be written in this way and should not be simplified to NADH$_2$.

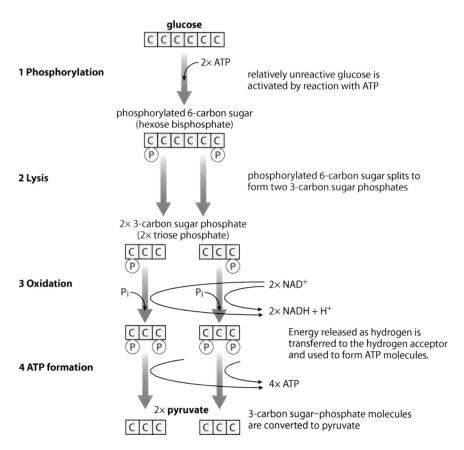

Figure 8.10 The stages of glycolysis. Note that for each molecule of glucose, two molecules of ATP are used and four are formed, so there is a net gain of two ATPs.

Exam tip
Think of your own acronym such as **P**eople **L**ove **O**utdoor **A**ctivities to help you recall the steps in glycolysis.

Phosphorylation

Phosphorylation of glucose is the first stage in its breakdown and involves the addition of phosphate groups from ATP. This turns the relatively unreactive glucose into a more unstable, phosphorylated compound, which can be split to form two three-carbon sugars.

Energy coupling

Energy coupling involves a sequence of reactions in which energy from an energy-releasing process is used to drive an energy-requiring process. Phosphorylation is an example of energy coupling – the transport of a phosphate group from ATP to a reactant molecule in the coupled reaction supplies energy for that reaction. The reactant molecule becomes a phosphorylated intermediate, an unstable molecule compared to the unphosphorylated state.

Bisphosphate and diphosphate

Although both these molecules contain two phosphate groups, the phosphates are joined in different ways. In a diphosphate molecule such as ADP, the two phosphates are joined to each other; in a bisphosphate molecule such as hexose bisphosphate, each phosphate is joined to a different part of the hexose molecule.

Test yourself

3 List **three** ways in which a substance can be reduced.
4 State the molecule used to phosphorylate glucose at the start of glycolysis.
5 State the name of the process that splits the hexose bisphosphate molecule into two triose phosphate molecules.

The link reaction and Krebs cycle

If oxygen is present, pyruvate formed during glycolysis moves into the mitochondrial matrix by facilitated diffusion. The structure of a mitochondrion is shown in Figures **8.11** and **8.12**.

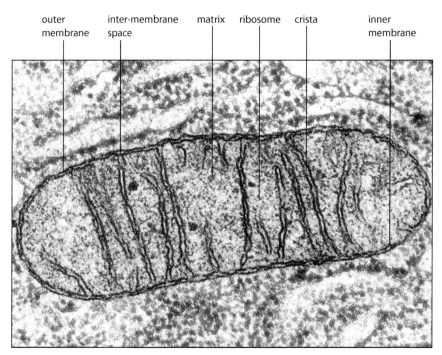

Figure 8.11 Coloured electron micrograph of a mitochondrion (×72 000).

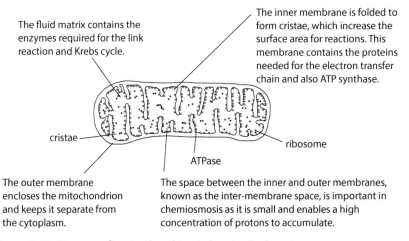

Figure 8.12 Diagram of a mitochondrion in longitudinal section.

Electron tomography and mitochondria

Electron tomography enables a 3D image of a mitochondrion to be built up from a series of two-dimensional images each 0.5 µm thick. The structure is viewed from many different angles and the 3D structure is calculated. Different membranes can be traced and changes that take place in active or even dying mitochondria can be followed.

The link reaction and the Krebs cycle pathways occur in the mitochondrial matrix (Figure **8.13**).

1. The link reaction converts pyruvate to acetyl CoA using coenzyme A, and a carbon atom is removed as carbon dioxide. This is called a **decarboxylation reaction**. At the same time as the carbon dioxide is removed, pyruvate is oxidised by the removal of hydrogen. The hydrogen atoms are removed by NAD^+ to form $NADH + H^+$.
2. Acetyl CoA now enters the **Krebs cycle** to continue the processes of aerobic respiration. Immediately, the coenzyme A is removed to be recycled. The acetyl component of the acetyl CoA combines with a four-carbon compound to form the six-carbon compound, citrate.
3. 4 The acetyl (two-carbon) groups are dehydrogenated to release four pairs of hydrogen atoms and decarboxylated to form two molecules of carbon dioxide so that the two carbons that enter with acetyl CoA leave as carbon dioxide.
5. One molecule of ATP is formed.
6. Hydrogen is removed during oxidation reactions to the two hydrogen carriers NAD^+ and FAD^+.
7. Since the Krebs cycle is a cyclic process, what enters must eventually leave so that the cycle begins and ends with the same substances.

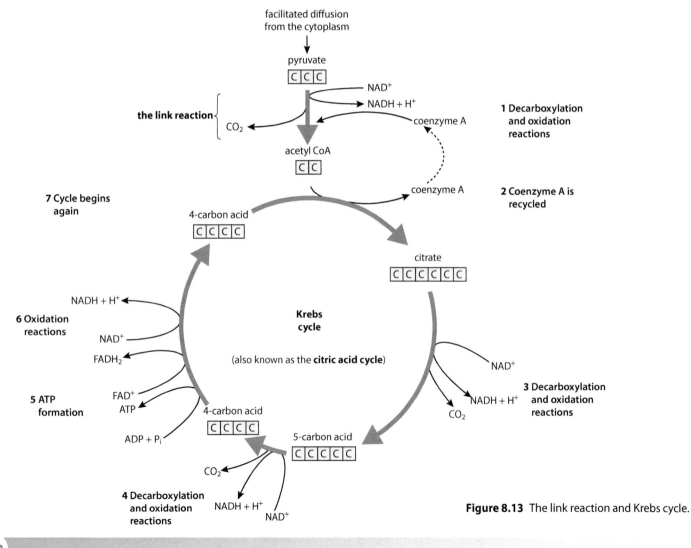

Figure 8.13 The link reaction and Krebs cycle.

Because each molecule of glucose forms two molecules of pyruvate during glycolysis, each glucose molecule requires two link reactions and two rotations of the Krebs cycle. Thus, when working out the products of the cycle we must consider two sets of products. So, to summarise, the products of the link reaction and Krebs cycle, per glucose molecule, are:
- 8 molecules of NADH + H$^+$
- 2 molecules of FADH$_2$
- 2 molecules of ATP
- 6 molecules of CO$_2$.

Note that the correct method to show reduced FAD is FADH$_2$.

The electron transport chain, oxidative phosphorylation and chemiosmosis

Most of the ATP produced from glucose breakdown occurs in the last phase of respiration at the end of the **electron transport chain** (ETC). Reactions take place on the inner mitochondrial membrane of the cristae and in the inter-membrane space between the inner and outer membranes (Figures **8.11** and **8.12**). The inner membrane holds molecules called **electron carriers**, which pick up electrons and pass them from one to another in a series of oxidations and reductions. The pathway is called the electron transport chain because electrons from hydrogen are moved along it. Just as the inner lining of the small intestine is folded to increase its surface area to absorb food, so the inner mitochondrial membrane is highly folded into cristae to increase its surface area. The cristae provide a large area for the protein molecules used in the electron transport chain. Several protein molecules are electron carriers and the three key ones are shown in Figure **8.14**.

Oxidative phosphorylation the formation of ATP in the mitochondria using energy released by the oxidation of glucose during respiration.

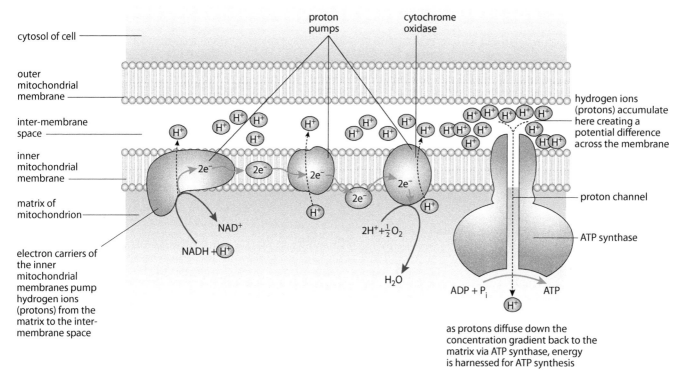

Figure 8.14 The electron transport chain showing oxidative phosphorylation and chemiosmosis.

Chemiosmosis

Osmosis is the passive flow of water molecules down a concentration gradient through a partially permeable membrane. Chemiosmosis is similar but instead of water moving, it is protons that pass down a concentration gradient.

As Table **8.3** shows, the net production of ATP from one molecule of glucose is, in theory, 36. Biochemists have discovered that the actual production is closer to 30 ATPs and propose that this discrepancy occurs because some protons are used to transfer ATP from the matrix to the cytoplasm. There are also losses such as the cost of moving pyruvate, phosphate and ADP (for ATP synthesis) into the mitochondria.

Electrons from NADH + H^+ are transferred onto the first electron carrier. As they pass through the carrier, they lose energy and this is used to pump a proton (H^+) from the matrix to the inter-membrane space, lowering the pH of the space. The electrons are then transferred to two further carriers and the process is repeated. As the electrons from one NADH + H^+ pass along the chain, a total of nine protons are pumped into the inter-membrane space. At the end of the chain, the electrons are combined with protons and oxygen atoms to make water, in the oxidative part of **oxidative phosphorylation**, completing the release of energy from the oxidation of glucose to produce ATP. The formation of water ensures that the H^+ gradient is maintained.

The space between the membranes is very narrow and allows for a rapid increase in the concentration of the protons that are pumped into it during the electron transfer reactions. The protons in the inter-membrane space create a concentration gradient between the space and the matrix. These protons can now flow passively down this concentration gradient back into the matrix, through a very large integral protein. This is called **chemiosmosis**. The large protein contains the enzyme **ATP synthase**, which joins ADP and P_i to form ATP. Three protons flowing through this enzyme result in one ATP being formed. Since the electrons from one NADH + H^+ pump nine protons into the inter-membrane space, each NADH + H^+ results in the formation of three ATP. This is the phosphorylation part of oxidative phosphorylation.

$FADH_2$ also supplies electrons to the electron transport chain but further down the chain than NADH + H^+, missing the first protein pump. $FADH_2$ allows the production of just two ATPs.

Overall ATP production during aerobic respiration

Stage		ATP use	ATP yield
glycolysis	2 ATP used at the start	−2 ATP	
	2 NADH + H^+		+4 ATP
	ATP formation		+4 ATP
link reaction	2 NADH + H^+		+6 ATP
Krebs cycle	ATP formation		+2 ATP
	6 NADH + H^+		+18 ATP
	2 $FADH_2$		+4 ATP
net energy yield			+36 ATP

Table 8.3 Together, glycolysis, the link reaction and the Krebs cycle yield 36 ATP molecules for each molecule of glucose broken down by aerobic respiration.

Nature of science

Paradigm shift – chemiosmosis theory required a significant change of view

The chemiosmosis hypothesis was proposed in 1961 by Peter Mitchell (1920–1992) to explain how the mitochondria convert ADP to ATP. At the start of the 1960s, scientists did not understand the exact mechanisms by which electron transfer is coupled to ATP synthesis. Various hypotheses current at the time proposed a direct chemical relationship between oxidising and phosphorylating enzymes and proposed that a high-energy intermediate compound was formed. Mitchell's theory was completely new and proposed an indirect interaction between these enzymes with no intermediate compound. He suggested that ATP synthesis is driven by a reverse flow of protons down a concentration gradient, the so-called the 'chemiosmotic theory'. This theory was first received with scepticism as his work was considered to be radical and outside the popularly held view. Mitchell struggled to persuade his contemporaries to reject the more accepted theories because his theory used a completely different approach. After several years of research, he published detailed evidence to support his theory, both in a pamphlet in 1966 and also in further publications in 1968, which were known as 'the little grey books' because of their bland covers. Eventually in the early 1970s, Mitchell's chemiosmosis theory gained scientific acceptance, and scientists conceded that no high-energy intermediate compounds were likely to be found. Mitchell was awarded the Nobel Prize for Chemistry in 1978.

 Despite Peter Mitchell's strong evidence for chemiosmosis, which falsified previous theories, he struggled to have his work accepted.

Question to consider

- Why is it often difficult for a paradigm shift to gain acceptance?

Test yourself

6. State the precise sites of the link reaction and the reactions of Krebs cycle.
7. State the precise site of the reactions of the electron transport chain.
8. State the name of the molecule that enters the Krebs cycle.
9. Explain the purpose of the folding of the inner mitochondrial membrane.
10. Explain what happens to the pH of the inter-membrane space as electrons move along the ETC.
11. State the name of the molecule that the protons pass through going from the inter-membrane space to the matrix.

8.3 Photosynthesis

Learning objectives

You should understand that:
- Light-dependent reactions occur in the inter membrane space of the thylakoids of a chloroplast.
- Light-independent reactions occur in the stroma of a chloroplast.
- Light-dependent reactions lead to the production of reduced $NADP^+$ ($NADPH + H^+$) and ATP.
- As photosystems absorb light, excited (high-energy) electrons are generated.
- Photolysis of water generates electrons, which are used in the light-dependent reactions.
- Excited electrons are transferred between carriers in the thylakoid membranes.
- Excited electrons from photosystem II are used to generate a proton gradient.
- ATP synthase in thylakoids generates ATP using this proton gradient.
- Excited electrons from photosystem I reduce $NADP^+$ to $NADPH + H^+$.
- A carboxylase catalyses the carboxylation of ribulose bisphosphate (RuBP) in the light-independent reactions.
- Reduced $NADP^+$ ($NADPH + H^+$) and ATP are used to reduce glycerate 3-phosphate to triose phosphate.
- Triose phosphate is used to produce carbohydrates and regenerate RuBP.
- ATP is used to re-form RuBP.
- The structure of a chloroplast is closely linked to its function.

The light-dependent and light-independent reactions

Both of these sets of reactions are part of photosynthesis and can only occur when there is sufficient light. Light-dependent reactions can only take place in light, and although light-independent reactions do not require light directly – and *can* take place when it is dark – they do require the products of the light-dependent reactions.

The reactions of photosynthesis

Photosynthesis is the process by which light energy is harvested and stored as chemical energy, primarily in sugars but also in other organic molecules such as lipids. It occurs in green plants, algae and some bacteria. All these organisms are known as autotrophs, which means they can make their own food.

Photosynthesis can be divided into two parts:
- the light-dependent reactions
- the light-independent reactions.

The light-dependent reactions produce compounds that are used in the light-independent reactions.

Both the light-dependent and the light-independent reactions take place in the chloroplasts of plant cells (Figures **8.15** and **8.16**). The stroma contains the enzymes required for the light-independent reactions and the stacks of thylakoid membranes increase the surface area for the light-dependent reactions.

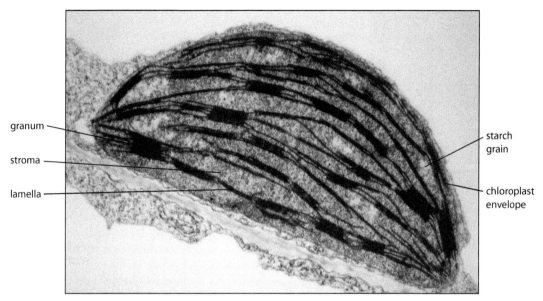

Figure 8.15 Coloured electron micrograph of a chloroplast (×20 000).

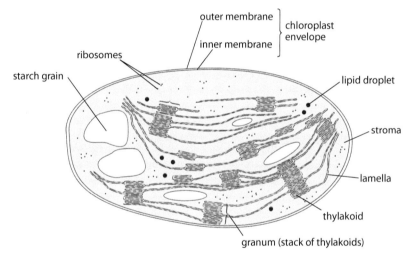

Figure 8.16 Diagram of a chloroplast.

The light-dependent reactions

The **light-dependent reactions** occur on the **thylakoid membranes** of the chloroplast and are powered by light energy from the Sun. Each thylakoid is a flattened sac so the space in the middle is narrow. The thylakoid membranes form stacks called **grana**, which may be joined together by inter-granal membranes. Light is absorbed by photosynthetic pigments such as chlorophyll, which are found on the granal membranes. There are several pigments found in plants and each one absorbs light of a slightly different wavelength. The pigments are associated with proteins that are involved in electron transport, proton pumping and chemiosmosis.

The photosynthetic pigments are combined into two complex groups called **photosystems I and II**, which absorb the light energy and use this to boost electrons to a higher energy level so that they become 'excited', as shown in Figure **8.17**.

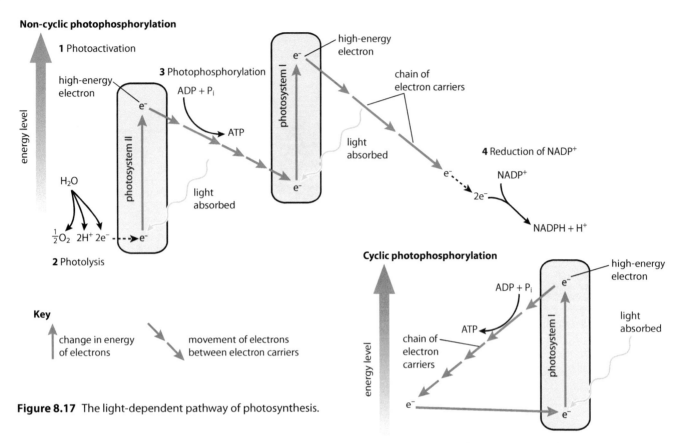

Figure 8.17 The light-dependent pathway of photosynthesis.

Exam tip

$NADP^+$ is very similar to NAD^+ – it simply has a phosphate group attached. An easy way to remember that photosynthesis uses $NADP^+$ is to note that they both have a letter 'P'.

1. The first step in the light-dependent reaction is the **photoactivation** of photosystem II. Pigment molecules in the photosystem absorb light energy and boost electrons in a molecule of chlorophyll to a higher energy level. The electrons are accepted by a carrier protein molecule at the start of the electron transport chain.
2. Photosystem II has to replace these lost electrons and it does this by taking them from water. Water is split into electrons, protons (hydrogen ions) and an oxygen atom. Since the splitting is brought about by light energy, it is called photolysis. The oxygen is released as an excretory product.
3. Excited electrons travel along the electron transport chain into photosystem I. As they do this, they lose energy but this is used to pump protons into the thylakoid interior (in a similar way as occurs in the electron transport chain in the mitochondrion). The thylakoid interior is small and so a proton concentration gradient builds up quickly. The protons then flow out through a large channel protein, almost identical to the one in mitochondria, which contains the enzyme ATP synthase. This time though, the formation of ATP is called photophosphorylation and it occurs between photosystems II and I (Figure 8.18).
4. Absorption of light energy causes photoactivation in photosystem I, boosting more electrons to an even higher energy level. The electrons that arrive from photosystem II replace those that are displaced. The electrons at the higher energy level are combined with protons in the hydrogen carrier $NADP^+$ to form $NADPH + H^+$.

The two products of the light-dependent reaction, ATP and $NADPH + H^+$, are used to drive the light-independent reaction.

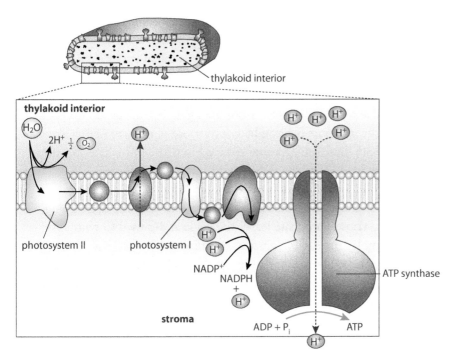

Figure 8.18 Chemiosmosis in photosynthesis.

Cyclic and non-cyclic photophosphorylation

When ATP is produced using energy from excited electrons flowing from photosystem II through photosynthesis I and on to NADP$^+$, the process is called **non-cyclic photophosphorylation**.

When light is not a limiting factor, the light-independent reactions may proceed more slowly than the light-dependent reaction, so that the supply of NADP$^+$ runs out. This means the electrons boosted up from photosystem I have no acceptor available to take them. They are sent back to photosystem II and rejoin the electron transport chain near the start, generating more ATP. This alternative pathway is called **cyclic photophosphorylation** (Figure **8.17**) – it produces neither O$_2$ nor NADPH + H$^+$.

The light-independent reactions

The light-independent reactions occur in the stroma of the chloroplast and comprise a cyclic pathway called the Calvin cycle. The pathway is shown in Figure **8.19**. ATP and NADPH + H$^+$ formed during the light-dependent stage supply energy and reducing power for the Calvin cycle. The final product of the cycle is carbohydrate.

During each turn of the Calvin cycle, one molecule of carbon dioxide is used so Figure **8.19** shows three cycles combined together. As this is a cycle, what goes in must leave, so three carbons enter in three molecules of carbon dioxide and three carbons leave in one molecule of triose phosphate, which can be used to form glucose or other organic compounds.

1. At the start of the cycle, the acceptor molecule ribulose bisphosphate (RuBP) combines with incoming carbon dioxide from the air to form glycerate 3-phosphate (GP). This reaction is called **carbon fixation**. It is catalysed by **RuBP carboxylase**, an enzyme that is sometimes called **Rubisco**.
2. The ATP and NADPH + H$^+$ from the light-dependent reaction convert the glycerate 3-phosphate into triose phosphate (TP). Glycerate 3-phosphate therefore becomes reduced to triose phosphate. No more phosphate is added so the only input from ATP is energy.
3. Six molecules of triose phosphate are produced but only five are needed to reform the ribulose bisphosphate to keep the cycle going. The extra triose phosphate leaves the cycle and is used to synthesise organic molecules such as glucose or amino acids.
4. Since the triose phosphate that leaves the cycle takes a phosphate with it, this is replaced in the cycle with a phosphate from ATP, as the five remaining triose phosphates are converted back to three ribulose bisphosphate molecules, and the cycle begins again.

Six 'turns' of the Calvin cycle produces two triose phosphate molecules, which can be combined to form the final product, glucose. Some triose phosphate molecules will follow other pathways to make other organic carbohydrate molecules, such as sucrose or cellulose, or other molecules that the plant needs, such as amino acids, fatty acids or vitamins.

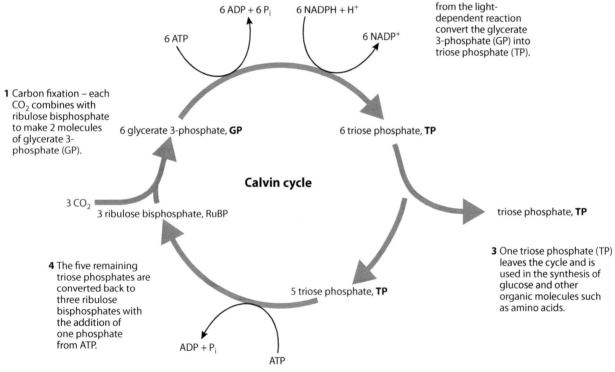

Figure 8.19 The light-independent pathway of photosynthesis – the Calvin cycle.

Nature of science

Scientific advance follows technical innovation – Melvin Calvin's experiments with ^{14}C

Melvin Calvin (1911–1997) was awarded a Nobel Prize in 1961 for experiments leading to our understanding of the Calvin cycle. He used the green alga *Chlorella* and radioactively labelled carbon dioxide, $^{14}CO_2$, to follow the path of carbon in photosynthesis.

The *Chlorella* was exposed to light – in a flat, round flask resembling a lollipop – and photosynthesis stabilised at a steady rate. Radioactive carbon dioxide was added for different periods of time, from a second to a few minutes, so that different products of photosynthesis became labelled. Then the *Chlorella* was killed right away with boiling ethanol so that all enzymatic reactions stopped. The labelled compounds were separated from one another using chromatography (Figure **8.20**). As different compounds contained different amounts of radioactivity, Calvin and his team were able to identify glycerate 3-phosphate and other sugar phosphates as intermediates that form at different stages in the cycle (Figure **8.19**).

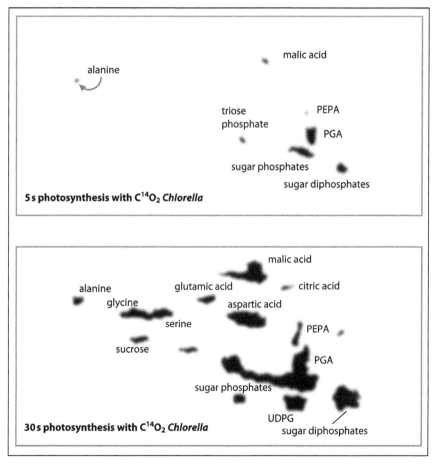

Figure 8.20 Representations of autoradiograms showing the labelling of carbon compounds in the alga *Chlorella* after exposure to $^{14}CO_2$. After 5 s, and after 30 s, the reaction was terminated and the labelled compounds in the cell homogenates were separated by paper chromatography. Notice that glycerate 3-phosphate (PGA) appears very quickly, because it is the first stable intermediate of the Calvin cycle.

By exposing the cells to $^{14}CO_2$ for progressively shorter periods of time, Calvin identified glycerate 3-phosphate as the first stable intermediate in the process and subsequently other labelled sugar phosphates were found to be produced as a result of its reduction.

From the distribution of ^{14}C in all the labelled compounds, ribulose bisphosphate was eventually identified as the acceptor of carbon dioxide. This final piece of the cycle explained the labelling patterns of the all the other intermediates.

Creative endeavour

Occasionally, the answer to a question about an unknown biochemical pathway is discovered as a result of designing an investigation that is simple and elegant. To find the steps involved in the light-independent reactions, a team led by Melvin Calvin designed what he called the 'lollipop' apparatus (Figure **8.21**).

Question to consider

- To what extent could an elegant scientific protocol be considered a work of art?

Figure 8.21 Calvin's apparatus for investigating the light-independent reactions of photosynthesis.

Test yourself

12 State the site of the light-independent reaction
13 State the colour of the light spectrum that is not absorbed by plants.
14 Explain what the light energy is used for when it is absorbed by the photosystems.
15 State the name of the waste product from the light-dependent reaction.
16 State the names of **two** products from the light-dependent reaction that are needed for the light-independent reactions.
17 State the name of the starting acceptor molecule in the Calvin cycle, which reacts with carbon dioxide.

Exam-style questions

1 Which of the following statements is correct about enzymes?

    A Most enzymes use the lock-and-key model of substrate interaction as it is the most stable.
    B In allosteric control of metabolic pathways, a product within the pathway can act as a non-competitive inhibitor of an enzyme earlier in the pathway.
    C Increasing the concentration of substrate has no effect on the rate of a reaction being inhibited by a competitive inhibitor.
    D Competitive inhibitors bind to an allosteric site and alter the shape of the active site. [1]

2 Which of the following statements is correct?

    A Oxidation can involve the removal of oxygen from a compound.
    B The solution inside a mitochondrion is called the matrix and the solution inside a chloroplast is called the stroma.
    C The folds of the inner membrane of a mitochondrion are called grana.
    D The photosynthetic pigments in a chloroplast are found on the cristae. [1]

3 Which of the following statements about glycolysis is correct?

    A It is anaerobic, occurs in the cytoplasm and includes at least one phosphorylation reaction.
    B It is aerobic and includes a lysis reaction.
    C In the final stages, two molecules of ATP are used in the formation of pyruvate.
    D It is aerobic and does not include a lysis reaction. [1]

4 Which of the following statements about the link reaction and Krebs cycle is correct?

    A Pyruvate in the mitochondrion matrix is oxidised to acetyl CoA.
    B At the end of the link reaction, coenzyme A is recycled back into the cytoplasm to combine with another pyruvate molecule.
    C During one rotation of the Krebs cycle there are three decarboxylation reactions.
    D The link reaction occurs in the mitochondrial matrix and Krebs cycle occurs on the mitochondrial cristae. [1]

5 Which of the following statements about the electron transport chain is correct?

    A As electrons flow along the chain, protons are pumped into the mitochondrial matrix.
    B ATP is formed as electrons flow through the enzyme ATP synthase.
    C The flow of electrons along the chain causes the pH of the inter-membrane space to increase.
    D When protons diffuse through ATP synthase into the matrix, ATP is formed. [1]

6 Which of the following statements about the light-dependent reaction is correct?

    A Photoactivation of photosystem I causes photolysis.
    B Cyclic photophosphorylation involves both photosystems I and II.
    C Electrons from the photolysis of water are transferred to $NADP^+$ from photosystem I.
    D $NADP^+$ becomes oxidised when it combines with electrons and protons. [1]

7 Which of the following statements about the light-independent reaction is correct?

    A The enzyme RuBP carboxylase is used to convert ribulose bisphosphate into triose phosphate.
    B The conversion of glycerate 3-phosphate to triose phosphate requires ATP and $NADP^+$ from the light-dependent reaction.
    C At the start of the Calvin cycle, carbon dioxide combines with glycerate 3-phosphate to form ribulose bisphosphate.
    D The formation of ribulose bisphosphate from triose phosphate requires a phosphorylation reaction. [1]

8 Which of the following statements is correct?

    A A limiting factor is one that determines what the end-product of a reaction is.
    B A limiting factor is one that determines the rate of a reaction.
    C Limiting factors for photosynthesis include light intensity, carbon dioxide concentration and glucose concentration.
    D An absorption spectrum can be obtained by measuring the release of oxygen gas from pondweed at different wavelengths of light. [1]

9 Explain the reasons for:

  a many grana in the chloroplast [2]

  b poor growth of plants growing beneath trees [2]

  c large amounts of RuBP carboxylase in the chloroplast. [2]

10 Explain the process of aerobic respiration in a cell starting at the end of glycolysis and including oxidative phosphorylation. [8]

11 Explain how the light-independent reaction of photosynthesis relies on the light-dependent reaction. [5]

12 **a** Draw an annotated diagram of a chloroplast and indicate how the structure of the chloroplast is adapted to the function it performs. [5]

  b Describe how solar energy is converted into the chemical energy of ATP during the light-dependent stage of photosynthesis. [5]

  c Name the products of the light-dependent stage and state their importance to biological systems. [4]

Plant biology (HL) 9

Introduction

Plants are a vital part of almost every ecosystem. As autotrophs, they start food chains and produce carbohydrates and other organic molecules that are needed by heterotrophs. The oxygen they release and the carbon dioxide they absorb are important in maintaining the balance of gases in the atmosphere. When plants die, they contribute to the formation of humus, an important part of the structure of soil.

9.1 Transport in the xylem of plants

Transpiration

Transpiration is the loss of water vapour from the leaves and stems of plants. Water is absorbed into the roots, travels up the stem in the xylem vessels in the vascular bundles to the leaves, and is lost by evaporation through stomata, which open to allow the exchange of oxygen and carbon dioxide in the leaf (Figure 9.1).

Most plants grow in areas where the amount of water in the air, the humidity, is less than in the leaves. During the day, water vapour leaves the air spaces in the spongy mesophyll through open stomata in the lower epidermis of the leaf, and in the stem. The evaporating water is drawn from the xylem in the vascular bundles in the leaf and stem. The vascular bundles are continuous with those in the roots so a column of water is formed connecting the roots, stem, leaves and air spaces. This is known as the **transpiration stream**.

Transpiration also carries minerals through the plant, and serves to cool leaves in warm conditions.

Cohesion–tension theory

The movement of water in the xylem can be explained by the cohesion–tension theory.

1. Loss of water vapour from the stomata in the leaves results in 'tension' or negative pressure in the xylem vessels.
2. Water vapour enters the air spaces in the leaf from the xylem vessels to replace water lost through the stomata.
3. Continuous columns of water are drawn up the xylem due to cohesion among water molecules in the xylem and forces of adhesion between the water molecules and the xylem vessel walls. Cohesion is due to hydrogen bonding among water molecules and adhesion is caused by the hydrogen bonds between water molecules and molecules in the walls of the xylem vessels (Subtopic **2.2**).
4. The tension in the xylem is strong due to loss of water and there would be a tendency for xylem vessels to collapse inwards. The thickening provided by lignin prevents this happening (Figure **9.2**).

Learning objectives

You should understand that:
- Transpiration occurs as a consequence of gas exchange in the leaf.
- Water is transported from the roots to the leaves to replace losses from transpiration.
- The structure of the xylem and cohesion among water molecules allow transport to occur under tension.
- Evaporation and adhesion between water molecules and the surfaces they contact lead to tension forces in the walls of plant cells.
- As minerals are taken up by active transport by the roots, water enters by osmosis.

Transpiration the loss of water vapour from the aerial parts of plants (the leaves and stems)

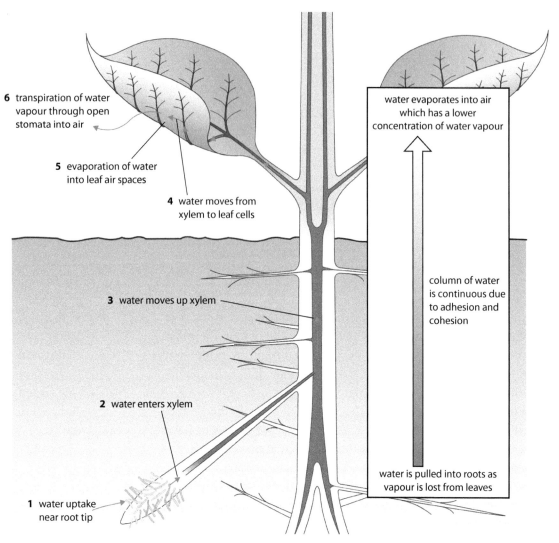

Figure 9.1 The movements of water through a plant: overall, water moves from the soil to the air (from where there is more water to where there is less water).

5 Water is drawn in from the cortex in the roots to replace water that is lost in transpiration.
6 The tension caused by transpiration also causes water to be drawn into the roots from the soil.

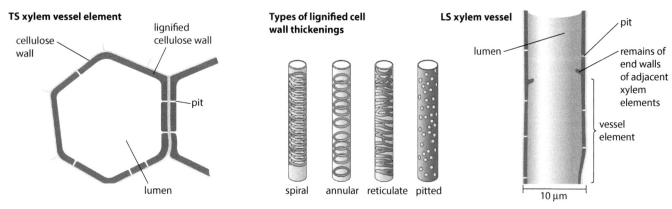

Figure 9.2 Xylem vessels are not alive and have no plasma membrane, so water can easily move in and out of them.

Transpiration is largely controlled by the pairs of guard cells that surround each stoma. **Guard cells** have unevenly shaped cell walls with more cellulose on the side adjacent to the stoma. This inner part of the cell wall is less elastic, so that when guard cells take up water and become turgid, they take on a sausage-like shape and an opening – the **stoma** – is formed between them (Figure **9.3**). When the guard cells lose water, the cell walls relax and the stoma closes.

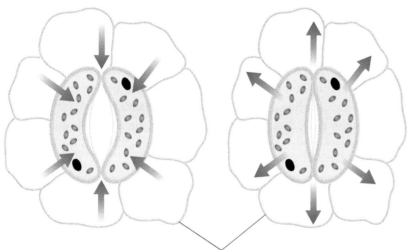

Figure 9.3 The opening and closing of stomata. Gases can diffuse in and out of open stomata. When stomata are closed, water loss is minimised.

The opening and closing of stomata is controlled by the concentration of potassium ions. In darkness, these ions move out of the guard cells into surrounding cells. In light conditions, potassium ions are actively pumped into the vacuoles of guard cells. This creates an increased solute concentration so that water enters by osmosis, making the cells turgid and opening the stomata. A plant hormone called abscisic acid, produced in the roots during times of drought, affects potassium ion movement in guard cells. When abscisic acid is present, potassium ions leak out and water follows by osmosis. This means that the guard cells lose turgor and stomata close, thus conserving water.

Factors affecting transpiration

Several abiotic environmental factors (notably light, temperature, humidity and wind speed) influence the rate of transpiration in plants.
- Light affects transpiration directly by controlling the opening and closing of stomata. As light intensity increases, stomata open, speeding up the rate of transpiration. In darkness, stomata close, thus restricting transpiration.
- Temperature affects transpiration because heat energy is needed for the evaporation of water. As the temperature rises, the rate of transpiration also rises as water evaporates from the air spaces in the spongy mesophyll and diffuses out of the stomata.

- An increase in atmospheric humidity reduces the rate of transpiration. Air in the mesophyll air spaces tends to be saturated with water vapour so if atmospheric air becomes more humid, the concentration gradient between the air space and the atmosphere is reduced and transpiration is slowed down.
- An increase in wind speed increases the rate of transpiration because it blows away the air just outside the stomata, which is saturated with water vapour. Reduced humidity near the stomata enables water vapour to diffuse more readily from the spongy mesophyll, where the air is very humid, to the air just outside the leaf, which has lower humidity.

Transpiration in xerophytes

Xerophytes are plants that live in arid climates – an example is shown in Figure **9.4**. Some xerophytes grow in areas where there is very little rainfall all the year, while others live in places where rainfall is intense but short lived, with long, dry periods for the rest of the year. In both cases, plant species have evolved specialisations that enable them to survive shortages of water by reducing water loss.

Marram grass is well adapted to survive in the dry conditions found in sand dunes. Its leaves are rolled into tube-like shapes, which are protected on the outside by a thick waxy cuticle. Stomata of marram grass are protected deep inside pits (Figure **9.5**) which themselves are rolled up inside the leaf. A lining of hairs on the inner side of the leaf keeps humid air trapped inside the rolled-up leaves. Hairs prevent water loss by diffusion. When the humidity of the air rises and water loss is not a problem, marram grass is able to unroll its leaves using specialised hinge cells and maximise photosynthesis. When the leaf is unrolled, leaf hairs help conserve a supply of water by trapping moist air. This air remains inside the leaf when it rolls up again.

The Crassulaceae, a group of succulent plants, have evolved a mechanism allowing them to keep their stomata closed during the heat of the day, to reduce water loss. At night, the stomata open and carbon dioxide diffuses into the leaf. It is fixed temporarily in cells, and then is released for photosynthesis during the day when the stomata are closed. This is called crassulacean acid metabolism (CAM).

Halophytes ('salt plants')

Halophytes are plants that live on salt marshes or mud flats. Their roots are surrounded by water that is salty and has a higher concentration of salt than the plants' cells. Halophytes such as glasswort (*Salicornia europaea*) (Figure **9.6**) actively absorb salt into their roots so that the concentration in root tissue is higher than that of the water. This ensures that water enters the roots, but the excess salt could become toxic if it was not removed. Some halophytes dispose of the excess salt by storing it, while others secrete it from salt glands on their leaves.

Figure 9.4 *Opuntia*, or prickly pear, is a cactus with flattened photosynthetic stems that store water. The leaves are modified as spines, and this minimises water loss by reducing the surface area from which transpiration can take place. The spines also protect the plant from being eaten by animals.

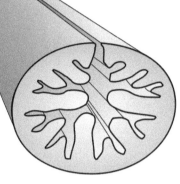

The leaves can roll up, exposing a tough, waterproof cuticle to the air outside the leaf.

Marram grass is one of the few plants that can thrive in mobile sand dunes, surviving the very dry conditions that are found there.

Specialised cells – hinge cells – cause the opening and closing of the marram grass leaf in response to air humidity. Leaf rolling traps air inside the rolled leaf. This air can remain humid even is the air outside is very dry.

The stomata are found deep in the grooves and open into the enclosed humid air space inside a rolled leaf. Water has to diffuse a long way before it reaches moving air outside the leaf, slowing water loss.

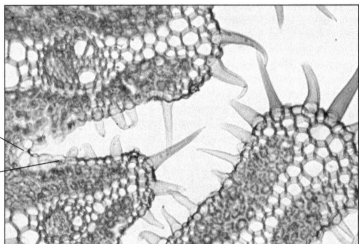

A coloured light micrograph of a transverse section through a marram grass leaf (×137).

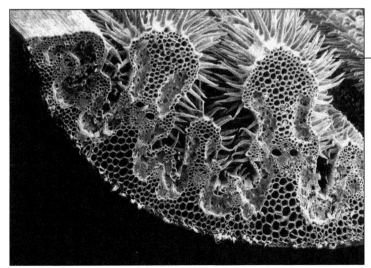

Hairs help to keep the humid air trapped inside a rolled leaf. When the lead is unrolled, the hairs help to trap a thick layer of moist air close to the leaf surface, reducing air movement. Water vapour has to diffuse through this layer before it can be carried away in air movements. The thicker the layer, the more slowly water is lost by transpiration.

A stained scanning electron micrograph of a transverse section through part of a rolled leaf of marram grass (×90).

Figure 9.5 Adaptations of the xerophyte marram grass (*Ammophila arenaria*) to dry conditions.

Figure 9.6 *Salicornia europaea* grows in saline water and actively absorbs salt into its roots.

Roots

Roots are responsible for absorbing water and mineral ions from the soil. Many plants develop an extensive, branching root system in order to increase the surface area of root in contact with the soil. In addition, as new roots grow, numerous root hairs develop to increase the surface area even more (Figure **9.7**). Root hairs are temporary and die away to be replaced by new ones near the growing tip.

Plants require a number of minerals to make a variety of substances necessary for growth. A few of these are listed in Table **9.1**.

Mineral ion	Importance
calcium	constituent of cell walls
magnesium	needed to make chlorophyll
iron	required as a cofactor for many enzymes

Table 9.1 How plants use some important mineral ions.

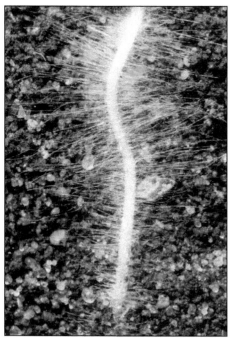

Figure 9.7 A root of a young radish showing the root hairs.

Minerals are present in the soil as salts – for example, calcium occurs in the form of calcium carbonate. These dissolve in soil water and the dissolved ions can move into root cells in different ways.

- Dissolved minerals may move into the root by **mass flow** of water carrying the ions, or by facilitated diffusion of ions from the soil water into root hairs, down their concentration gradient (Figure **9.8**). Both these processes are passive – that is, they do not require energy in the form of ATP.

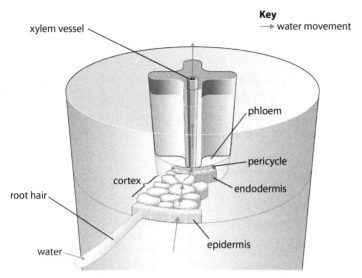

Figure 9.8 The pathway of water movement from root hair to xylem. The water may carry dissolved mineral ions.

- Where the concentration of a mineral is lower in the soil water than in plant cells, **active transport** is needed to take it up. Potassium, nitrate and phosphate are usually absorbed by active transport. Root hair cells contain mitochondria to provide ATP and most roots can only take in minerals if oxygen is available for aerobic respiration, to provide sufficient ATP. Experiments have shown that potassium ions stop moving into root cells from the soil when potassium cyanide is added. Cyanide is a potent blocker of respiration as it inhibits enzyme action, and so it prevents active transport.
- In other cases, there may be a close association, known as a **mutualistic relationship**, between the roots and a fungus (a mycorrhiza). Roots become covered with an extensive network of hyphae, which increase the surface area for absorption of both water and minerals. Minerals can pass directly from fungal hyphae to root cells.
- Active uptake of minerals into roots leads to an increase in the solute concentration inside root cells. This in turn causes the absorption of water by osmosis. The water then travels to the leaves in the transpiration stream.

Modelling water transport in plants

Some of the principles of water transport in plants can be demonstrated with simple experiments.

Root pressure

As water is absorbed rapidly in a root, it creates a strong upward pressure in the xylem. To demonstrate this, a stem from a potted plant can be cut off about 1 cm above the soil. A narrow glass tube approximately 1 m long is then fixed to the stem using rubber tubing and supported vertically in a stand. The tube is half filled with water, covered by a thin layer of oil to prevent evaporation, and the surface level is marked. After the soil is watered, and the apparatus left at room temperature, water is observed rising up the tube as a result of root pressure (Figure **9.9**).

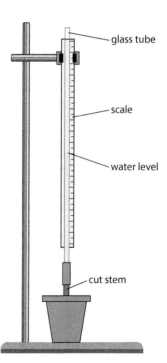

Figure 9.9 Apparatus used to demonstrate root pressure.

Support and the stem

The stem of a plant connects the roots to the leaves, and xylem within it transports water. A stem is supported by cellulose cell walls, cell turgor and by thickening with lignin. All plant cells are surrounded by a firm **cellulose** cell wall (Subtopic **2.3**), which also contains hemicelluloses and pectin. Over time, some cells may thicken their cell walls with other carbohydrates such as **lignin**, which strengthen the cell. The large fluid-filled vacuole exerts pressure on the cell wall, so plant cells become rigid and press on adjacent cells. Cells in this condition are said to be **turgid**, and each exerts **turgor pressure** on surrounding cells. Turgor pressure is sufficient to support leaves and new, soft tissue but not the stem of a tall plant. In tall plants, **xylem tissue** not only carries water but also provides support, as shown in Figure **9.2**. The xylem contains elongated cells, which are hollow, and become thickened with lignin forming a 'backbone' to support the stem. Lignin is a complex substance that is very hard and resistant to decay. Perennial plants like trees lay down more lignin each year, forming wood.

Cohesion of water molecules

The apparatus shown in Figure **9.10** can help demonstrate how water molecules 'stick' together and the importance of this for the transpiration stream. The porous pot allows water to evaporate freely from its surface, which causes water to be drawn up the tube to replace it – and the level of water in the beaker to fall. For comparison, a leafy shoot connected to similar apparatus will show a similar loss of water. Since water molecules are held together by cohesive forces, they move together in a continuous column of water up the tube.

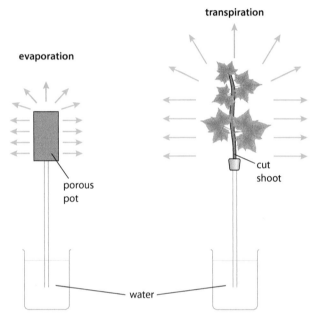

Figure 9.10 Apparatus for a porous pot experiment.

A simple potometer to estimate the rate of transpiration

The apparatus shown in Figure **9.11** is a simple potometer, which can be used to measure the rate of water loss from a leafy shoot in different environmental conditions, such as temperature, humidity and light. The apparatus is completely filled with water and the shoot must be cut under water so that no air bubbles enter the xylem. The movement of the meniscus in the glass tube can be timed to estimate rate of transpiration. The apparatus is reset using the tap and reservoir of water.

Nature of science

Using theories to explain natural phenomena – modelling water movement in plants

As the experiments above demonstrate, it is often possible to represent living processes as models. A porous pot behaves in similar way to a leaf as it loses water by evaporation, glass tubing allows us to see a column of water moving in a similar way to that in the xylem and a strip of filter paper will draw up water by capillary action just as a plant does. Models such as these have limitations but they have contributed to our understanding of real processes.

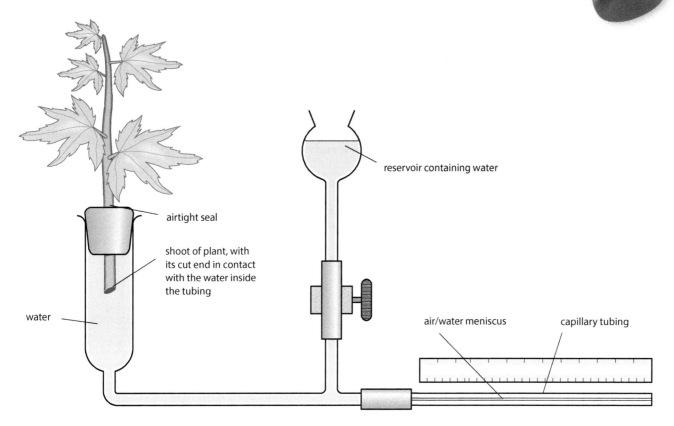

Figure 9.11 A simple potometer.

Test yourself

1. List **three** ways in which a xerophyte is adapted to survive in dry conditions.
2. Explain how mineral ions enter a plant root.

9.2 Transport in the phloem of plants

Translocation in the phloem

Translocation is the movement of organic molecules through the phloem tissue of plants. The phloem consists of two types of living cell: sieve tube cells, which are perforated to allow the movement of solutes through them, and companion cells, which are connected to the sieve tube cells as shown in Figure **9.12**.

Whereas the xylem carries water and mineral salts in one direction from the roots to the leaves, the phloem can transport materials in either direction through the plant. Translocation moves materials from a **source**, where they are made or stored, to a **sink**, where they are used, as shown in Figure **9.13**.

The products of photosynthesis, including sugars and amino acids, move from leaf cells, which are a source, into the phloem. Once in the

Exam tip
Make sure you can describe using a potometer to investigate effect of wind, temperature or humidity on transpiration.

Learning objectives

You should understand that:
- Plants transport organic compounds from a source to a sink.
- The incompressibility of water allows plants to transport substances along hydrostatic pressure gradients.
- Active transport is used to load organic compounds into the sieve tubes of the phloem at the source.
- High concentrations of solutes in the phloem at a source lead to water uptake by osmosis.
- High hydrostatic pressure causes the phloem to flow toward a sink.

Storage structures such as seeds and bulbs are sinks during the growing season but may also act as sources when they begin to sprout.

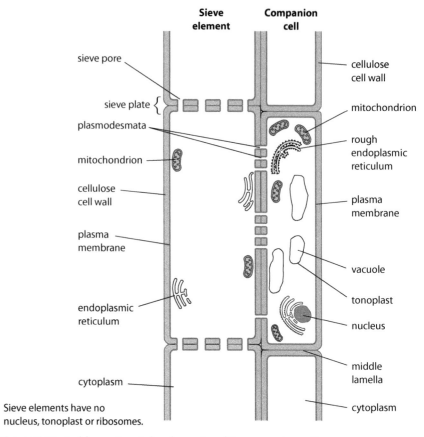

Figure 9.12 A phloem sieve tube element and its companion cell.

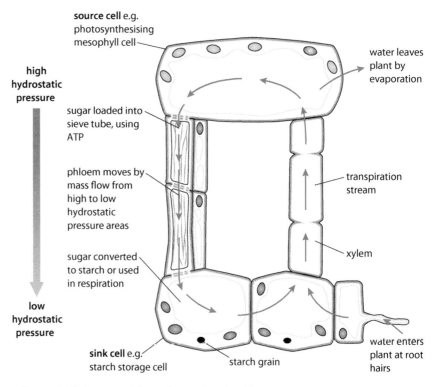

Figure 9.13 Sources, sinks and mass flow in phloem.

phloem, they are translocated to sink regions, such as growing tissue in the meristems of roots, buds and stems, or storage organs like fruits and seeds.

All the materials that are moved by translocation are dissolved in water to form a solution called 'sap', which also carries plant hormones. Sugar is usually carried as sucrose, which enters and leaves the phloem by active transport using energy provided by the companion cells. High concentrations of sucrose and other solutes in the phloem at sources, such as the leaves, lead to the uptake of water by osmosis. Once materials have entered the phloem, they move passively throughout the plant towards sinks by **mass flow** as a result of raised hydrostatic pressure in the phloem.

Hydrostatic pressure is defined as the pressure exerted by a liquid. In plants, a hydrostatic pressure gradient occurs in the phloem as sugar is loaded and unloaded at sources and sinks. In the leaves, sugar enters the phloem and is followed by water entering by osmosis. This creates a pressure gradient, which 'pushes' the content of the phloem to other parts of the plant. At a sink, the hydrostatic pressure is lower because sugar is removed from the phloem and converted to starch or used in respiration. In this way, the solute concentration of the phloem is reduced and the hydrostatic pressure gradient is maintained.

In a physical or mechanical situation, the hydrostatic pressure in a column of liquid depends on gravity, and on the height of the column – the greater the height of liquid above a certain point, the greater the hydrostatic pressure at that point.

Tissues in the root and stem

Vascular bundles in the roots, stems and leaves of plants are made up of both xylem and phloem. Xylem tissue is made up of long series of cells joined end to end, which form a continuous fine tube once the cells have stopped growing and their end walls have broken down. The phloem, on the other hand, is composed of living cells with perforated end walls known as sieve plates. When viewed under a microscope, the xylem and phloem can easily be distinguished in both cross sections and longitudinal sections through stems (Figure **9.14**). Table **9.2** summarises the relationships between structure and function for xylem and phloem tissues.

Xylem	Phloem
composed of a column of dead cells, once mature – cell end walls removed	composed of a column of living cells with perforated walls between them
continuous tube of cells enables an unbroken column of water (held together by cohesive forces) to move inside the xylem	living cells enable substances to be loaded by active transport
thickened with lignin to withstand negative pressure as water vapour is lost in transpiration	associated with companion cells which carry out cell functions and supply energy for active transport into the phloem
transports water and minerals passively from roots to leaves	transports sugars, amino acids, hormones to all parts of the plant by mass flow

Table 9.2 Structure and function in the xylem and phloem.

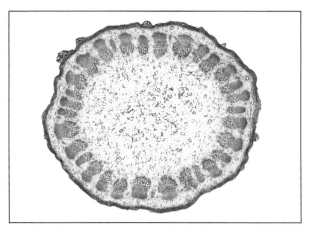
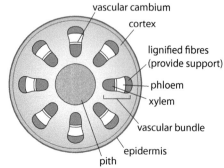

Tissue map of a transverse section through a young stem to show the distribution of tissues.

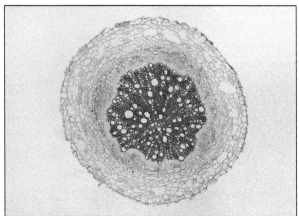
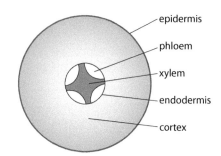

Tissue map of transverse section through a root. The epidermis protects the outer layer of the root. The cortex carries water to the xylem, which transports it to the rest of the plant. The phloem carries nutrients and minerals around the plant.

Figure 9.14 Stained micrographs showing transverse sections through a stem (×4.2) and through a root (×20), with diagrams showing the distribution of vascular tissue.

Nature of science

Scientific advance follows technical innovation – radioactive labelling

Details of the substances transported in the phloem were elucidated as radioactively labelled isotopes enabled scientists to follow the route of various substances through the plant. Radioactive isotopes became more widely available from the 1950s onwards and by the 1970s labelled carbon dioxide was being used in experiments into transport in the phloem.

Carbon dioxide labelled with ^{14}C can be introduced to the leaves, which incorporate it into sugars and other products. As the labelled substances move into the phloem, they can be followed with the help of aphids. Aphids are small insects that use needle-like mouthparts known as stylets to puncture a single sieve tube element of the phloem and feed on the contents (Figure **9.15**). High pressure in the phloem causes the sap to enter the insect's digestive system through its stylet. If an insect that has attached itself to a phloem element is anesthetised and its stylet then severed, the sap will exude from the cut end of the stylet, and it is possible to analyse the contents, providing an accurate record of the substances in the phloem and how far they have travelled.

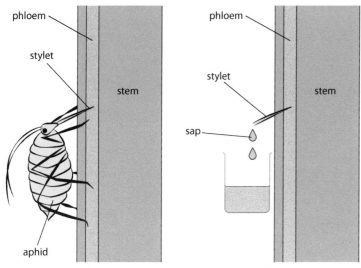

Figure 9.15 Aphid stylets can be used in the analysis of phloem.

Test yourself

3 Draw a plan of the structure of a typical stem to show the arrangement of tissues inside.
4 Compare the structure of xylem and phloem.
5 State the substances that are carried in the xylem.
6 List the factors that affect the rate of transpiration.
7 Explain why hydrostatic pressure is important in translocation.

Exam tip
Practise drawing xylem and phloem side by side to compare the structures.

9.3 Growth in plants

Meristems

Meristems are the growing parts of a flowering plant where cells may divide by mitosis throughout the life of the plant. In a dicotyledonous plant (Subtopic **9.4**), there are two types of meristem – the **apical meristems** found at the tip of the root and the shoot, and the **lateral meristems** found in the vascular bundles of the stem. Growth in the main stem occurs at the apical or primary meristem, as cells divide by mitosis. New tissue forms the cells needed to elongate the stem and to form leaves at the growing tip (Figure **9.16**). This allows a plant to grow upwards from the soil and towards light so that its leaves can obtain sufficient light for photosynthesis. In the root, growth of the apical meristem extends the root into the soil.

Many plants also grow at lateral meristems in the vascular **cambium**. This makes stems and roots thicker, and is known as **secondary growth**. Side growth may develop from the main stem as shoots or branches to take advantage of favourable conditions and to avoid competition from other plants.

If plants are damaged at the apical meristems, often the first lateral meristem is induced to grow and take over the role of the apical

Learning objectives

You should understand that:
- Cells in the meristems of plants are undifferentiated and allow indeterminate growth.
- Mitosis and cell division in the shoot provide cells needed for elongation of the stem and the development of leaves.
- Growth in the shoot apex is controlled by plant hormones.
- Plants respond to the environment with growth movements known as tropisms.
- Concentration gradients of auxin in plant tissue are set up by auxin efflux pumps.
- Auxin changes the pattern of gene expression to influence cell growth rates.

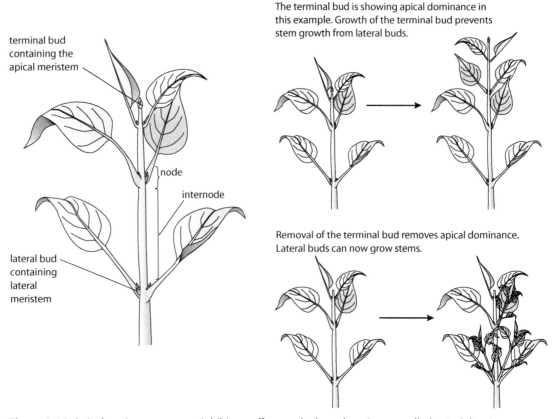

Figure 9.16 Apical meristems exert an inhibitory effect on the lateral meristems, called apical dominance.

meristem. If a flower on the apical meristem is cut off, this will induce the lateral meristems to switch to producing flowers.

Micropropagation

Cells in the meristems of plants are undifferentiated and so small pieces of tissue from meristems can be used to grow whole plants in a technique known as **micropropagation** (Figure 9.17). A few cells or small pieces of tissue are grown in sterile conditions on special agar growth medium, which contains an energy source such as sucrose, and minerals, as well as plant hormones and other substances such as amino acids or vitamins. The tissue will differentiate into a range of cells, depending on the medium used.

Micropropagation and plant tissue culture are important in horticulture and conservation. Many identical (cloned) plants can be grown from a single cell or small sample of tissue. Botanic gardens propagate rare and endangered plants or those that are difficult to grow. In horticulture, large numbers of ready-rooted, disease-resistant plants can be produced in a relatively short period of time, which saves time and money.

Plant hormones and the growth of shoots

Plants produce **growth-regulating substances** that act to control growth and development. These substances are sometimes called plant hormones.

The existence of these growth substances was noted by Charles Darwin in a report of his experiments that he published in 1880. He

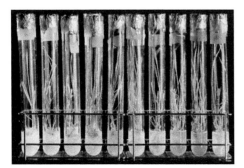

Figure 9.17 Cereal plants being grown from tissue culture.

Plant communications

Plants use chemicals such as auxin to coordinate the activities of their cells and communicate internally. Externally, plants produce substances such as ethylene to control ripening of fruits. Ethylene accelerates the normal process of fruit maturation and senescence (dying or becoming dormant) and causes ripening of bananas, mangoes, apples and pineapples.

Question to consider

- To what extent can plants be said to have 'language'?

observed that oat shoots grew towards light because of some 'influence', which he proposed was transmitted from the shoot tip to the area immediately below. We now know that the substance that causes shoots to bend towards the light is auxin and the response it causes is called phototropism. Auxin is found in the embryos of seeds and in apical meristems, where it controls several growth responses, or tropisms.

The exact mechanism of auxin action is still the subject of much research. It has been suggested that auxin is redistributed to the side of a shoot tip that is away from a light source. The uneven distribution of auxin allows cell elongation on the shaded side of a shoot, which in turn causes bending towards light (Figure 9.18).

Auxins influence and coordinate the development of plants by influencing the expression of genes. Auxin may activate certain genes, leading to a rapid response, or it may inhibit other sets of genes. Auxin affects both cell division and cell enlargement. In a shoot, auxin promotes the elongation of cells at the shoot apex. It seems to act by stimulating factors such as elastins, which loosen the bonds between cellulose fibres in cell walls, thereby making their

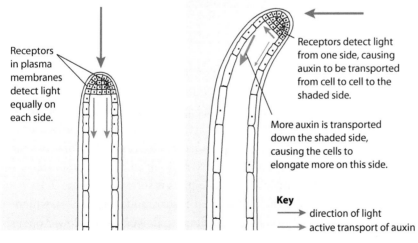

Figure 9.18 Some proteins in the plasma membranes of certain cells in plant shoots are sensitive to light. When light falls on them, they cause auxin to be transported to the shaded side of the shoot, which in turn causes the shoot to bend towards the light.

Tropisms

Tropic responses are growth responses that occur as a result of an external stimulus, and are directional. The direction of the stimulus determines the direction of the response. Plants respond to light (phototropism), gravity (geotropism), chemicals (chemotropism) and in some cases to the presence of water (hydrotropism). The response may either be positive, so that part of the plant grows towards the stimulus, or negative, so that it turns away from it. Plants also respond to stimuli with **nastic responses** but these are non-directional. For example, the opening and closing of flowers in response to light is a photonasty.

structure more flexible. Auxins also interact with other plant hormones such as cytokinins to influence the general pattern of development.

Auxin is not produced in all cells and has to be moved to where it is needed. In general, auxin moves from shoots towards roots but it can move for short distances, such as from one side of a shoot to another, by a method known as polar auxin transport. Movement can only occur in one direction and is controlled by **auxin efflux pumps**, which are special protein channels unevenly distributed in the plasma membranes of the plant's cells. Pin-formed or PIN proteins transport the hormone and control its direction of moment from cell to cell. In certain places, high concentrations of auxin are built up. As well as influencing tropisms, this build-up is important in developing roots and shoots where the hormone organises the development of important structures.

Nature of science

Scientific advance follows technical innovation – identifying plant hormones

Even though the existence of auxins was known to Darwin more than a century ago, it was not until analytical techniques improved that small quantities of plant growth substances could be identified. Advances in these techniques and in our ability to interpret the results of analysis have enabled us to understand more about the effect of plant hormones on gene expression.

? Test yourself

8 Outline the role of auxin in phototropism.
9 State the role of apical meristems in the growth of a plant.
10 Outline the importance of auxin efflux pumps.

9.4 Reproduction in plants

Learning objectives

You should understand that:
- Flowering of a plant involves a change in gene expression in the shoot apex.
- Flowering in many plants is induced in response to the duration of light and dark periods.
- Plant reproductive success depends on pollination, fertilisation and dispersal of seeds.
- Most sexually reproducing plants use mutualistic relationships with pollinators to aid reproduction.

Flowers

There is enormous variety in the shapes and structures of flowers. Some are male, some are female, others have both male and female parts. For this reason, there is a huge variety in the appearance of flowers that we see. Flowers also differ in the way they are pollinated. Animal-pollinated flowers need structures to attract a pollinator, whereas flowers that use the wind to carry their pollen do not. An example of an animal-pollinated flower containing both male and female parts is shown in Figure **9.19**.

When the flower is in the bud stage, it is surrounded by sepals, which fold around the bud to protect the developing flower. As the flower opens, the sepals become insignificant and resemble small dried petals below the flower.

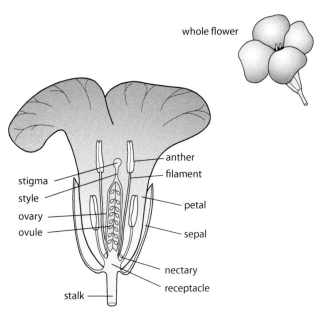

Figure 9.19 Half-flower of wallflower (*Cheiranthus cheiri*). The flower is about 2.5 cm in diameter. It is pollinated by bees and hoverflies. Its petals are usually brightly coloured and fragrant.

Petals of animal-pollinated flowers are often brightly coloured to attract insects or other animals that may visit. Many have nectar guides, which are markings on the petals that tempt pollinators deep into the flower.

Pollen, containing the male gametes, is produced in the anthers, which are held up on long filaments in many flowers, so that as pollinators enter they brush past the anthers and are dusted with pollen.

The female organs are the stigma, style and ovary. The stigma receives pollen grains, which arrive with pollinators as they delve into a flower to obtain nectar. The sticky stigma has sugars present on its surface that cause pollen grains to germinate.

Pollination and fertilisation

Pollination is the transfer of pollen (containing the male gametes) from the anther to the stigma. Pollen may be carried by insects, other animals (such as birds, bats or mice), or by the wind. If pollen travels from the anther of one plant to the stigma of another plant, the process is known as **cross-pollination**. If pollen is deposited on the stigma of the same plant that produced it, **self-pollination** occurs. Self-pollination produces less genetic variation than cross-pollination.

Most flowering plants use a mutualistic relationship with a pollinator to enable them to reproduce. Both the plant and pollinator benefit in this type of relationship. For example, a bee that visits a flower benefits by receiving food in the form of nectar, while the flower benefits as it receives pollen to fertilise its ovules. The pollen is carried on the bee's body from flowers it has already visited.

Fertilisation occurs when male and female gametes fuse to form a zygote. This occurs in the ovule of the flower. When pollen grains from a plant of the right species arrive on the stigma, they germinate and each

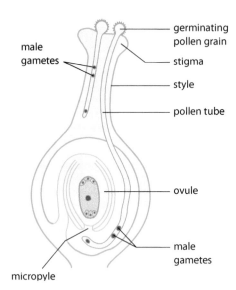

Figure 9.20 Fertilisation of an ovule in the ovary of a plant.

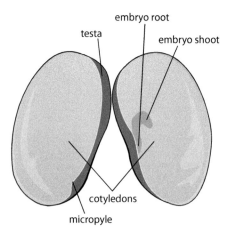

Figure 9.21 Two halves of a broad bean seed showing the main parts of a dicotyledonous plant seed.

produce a **pollen tube**, which grows down the style to the ovary (Figure 9.20). The tube enters the ovary and a pollen nucleus passes down the tube to fuse with and fertilise the nucleus of the female gamete in the ovule.

Seed structure and dispersal

Fertilised ovules develop over time into **seeds**, which protect the developing embryo inside. Seeds are held within a seed pod, fruit or nut, which can be dispersed to new locations so that when they germinate, the new plants that develop do not compete with their parents.

Plants have evolved many ingenious means by which to bring about **seed dispersal**. A few are listed below.

- Some seed pods, such as those in the pea family, mature and dry out, so they eventually snap, causing the seed pod to open quite suddenly, ejecting the seeds some distance from the parent plant.
- Fruits containing seeds are frequently eaten by birds or animals. A tasty fruit tempts an animal to eat and digest it, but the tough seeds inside it pass through the digestive tract and emerge in the feces. These seeds may appear long distances from the parent plant and are contained within a rich fertiliser.
- Nuts are collected by animals like squirrels, which bury them as a reserve of food for the winter. They may bury several groups of nuts and fail to dig them all up during the winter. The nuts remain in the soil ready to germinate when conditions are favourable.

Seeds have all the necessary components to ensure successful germination and the growth of a new plant. Within every seed is an embryo root and shoot, ready to develop when the time is right. Once a seed has been formed in the ovary, it loses water so that it can enter a dormant phase and not develop further until conditions for growth are favourable.

Inside their seeds, dicotyledonous plants have two seed leaves, or **cotyledons**, which store food reserves needed for germination (Figure 9.21). The cotyledons are surrounded by a hard protective seed coat called the **testa**. Many seeds have to endure quite harsh environmental conditions, so the testa protects the delicate tissues inside. In the wall of the testa is a pore called the **micropyle** through which water is absorbed to begin the process of germination.

Germination

Germination is the development of the seed into a new plant (Figure 9.22). A dormant seed needs three vital factors to be in place for germination to occur.

- **Temperature** – A suitable temperature is essential for the enzymes in a seed to become active. They cannot work in cold conditions, and very high temperatures also inhibit their activity. Many seeds remain dormant until the temperature is at a particular level so that they germinate when the seedling will have the best chance of survival.
- **Water** – Most seeds contain only about 10% water, so water must be taken in to start the germination process. Water rehydrates the seed and

the enzymes contained within it. The enzymes break down food stores to provide energy for the emerging root and stem.

- **Oxygen** – This is essential to provide energy for aerobic respiration. Germination begins as water is absorbed by the seed in a process known as **imbibition**. Water enters through the micropyle of the testa.

Water rehydrates stored food reserves in the seed and, in a starchy seed such as a barley grain, it triggers the embryo plant to release a plant growth hormone called **gibberellin** (Figure 9.23). The gibberellin in turn stimulates the synthesis of **amylase** by the cells in the outer **aleurone layer** of the seed. The amylase hydrolyses starch molecules in the **endosperm** (food store), converting them to soluble maltose molecules. These are converted to glucose and are transported to the embryo, providing a source of carbohydrate that can be respired to provide energy as the **radicle** (embryo root) and **plumule** (embryo shoot) begin to grow, or used to produce other materials needed for growth, such as cellulose.

Absorption of water by the seed splits the testa, so that the radicle and plumule can emerge and grow. When the leaves of the seedling have grown above ground, they can begin to photosynthesise and take over from the food store in the seed in supplying the needs of the growing plant.

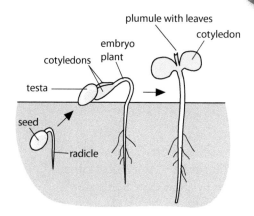

Figure 9.22 Germination and early growth in a dicotyledonous plant.

Control of flowering

At certain times of the year, shoot meristems produce flowers. Flowering involves a change in the expression of genes in the cells of the shoot apex. New proteins are produced and the cells and tissues that make up flowers rather than leaves begin to develop from the meristems. Nearly 100 years ago, plant scientists discovered that light influences the timing of flowering in many plants, although they did not know that the reason for this is a change in gene expression. In the middle of the 20th century, scientists showed that the period of darkness is the critical factor in controlling flowering. **Photoperiodism** is the term given to plants' responses to the relative periods of darkness and light.

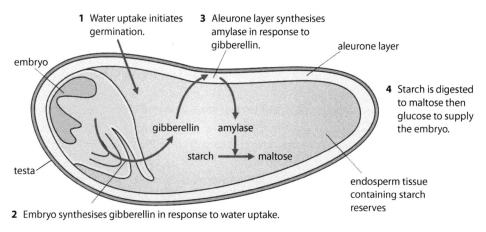

Figure 9.23 Longitudinal section through a barley seed, showing how secretion of gibberellin by the embryo results in the mobilisation of starch reserves during germination.

- **Long-day plants** flower when days are longest and the nights are short. A long-day plant requires less than a certain number of hours of darkness in each 24-hour period to induce flowering. In the northern hemisphere, these plants flower during late spring or early summer (April to July) as days become longer and periods of darkness decrease. This situation is reversed in the southern hemisphere, where long-day plants flower between September and December. Examples of long-day plants are carnations, clover and ryegrass.
- **Short-day plants** flower as nights become longer and days are shorter. In the northern hemisphere this is during the late summer or autumn (fall). They require a continuous period of darkness before flowering and the dark period must be longer than a critical length. The length of the dark period differs between species. Examples of short-day plants are chrysanthemum, poinsettia, coffee and tobacco, which require 10–11 hours of darkness in a 24-hour period before they will flower. Short-day plants cannot flower if the period of darkness is interrupted by a pulse of artificial light shone on them even for just a few minutes.

So-called 'short-day' plants are stimulated by long nights rather than by short days. Some short-day plants, such as poinsettias or chrysanthemums, are commercially produced so that they flower at any time of year. Horticulturalists grow them in shaded glasshouses where an extended period of darkness induces them to produce flowers.

Plants respond to periods of darkness using a leaf pigment called **phytochrome**. This pigment can exist in two inter-convertible forms – inactive P_r and active P_{fr}.

Active P_{fr} is produced from inactive P_r during daylight hours. The conversion occurs rapidly in response to an increase in red light (wavelength 660 nm), which is absorbed by P_r. In darkness, P_{fr} reverts slowly to the more stable P_r.

$$P_r \underset{\text{far-red light (730 nm)}}{\overset{\text{red light (660 nm)}}{\rightleftharpoons}} P_{fr} \rightarrow \text{response}$$

$$\text{darkness}$$

During long days, increased amounts of P_{fr} promote flowering in long-day plants. Flowering in short-day plants is inhibited by P_{fr}, but during long nights, sufficient P_{fr} is removed to allow them to flower.

The exact mechanism of phytochrome action is not fully understood but it is thought that it causes changes in the expression of genes concerned with flowering.

Recent research by German scientists found that 87 of 115 common crop plants worldwide depend on bees or other animal pollinators for their reproduction. Our dependence on such pollinators is an important reason why bee keepers and farmers have been very concerned, in recent years, about the death of bees due to Colony Collapse Disorder. Whole colonies have been wiped out and various causes including the varroa mite, habitat loss, pesticides and poor nutrition have been blamed. In 2013, the European Union passed legislation to restrict the use of certain neonicotinoid pesticides in the hope of restoring bee populations. In the USA, farmers have been renting hives of bees from across the country to pollinate their crops and make up for the shortage of bees.

Nature of science

Paradigm shift – a new approach to conservation

In the past, conservation efforts have often focused on saving one or a few species, such as a rare orchid or bird. Today it is far more common for conservationists to adopt a systems approach which aims to protect entire ecosystems rather than separate species (see Option C). This paradigm shift has occurred because studies have revealed how vital the interdependence of species is. For example, scientists have discovered that the majority of the quarter of a million species of flowering plants on Earth depend on pollinators for their reproduction. These relationships are crucial for plant survival and conserving a single species cannot happen in isolation.

Test yourself

11 Explain why temperature is an important factor in the process of germination.
12 Distinguish between 'pollination' and 'fertilisation'.
13 State the name of the male parts of a flower.
14 Outline why it is important for seeds to be dispersed.

Exam-style questions

1 Which of the following statements concerning transport in plants is correct?

    A A plant loses support if its cells increase their turgor.
    B Fungal hyphae can play an important part in the uptake of mineral ions into the root of a plant.
    C Mineral ions can only move into plant root hairs as a result of active transport.
    D Transpiration is the flow of water through the phloem tissue of a plant. [1]

2 Which of the following statements concerning transport in plants is correct?

    A Cohesion is important in maintaining a column of water in xylem vessels when transpiration causes transpiration pull.
    B Plant hormones cause an increase in transpiration.
    C An increase in humidity will lead to an increase in transpiration.
    D Phloem is involved in the transpiration of sugar. [1]

3 Which of the following statements concerning reproduction in plants is correct?

    A Transfer of pollen from the stigma to the anther results in pollination.
    B In a dicotyledonous seed, the micropyle is part of the embryo root.
    C During germination, production of gibberellin in the cotyledons opens the micropyle to allow water absorption.
    D In a long-day plant, flowering is promoted as a result of an increase in the proportion of P_{fr} compared to P_r. [1]

4   Explain the role of auxin in phototropism. [3]

5   Outline the movement of water in plants from root to leaf, including the effects of environmental abiotic factors on the rate of transport. [6]

6   Explain how manipulating day length is used by commercial flower growers. [6]

7   Describe the metabolic events of germination in a starchy seed. [4]

8   Outline the role of the phloem in the active translocation of food molecules. [6]

9   In 1893, Joseph Boehm (who proposed the cohesion–tension theory) constructed the apparatus shown in the diagram below to model the movement of water through plants.

If air is drawn out of the tube of mercury so that a vacuum is created, mercury inside the tube will rise to 76 cm. In his experiment Bohm forced water into the capillary tube by placing the porous clay vessel in boiling water (diagram **A**). The water formed a continuous stream with the mercury. The beaker of hot water was then removed and the porous clay pot exposed to the air so that water could evaporate from it (diagram **B**).

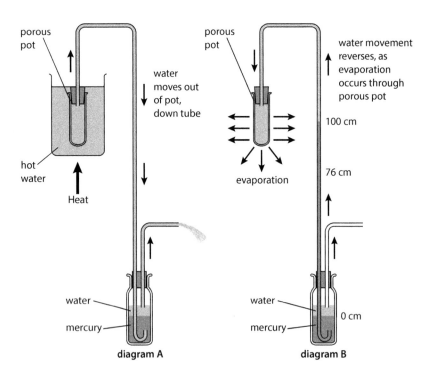

a   Describe the conclusions that could be drawn from this experiment. [2]

Bohm discovered that if an air bubble entered the apparatus anywhere, the height of the mercury fell to around 76 cm.

b   Outline the forces which hold the column of water and mercury together. [2]

c   Suggest why the mercury did not always fall to exactly 76 cm if an air bubble entered the apparatus. [1]

10 The graph shows the result of an experiment in which roots and shoots were exposed to different concentrations of auxin.

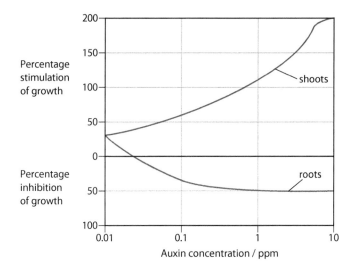

a   Compare the effect of different concentrations of auxin on roots and shoots. [4]

The diagrams show the structure of auxin and a herbicide called 2,4-D. The herbicide has the same effect on cell walls as auxin but it is not broken down by plants.

b   Outline the effect of auxin on plant cell walls and suggest why the herbicide has a similar effect. [2]

When the herbicide is used on plants they grow rapidly but in a distorted form and eventually die.

c   Suggest why the plants grow rapidly but are distorted. [2]

10 Genetics and evolution (HL)

Introduction

Topic **3** dealt with meiosis, how gametes are produced and how the genes they carry affect human characteristics. Single genes and monohybrid genetic crosses produce variation, and multiple alleles, such as those that control blood groups, increase the possible variety still further. Here, we consider dihybrid crosses – the simultaneous inheritance of two pairs of characteristics, which involve more than one gene. We see how the processes of random assortment and crossing over of homologous chromosomes result in enormous variety among individuals. Very few characteristics are controlled by single genes and when a number of genes control a characteristic, the phenotype is determined by their combined effect. These groups of genes are known as polygenes, and help produce the variety that is essential for evolution.

10.1 Meiosis

Functions of meiosis

Meiosis is essential for gamete production in all organisms that reproduce sexually because at fertilisation, when two haploid gametes meet, the chromosome number of the zygote becomes the diploid number for the organism. Remember that chromosomes replicate during interphase before meiosis begins.

Meiosis has two functions:

- **Halving the chromosome number** – Meiosis consists of two nuclear divisions (meiosis I and II, Subtopic **3.3**) but the chromosomes replicate only once, so the four resulting daughter cells each have half the chromosome number of the parent cell.
- **Producing genetic variety** – This occurs in two ways: through crossing over during prophase I, and through random assortment during metaphase I. In addition, random fertilisation also produces variety since any gamete has an equal chance of combining with any other gamete.

Behaviour of chromosomes during meiosis

Crossing over

During prophase I, chromosomes shorten and coil. Homologous pairs of chromosomes come together to form a bivalent so that maternal and paternal chromosomes are next to one another (Figure **10.1**). Homologous chromosomes contain the same genes, but since they came from different parents, they can have different alleles. As they line up together, the non-sister chromatids may touch and break. The two

Learning objectives

You should understand that:
- Chromosomes are replicated in interphase before meiosis.
- Crossing over is the exchange of genetic material between non-sister homologous chromatids.
- Crossing over produces new combinations of alleles in the chromosomes of the haploid cells which are produced as a result of meiosis.
- Formation of chiasmata between non-sister chromatids in a bivalent can result in an exchange of alleles.
- Homologous chromosomes separate in meiosis I.
- Sister chromatids separate in meiosis II.
- Independent assortment of genes is due to the random orientation of homologous pairs of chromosomes during meiosis I.

Look back at Figure **3.7** to remind yourself of the stages of meiosis. Note that homologous chromosomes separate in the first division of meiosis while sister chromatids separate in meiosis II.

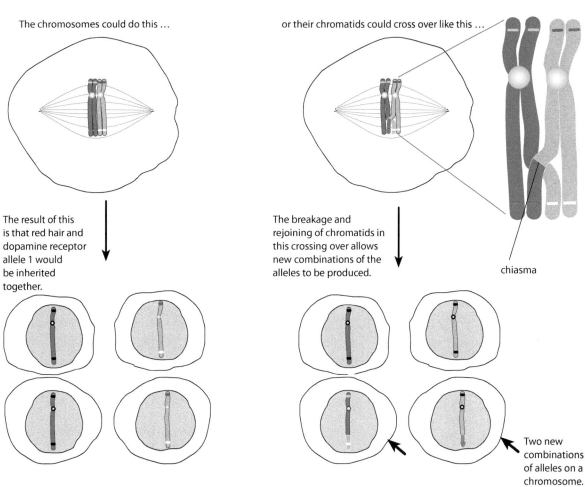

Figure 10.1 How crossing over produces variation. If a single cross-over occurs in a pair of chromosomes, four different daughter chromatids are produced instead of two.

segments may then re-join at the corresponding position on the other chromatid. In this way, chromatids are formed that are a mixture of paternal and maternal alleles. The region where this happens is called a **chiasma** (plural **chiasmata**).

The point at which a chiasma forms is largely random. However, not all chromatids will form chiasmata and they never occur at all on some chromosomes. In this way, some chromatids will retain their full complement of paternal or maternal alleles. This is shown in Figure **10.1** for two pairs of alleles (for a dopamine receptor gene and a hair colour gene) on human chromosome 4.

Crossing over does not occur between the X and Y chromosomes. This is because the two chromosomes are very different sizes and therefore do not sit alongside each other for their full length in the same way as other homologous pairs. The fact that there is no crossing over between the sex chromosomes is advantageous because it means that genes that determine sex remain on the appropriate chromosome.

Random orientation of chromosomes

During metaphase I, the bivalents line up on the equator and spindle microtubules become attached to their centromeres. However, the way in which they line up is random. This is shown in Figure **10.2**, which illustrates the possibilities for just two chromosomes, 4 and 7.

- The paternal chromosomes could both line up together on one side of the equator with the maternal ones on the other side, as shown on the left in Figure **10.2**. Two of the gametes that are produced, after the sister chromatids separate in meiosis II, then contain just paternal chromosomes while the other two contain just maternal chromosomes.
- Another possibility is that the chromosomes line up as shown on the right in Figure **10.2**, with maternal and paternal chromosomes on both sides of the equator. The end result here is that all four gametes contain a mixture of paternal and maternal chromosomes.

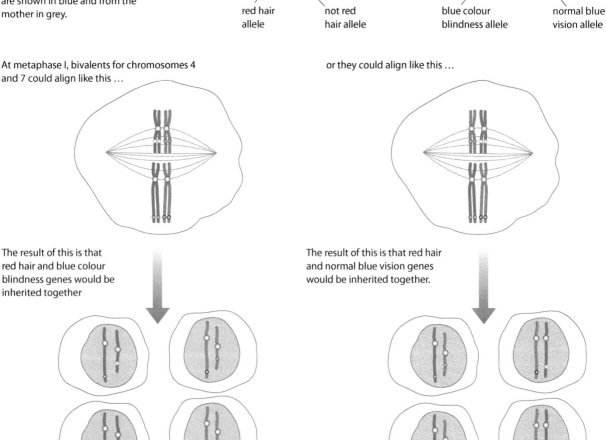

Figure 10.2 How independent assortment produces variation.

Look carefully at the eight gametes in Figure **10.2**. How many different gametes are there? Your answer should be that there are four different gametes. Now consider what happens if a third pair of chromosomes is added. The number of genetically different gametes will increase to eight.

The number of possible genetic combinations that can occur as a result of random orientation of chromosomes is vast. If only chromosome combinations in a haploid cell are considered, the figure is 2^n where n is the haploid number. For humans, $n = 23$ so the possible combinations are over 8 million for just one gamete. The formula can be used to work out the combinations in Figure **10.2** – for two pairs of chromosomes, $n = 2$, so the number of possible combinations $= 2^2 = 4$. With three pairs of chromosomes, the number of possible combinations $= 2^3 = 8$.

Meiosis and variety

The calculation of the number of genetic possibilities in gametes using the formula 2^n can only be an approximation. It does not account for the fact that there are two types of gametes – male and female. Also it does not include the extra variation that arises from crossing over. If a single cross-over occurs then the number of combinations is doubled, as shown in Figure **10.1**. Since several crossovers normally occur during a meiotic division, it is clear that the amount of genetic variety is almost infinite even in organisms that only have a small haploid number.

Meiosis and Mendel's law of independent assortment

Gregor Mendel's 'law of independent assortment' states that:

> When gametes are formed, the separation of one pair of alleles into the new cells is independent of the separation of any other pair of alleles.

Or:

> Either of a pair of alleles is equally likely to be inherited with either of another pair.

Random assortment during metaphase I can produce variety in the gametes as shown in Figure **10.2**. How this relates to Mendel's law of independent assortment for two pairs of chromosomes carrying the alleles **Aa** and **Bb** is shown in Figure **10.3**.

As a result of the random alignment, the **A** allele has an equal chance of being in a gamete with either the **B** allele or the **b** allele, and similarly the **a** allele has an equal chance of being in a gamete with either the **B** allele or the **b** allele. Thus, four types of gamete can be produced carrying the alleles **AB**, **Ab**, **aB** or **ab**. They will occur in the ratio 1 : 1 : 1 : 1.

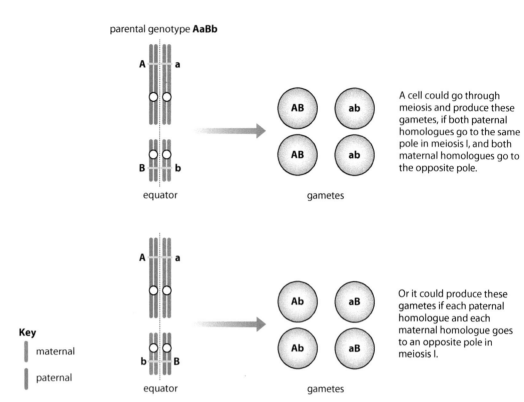

Figure 10.3 Meiosis and Mendel's law of independent assortment.

Nature of science

Careful observation – were Mendel's results 'too good to be true'?

In Mendel's time, statistical analyses were not routinely used to test the validity of scientific results. Analysis of the ratios that Mendel published in 1866 suggests that they may be too close to the expected 3 : 1 ratio. Some have suggested that Mendel selected the data to present or that, having obtained a 3 : 1 ratio in some experiments, he persisted with counting until he achieved his expectations in subsequent experiments.

Mendel considered seven genes in his monohybrid experimental crosses, and also carried out dihybrid crosses involving pairs of these same genes. Crosses involving a pair of genes on different chromosomes produce the ratios Mendel reported but he would have obtained unexpected results had he used linked genes – that is, genes that occur on the same chromosome. It is not likely that the seven genes Mendel investigated would each occur on a different chromosome by chance (the pea plant only has seven pairs of chromosomes) so perhaps some of his crosses did involve linked genes. He would not have been able to explain the results from such crosses since he did not know about chromosomes, so it may be that he only published results from crosses that met his expectations.

Later work by Thomas Hunt Morgan (1866–1945) involving meticulous observation and record keeping revealed anomalous data in dihybrid crosses similar to those which Mendel had performed. As a result, Hunt developed his ideas of gene linkage to explain the results (Subtopic **10.2**).

? Test yourself
1. Define 'crossing over'.
2. State when sister chromatids separate during meiosis.
3. Outline the reasons for independent assortment of genes.

10.2 Inheritance

Genotypes and gametes

Dihybrid crosses involve characteristics that are controlled by two genes. In a dihybrid genotype, there are two pairs of alleles to consider. Figure **10.3** shows what happens during meiosis in a parent with the genotype **AaBb**. Each pair of alleles can be combined in a mixture of dominant and recessive and some examples of genotypes and gametes are given in Table **10.1**.

Genotype	Gametes
AABB	all AB
aaBB	all aB
AaBB	AB and aB in a ratio of 1 : 1
Aabb	Ab and ab in a ratio of 1 : 1

Table 10.1 Some examples of diploid genotypes and the haploid gametes they produce.

There are more possible dihybrid genotypes than those shown in Table **10.1**, but each genotype can only produce either one type of gamete, or two, or four. No genotype produces three types of gamete and only the double heterozygous genotype (as in Figure **10.3**, for example) can produce four types of gamete. It is helpful to keep this in mind when working out the possible offspring of dihybrid crosses.

The dihybrid cross

Although the dihybrid cross involves two pairs of genes instead of just one, the principles of setting out a genetic cross diagram to predict the offspring that will be produced are exactly the same as for the monohybrid crosses you saw in Subtopic **3.4**. The genetic diagrams should include parental phenotypes, parental genotypes, gametes in circles and a Punnett grid for the F_1 or F_2 generation.

> **Parental generation** the original parent individuals in a series of experimental crosses
> **F1 generation** the offspring of the parental generation
> **F2 generation** the offspring of a cross between F1 individuals

Learning objectives

You should understand that:
- Gene loci on the same chromosome are said to be linked.
- Unlinked genes segregate independently as a result of meiosis.
- Variation in a species may be continuous or discrete.
- Phenotypes of polygenic characteristics tend to show continuous variation.
- The chi-squared test is used to determine whether the differences between observed and expected frequency distributions are statically significant.

Exam tip
Remind yourself of the correct way of writing dominant, recessive and co-dominant alleles in genetics problems.

Worked examples

10.1 Fur colour in mice is determined by a single gene. Brown fur is dominant to white. Ear size is also determined by a single gene. Rounded ears are dominant to pointed ears.

A mouse homozygous for brown fur and rounded ears was crossed with a white mouse with pointed ears.

Determine the possible phenotypes and genotypes of the offspring.

Step 1 Choose suitable letters to represent the alleles. Brown fur is dominant, so let **B** = brown fur and **b** = white fur. Rounded ears is dominant, so let **R** = rounded ears and **r** = pointed ears.

Step 2 The brown mouse with rounded ears is homozygous so its genotype must be **BBRR**.
Since white and pointed are recessive, the genotype of the white mouse with pointed ears must be **bbrr**.

Step 3 Set out the genetic diagram as shown.

Step 4 All the F_1 mice have brown fur and rounded ears.

parental phenotypes:	brown fur, rounded ears	white fur, pointed ears
parental genotypes:	**BBRR**	**bbrr**
gametes:	all (BR)	all (br)

Punnett grid for F_1:

	gametes from brown, round-eared parent
gametes from white, pointed-eared parent (br)	(BR) **BbRr** brown, rounded ears

10.2 Mendel carried out genetic studies with the garden pea. Tall plants are dominant to short, and green seed pods are dominant to yellow. A homozygous tall plant with yellow seed pods was crossed with a short plant homozygous for green seed pods. Determine the possible genotypes and phenotypes in the offspring.

Step 1 Tall is dominant to short, so **T** = tall and **t** = short. Green pod is dominant to yellow so **G** = green and **g** = yellow.

Step 2 Each parent has one dominant and one recessive characteristic but we are told the dominant characteristic is homozygous. The tall plant with yellow seed pods therefore has genotype **TTgg**, and the short, green-podded plant is **ttGG**.

Step 3 Set out the genetic diagram as shown.

Step 4 All the offspring are tall plants with green seed pods.

parental phenotypes:	tall, yellow seed pods	short, green seed pods
parental genotypes:	**TTgg**	**ttGG**
gametes:	all (Tg)	all (tG)

Punnett grid for F_1:

	gametes from tall, yellow-seeded parent
gametes from short, green-seeded parent (tG)	(Tg) **TtGg** tall, green seeds

It was fortunate that Mendel chose these two characteristics for his crosses. Seed pod colour and height are unlinked genes on different chromosomes. Had they been on the same chromosome and linked, the results would have been different.

10.3 One of the heterozygous F₁ mice with brown fur and rounded ears from the cross in Worked example **10.1** was crossed with a mouse with white fur and rounded ears. Some of the offspring had pointed ears. Deduce the genotype of the second mouse and state the phenotype ratio of the offspring.

Step 1 Use the same letters as in Worked example **10.1**: **B** = brown fur and **b** = white fur, **R** = rounded ears and **r** = pointed ears.

Step 2 The first mouse has the genotype **BbRr**.

We are told the second mouse is white, so it must have the alleles **rr**. It has rounded ears but we are not told if this is homozygous or heterozygous, so the alleles could be **RR** or **Rr**.

Reading on, we find that there are some offspring with pointed ears, so they must have the genotype **rr**. This means that the unknown parent genotype must have been heterozygous, **Rr**. If the parent was **RR**, no recessive allele would have been present so that a homozygous genotype could not occur in the offspring, and none of them would have had pointed ears.

Step 3 Having written down your reasoning, as above, now set out the usual genetic diagram.

parental phenotypes: brown, rounded ears   white, rounded ears

parental genotypes: **BbRr**   **bbRr**

gametes: (BR) (Br) (bR) (br)   (bR) (br)

Punnett grid for F₁:

gametes from white, rounded-eared parent	gametes from brown, rounded-eared parent			
	(BR)	(Br)	(bR)	(br)
(bR)	**BbRR** brown, rounded ears	**BbRr** brown, rounded ears	**bbRR** white, rounded ears	**bbRr** white, rounded ears
(br)	**BbRr** brown, rounded ears	**Bbrr** brown, pointed ears	**bbRr** white, rounded ears	**bbrr** white, pointed ears

Step 4 The phenotypes produced are:
3 brown fur, rounded ears
3 white fur, rounded ears
1 brown fur, pointed ears
1 white fur, pointed ears.
This produces a ratio of phenotypes of 3 : 3 : 1 : 1, which is an important Mendelian ratio.

Chromosomes and genes

The human genome contains between 25 000 and 30 000 genes but there are only 23 pairs of chromosomes. This means that each chromosome must carry very many genes. Chromosome 1 contains over 3000 genes but the much smaller chromosome 21 contains only around 400 genes. Any two genes on the same chromosome are said to be **linked**. Linked genes are usually passed on together. The genes on any chromosome form a linkage group, so a human has 23 linkage groups. The difference between unlinked and linked genes is shown in Figure **10.4**

Mendelian ratios

In monohybrid crosses, two ratios for the offspring of a genetic cross are possible.

The first is 1:1 if a heterozygous individual (**Aa**) and a homozygous recessive individual (**aa**) are crossed. The second is 3:1 when two heterozygous individuals are crossed (**Aa** × **Aa**). These are called Mendelian ratios.

Mendelian ratios also occur in dihybrid crosses, but with more gametes there are more possibilities. A heterozygous individual (**AaBb**) crossed with a homozygous recessive (**aabb**) produces a 1:1:1:1 ratio and the ratio produced by crossing two double heterozygous (**AaBb**) individuals is 9:3:3:1. The 3:3:1:1 ratio in Worked example 10.3 is another Mendelian ratio. It is helpful to be familiar with these ratios.

Linkage group all the genes that have their loci on a particular chromosome

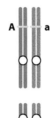

Genes **A** and **B** are on separate chromosomes and so are not linked. They will obey Mendel's law of independent assortment and be inherited independently.

Genes **D** and **G** are on the same chromosomes and so are linked. They will not follow Mendel's law of independent assortment. Genes **D** and **G** form a linkage group.

Figure 10.4 The difference between unlinked and linked genes.

Linkage and genes

If alleles are linked together on a chromosome, then it follows that they will be inherited together because during meiosis they will move together to the same pole as the cell divides. In genetics problems, dihybrid crosses involving linked genes do not produce Mendelian ratios. Linked genes do not follow Mendel's law of independent assortment – they are not inherited independently and can give a variety of different ratios.

Writing a linkage genotype

In the dihybrid crosses considered so far, genotypes have been written in the form **AABB**. With linked genes, a different notation has to be used because, although there are still four alleles to be considered, they are found on only one pair of chromosomes. The genotype is therefore always written as shown in Figure 10.5. The horizontal lines signify that the two genes occur on the same chromosome.

$$\frac{D \quad G}{d \quad g} \qquad \frac{D \quad g}{d \quad G}$$

Here, **D** is linked to **G** and **d** is linked to **g**. Therefore **D** and **G** are inherited together and **d** and **g** are inherited together.

Here, **D** is linked to **g** and **d** is linked to **G**. Therefore **D** and **g** are inherited together and **d** and **G** are inherited together.

Figure 10.5 This shows the two possible linkage patterns for the four alleles. The difference between the two linkage patterns makes a very big difference in the ratios of the phenotypes in the offspring of a cross.

Breaking the law

Mendel's law of independent assortment was found to have exceptions, which geneticist T. H. Morgan explained by using the idea of linked genes, described on page 325 at the end of Topic 10.2.

Questions to consider

- Is it correct to call Mendel's proposals a 'law'?
- What is the difference between a law and a theory in science?

Worked example

10.4 In the fruit fly, *Drosophila*, red eye colour is dominant to purple eyes and long wings is dominant to dumpy wings. These genes are linked on chromosome 2. A fly that was homozygous for red eyes and long wings was crossed with a fly that had purple eyes and dumpy wings. Determine the ratios of genotypes and phenotypes of the F_2 offspring by using a full genetic diagram.

Step 1 Red eye colour is dominant so **R** = red eye and **r** = purple eye. Long wings is dominant so **N** = long wings and **n** = dumpy wings.

Step 2 The fly with the dominant characteristics is homozygous and the other fly shows both recessive characteristics. The parental genotypes are:

$$\frac{R\ N}{R\ N} \text{ and } \frac{r\ n}{r\ n}$$

Step 3 Set out the genetic diagram as below.

	red eyes, long wings	purple eyes, dumpy wings
parental phenotypes:		
parental genotypes:	$\frac{R\ N}{R\ N}$	$\frac{r\ n}{r\ n}$
gametes:	all (RN)	all (rn)

Punnett grid for F_1:

	gametes from red-eyed, long-winged parent
gametes from purple-eyed, dumpy-winged parent (rn)	(RN) → $\frac{R\ N}{r\ n}$ red eyes, long wings

All the F_1 have red eyes and long wings. Now, the F_2 generation is obtained by crossing two offspring from the F_1 generation.

	red eyes, long wings	red eyes, long wings
parental phenotypes:		
parental genotypes:	$\frac{R\ N}{r\ n}$	$\frac{R\ N}{r\ n}$
gametes:	(RN) (rn)	(RN) (rn)

Punnett grid for F_2:

	gametes from red-eyed, long-winged parent	
gametes from red-eyed, long-winged parent	(RN)	(rn)
(RN)	$\frac{R\ N}{R\ N}$ red eyes, long wings	$\frac{R\ N}{r\ n}$ red eyes, long wings
(rn)	$\frac{R\ N}{r\ n}$ red eyes, long wings	$\frac{r\ n}{r\ n}$ purple eyes, dumpy wings

Step 4 In the F_2 generation, the ratio of phenotypes is:
3 red eye, long wing : 1 purple eye, dumpy wing.
Note that this 3 : 1 ratio is what you would expect in a monohybrid cross. The reason for this is that the two genes are linked and so there is only one pair of chromosomes involved, as in monohybrid crosses.

Figure 10.6 A fruit fly, *Drosophila melanogaster*, of 'wild type' with red eyes and long wings.

Drosophila

Drosophila is a genus of fruit fly that is commonly used in genetic experiments (Figure **10.6**). It breeds quickly, producing large numbers of offspring, and its genetic makeup has been well studied. (You can read about T. H. Morgan's work with *Drosophila* at the end of this Subtopic **10.2**.). Various alleles, controlling characteristics such as eye colour, wing shape and body colour, are used in genetic crosses. These features are easy to see and enable geneticists to deduce the genotypes of the flies from their appearance.

Linkage and crossing over

Figure **10.1** showed how crossing over creates genetic variety by exchanging parts of the maternal and paternal chromosomes. Figure **10.7** shows what happens to two closely linked alleles after a cross-over occurs between them.

Look back at the left-hand example in Figure **10.5**. Without crossing over, the parental gametes formed will be **DG** and **dg**. If a cross-over does take place (as shown by the red cross, below) then additional **recombinant** gametes **Dg** and **dG** will be formed.

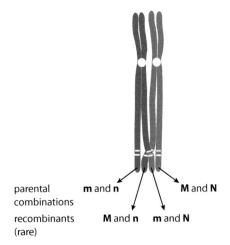

parental combinations   **m and n**   **M and N**
recombinants (rare)   **M and n**   **m and N**

Figure 10.7 A single chiasma has formed between two chromatids so crossing over of the alleles will take place to form the recombinant gametes. No crossing over has taken place with the other chromatids and so these will retain the parental combination of alleles.

Four types of gamete – **DG**, **dg**, **Dg** and **dG** – are possible, but there is a very significant difference in the numbers of each type that are formed. The chance of a chiasma forming between the two loci, which are close together, is very small. So the chance of forming the gametes **Dg** and **dG** is also very small. The majority of gametes therefore carry the alleles **DG** and **dg** and they will form in equal numbers. If a cross-over does takes place, for every **Dg** gamete there will be a **dG** gamete. The numbers of these two gametes will also be equal but very small. The geneticist T. H. Morgan made these observations in his work in the early part of the 20th century – you can read about his discoveries at the end of this subtopic.

Now look at the right-hand example in Figure **10.5**. What will be the allele combinations in gametes where crossing over has not taken place? What will be the allele combinations in gametes where crossing over has taken place, and which combinations will be present in greater numbers?

> **Recombinant** offspring in which the genetic information has been rearranged by crossing over so as to produce phenotypes that are different from those of the parents

Worked example

10.5 Grey body and red eyes are dominant to stripe body and cardinal eye in *Drosophila*. They are autosomal, linked genes on chromosome 3. Homozygous grey flies with red eyes were crossed with stripe flies with cardinal eyes. No crossing over occurred.

Then the F_1 flies were crossed with stripe, cardinal flies.
- If no crossing over occurs between the two loci, what phenotypes would be expected in the offspring of this second cross?
- If crossing over did occur between the loci, what phenotypes would be expected this time?

First cross, with no crossing over:

Step 1 Grey body is dominant, so **G** = grey body and **g** = stripe body. Red eye is dominant so **R** = red eye and **r** = cardinal eye.

Step 2 The fly with the dominant characteristics is homozygous and the other fly shows both recessive characteristics. The parental genotypes are $\frac{G\ R}{G\ R}$ and $\frac{g\ r}{g\ r}$.

Step 3 Set out the diagram as shown.

parental phenotypes:	grey body, red eyes	stripe body, cardinal eyes
parental genotypes:	$\frac{G\ R}{G\ R}$	$\frac{g\ r}{g\ r}$
gametes:	all (GR)	all (gr)

Punnett grid for F_1:

	gametes from grey-bodied, red-eyed parent
gametes from stripe-bodied, cardinal-eyed parent (gr)	$\frac{G\ R}{g\ r}$ grey body, red eyes

Step 4 All the F_1 flies have a grey body and red eyes.

Second cross, with no crossing over:

parental phenotypes:	grey body, red eyes	stripe body, cardinal eyes
parental genotypes:	$\frac{G\ R}{g\ r}$	$\frac{g\ r}{g\ r}$
gametes:	(GR) (gr)	all (gr)

Punnett grid for F_1:

	gametes from grey-bodied, red-eyed parent	
	(GR)	(gr)
gametes from stripe-bodied, cardinal-eyed parent (gr)	$\frac{G\ R}{g\ r}$ grey body, red eyes	$\frac{g\ r}{g\ r}$ stripe body, cardinal eyes

Second cross, with crossing over:

parental phenotypes: grey body, red eyes   stripe body, cardinal eyes

parental genotypes: $\dfrac{G\ R}{g\ r}$ $\dfrac{g\ r}{g\ r}$

gametes: (GR) (gr) (Gr)* (gR)*   all (gr)

Punnett grid for F_1:

	gametes from grey-bodied, red-eyed parent			
	(GR)	(gr)	(Gr)	(gR)
(gr)	$\dfrac{G\ R}{g\ r}$ grey body, red eyes	$\dfrac{g\ r}{g\ r}$ stripe body, cardinal eyes	$\dfrac{G\ r}{g\ r}$ grey body, cardinal eyes	$\dfrac{g\ R}{g\ r}$ stripe body, red eyes

(gametes from stripe-bodied cardinal-eyed parent)

Step 4 The four F_1 phenotypes are:
- grey body and red eyes
- stripe body and cardinal eyes
- grey body and cardinal eyes
- stripe body and red eyes

The 'grey, cardinal' and 'stripe, red' (shown in red type) flies are **recombinants** as they have a phenotype that is different from the parental phenotypes. These recombinant phenotypes will occur in approximately equal numbers. The parental phenotypes (shown in black type) will also be in approximately equal numbers among the offspring, but the recombinant phenotypes will be very few in number compared to the parental phenotypes.

Variation

Variation is defined as the differences that exist between individuals. We can identify two types – discrete variation (also known as discontinuous variation) and continuous variation.

Discrete (or discontinuous) variation is where clearly distinguishable categories can be identified and involves features that cannot be measured. Examples include blood groups and left- or right-handedness. This type of variation is controlled by the alleles of one or a few genes and it is not affected by environmental factors.

Continuous variation is not categorical. It is possible to make a range of measurements of continuous variation from one extreme to another. Human height, hand span and skin colour are all examples of continuous variations. The yield of wheat from crop plants or milk from dairy cattle is also continuously variable. These variations are controlled by genes but are also affected by environmental factors such as soil quality, weather and – in the case of the cattle – diet.

Continuous variation can be shown on a graph, which produces an even distribution of frequencies of the characteristic, called a **normal distribution**. The bell-shaped curves in Figure 10.8 represent continuous variation. Discrete variation, such as that observed in human blood groups, produces a bar graph, with each distinct category represented separately.

Exam tip
Linked alleles will always be represented like this
$\dfrac{T\ B}{t\ b}$
in examination papers

Polygenes

In the genetic examples considered so far, a particular characteristic is controlled by one gene, which can have different alleles at a specific locus on a pair of chromosomes. There is a clear difference between organisms with different alleles. An organism either has the characteristic or it does not – there are no intermediate forms. These phenotypes are examples of discrete (or discontinuous) variation.

Very few characteristics are controlled by single genes. Most are controlled by groups of genes, which together are known as **polygenes**. The genes that form polygenes are often unlinked – that is, they are located on different chromosomes. When two or more genes, each with multiple alleles, are responsible for a characteristic the number of possible phenotypes is greatly increased. Each gene separately may have little impact but their combined effect produces a whole variety of phenotypes. Unlinked polygenes result in a range of degrees of the characteristic from one extreme to another – that is, continuous variation.

Human skin colour

Human skin colour depends on the amount of the pigment melanin that is produced in the skin. Melanin synthesis is controlled by genes. The degree of pigmentation can range from the very dark skin of people originating from regions such as Namibia in southern Africa, through to the very pale skin of native Scandinavian people.

Melanin protects the skin from the harmful UV rays from the Sun. In parts of the world close to the equator, the Sun's rays are particularly intense so people need more protection from sunburn. Dark-skinned people have a high concentration of melanin, which protects them, while fair-skinned people have much less. Although skin colour is genetically determined, environmental factors also influence it. Fair-skinned people who are exposed to sunlight produce extra melanin and develop a protective suntan. Exposure to sunlight also allows vitamin D to be produced in the skin.

Several genes are involved in determining skin colour and they produce the almost continuous variation that can be seen in the global human population. In Figure **10.9** only three genes are shown. Each gene has two alleles. One allele, **M**, contributes to melanin production and the other, **m**, does not. These three genes, each with two alleles, give rise to seven possible skin tones – numbered 0 to 6 below. The number refers to the number of **M** alleles, which determine the level of skin pigmentation. A person with the genotype **MMMMMM** will have very dark skin (6), while a person with **mmmmmm** will have very pale skin (0). The Punnett grid shows the possible combinations of skin colour in children from two parents, both heterozygous for all three genes. The parents' phenotype is light brown skin (3).

Although in this simplified example there are only seven categories of pigmentation, nevertheless you can see that if the frequencies of the skin colour varieties in the Punnett grid are plotted on a histogram, as

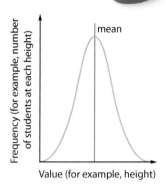

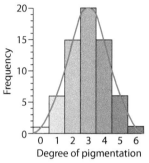

Figure 10.8 Human height demonstrates continuous variation – when frequency is plotted against height, a normal distribution is obtained (top). Frequency of skin variation is shown on bar graph variation.

Polygenic inheritance the inheritance of a characteristic that is controlled by two or more genes; polygenic inheritance accounts for the continuous variation that occurs in characteristics such as human height, body mass and skin colour

in Figure **10.8**, it produces a normal distribution. In the case of human skin colour, it is known that more than three genes are involved and the number of categories exceeds seven. The result is a wider distribution curve and more 'continuous' variation.

Many other examples of polygenic inheritance are known. In wheat plants, three genes interact to control the colour of seeds. Human height and body mass are also examples of polygenic characteristics, but both can be influenced by environmental factors such as nutrients in a person's diet or the quantity of food that they consume.

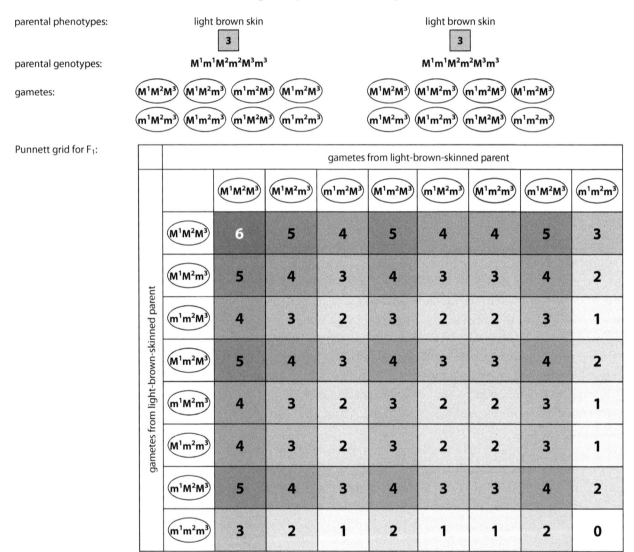

Figure 10.9 Punnett grid show the possible combinations of skin colour in children from two heterozygous parents.

The chi-squared test

The chi-squared (χ^2) test is a statistical test that can be used in cases where variation is discrete. In genetic or ecological investigations, the chi-squared test is useful to compare your observed results with the results that you would expect if your theory about how the system works were correct. The test tells you whether or not any difference between your observed results and your expected results is significant. If the difference *is* significant, that means the results do not fit well with your theory – so you may need to revise your theory. If the difference is *not* significant – that is, it's so small that it could have occurred by chance – then you can say that your results do support your theory.

The chi-squared value is calculated like this:

$$\chi^2 = \Sigma \frac{(O-E)^2}{E}$$

where Σ means 'the sum of', O = observed frequency and E = expected frequency.

The greater the value of chi-squared you calculate, the greater the difference between your observed and expected results. To find out whether the difference is significant or not, you must compare your chi-squared value with a table of 'critical values' like the one in Table **10.2**. The **null hypothesis** states that there is no significant difference between the observed and expected results – that is, that the results fit the expected pattern, and therefore support your theory.

The chi-squared test is used in genetics and in ecology to compare observed numbers of organisms to expected numbers. It can be used to test the outcome of monohybrid or dihybrid crosses – to see if the observed ratios fit the expected pattern. The worked examples that follow use data that were produced as a result of dihybrid crosses made by Mendel.

Degrees of freedom	p									
	0.995	0.99	0.975	0.95	0.90	0.10	0.05	0.025	0.01	0.005
1	---	---	0.001	0.004	0.016	2.706	3.841	5.024	6.635	7.879
2	0.010	0.020	0.051	0.103	0.211	4.605	5.991	7.378	9.210	10.597
3	0.072	0.115	0.216	0.352	0.584	6.251	7.815	9.348	11.345	12.838
4	0.207	0.297	0.484	0.711	1.064	7.779	9.488	11.143	13.277	14.860
5	0.412	0.554	0.831	1.145	1.610	9.236	11.070	12.833	15.086	16.750
6	0.676	0.872	1.237	1.635	2.204	10.645	12.592	14.449	16.812	18.548
7	0.989	1.239	1.690	2.167	2.833	12.017	14.067	16.013	18.475	20.278
8	1.344	1.646	2.180	2.733	3.490	13.362	15.507	17.535	20.090	21.955
9	1.735	2.088	2.700	3.325	4.168	14.684	16.919	19.023	21.666	23.589
10	2.156	2.558	3.247	3.940	4.865	15.987	18.307	20.483	23.209	25.188

Table 10.2 Critical values of the chi-squared distribution, showing how to read across the appropriate 'degrees of freedom' row to find the critical chi-squared value at the 0.05 level (*p*).

Worked examples

10.6 A monohybrid cross

If we consider a cross such as the ones Mendel carried out between pea plants, we might obtain data like this:
- Mendel grew 7324 pea plants, which were the result of a monohybrid cross between plants with green seeds and plants with yellow seeds.
- Among this F_1 generation, he observed 5474 plants with green seeds and 1850 plants with yellow seeds.

Step 1 Set the null hypothesis – this predicts the ratio of offspring of each phenotype according to Mendelian ratios for the cross. For a monohybrid cross, the expected ratio of phenotypes in the F_1 generation is 3 : 1. So here, if 7324 plants were counted, the expected ratio (E) would be 5493 : 1831 (which is 3 : 1).

Phenotype	Observation (O)	Expected (E)
green seeds	5474	$\frac{3}{4} \times 7324 = 5493$
yellow seeds	1850	$\frac{1}{4} \times 7324 = 1831$
Total	7324	7324

Step 2 We must now use the chi-squared equation to test whether the observed numbers differ significantly from our expectations. Calculate the value of chi-squared. This is the sum of the differences between each pair of observed and expected values, squared, and divided by the expected value:

$$\chi^2 = \Sigma \frac{(O - E)^2}{E}$$

$$= \frac{(5474 - 5493)^2}{5493} + \frac{(1850 - 1831)^2}{1831}$$

$$= 0.065 + 0.197$$

$$= 0.262$$

Step 3 Select the appropriate row in a table of critical values of chi-squared, like the one in Table **10.2**. To do this, we must calculate the '**degrees of freedom**', which is the number of categories among your results, minus 1. In this case there are two categories (green-seeded plants and yellow-seeded plants) so the degrees of freedom $2 - 1 = 1$. Look along the row of the table that corresponds to 1 degree of freedom.

Step 4 Find the critical chi-squared value at the 5% (0.05) significance level. In biology, the 5% significance level is used. This level means that the probability for rejecting the null hypothesis is 5% – that is, there is a 5% probability that the difference between the observed results and the expected values occurred purely by chance, and is not significant. If our calculated chi-squared value is less than the critical value at the 5% level, then the probability of obtaining the difference we observed by chance alone is greater than 5%, so we can accept the null hypothesis and have no reason to think that our results differ significantly from our expected values.

In this case the chi-squared (0.262) value is lower than the value in the table (3.841) so we accept the null hypothesis that there is no significant difference between the observed and expected results. In other words, our results support the theory on which the expected ratio of 3 : 1 was based.

10.7 A dihybrid cross using Mendel's data

In this dihybrid cross between heterozygous round and yellow-seeded pea plants (RrYy). A total of 556 seeds are counted, the expected ratios of phenotypes in the F_1 generation are 9:3:3:1.

Phenotype	Observed numbers (O)	Expected numbers (E)
round yellow seed	315	$\frac{9}{16} \times 556 = 321.72$
round green seeds	101	$\frac{3}{16} \times 556 = 104.25$
wrinkled yellow seeds	108	$\frac{3}{16} \times 556 = 104.25$
wrinkled green seeds	32	$\frac{1}{16} \times 556 = 34.75$
Total number of seeds	556	556

R = round
r = wrinkled
Y = yellow
y = green

To calculate the chi-squared value:

$$\chi^2 = \frac{(315 - 312.75)^2}{312.75} + \frac{(101 - 104.25)^2}{104.25} + \frac{(108 - 104.25)^2}{104.25} + \frac{(32 - 34.75)^2}{34.75}$$

$= 0.016 + 0.101 + 0.135 + 0.218$

$= 0.470$

Degrees of freedom: $4 - 1 = 3$

From Table **10.2**, the critical value at the 5% (0.05) level, for 3 degrees of freedom = 7.815
The observed value is less than the critical value so we can accept the null hypothesis that there is no significant difference between our observations and expectations. We can accept Mendel's conclusion that a 9:3:3:1 ratio is the result of this type of cross.

Nature of science

Looking for trends and discrepancies – the work of T. H. Morgan

Thomas Hunt Morgan (1866–1945) was a pioneering geneticist who studied the inheritance of mutations in the fruit fly *Drosophila melanogaster*. After the rediscovery of Mendelian genetics in 1900, Morgan worked to show that genes are carried on chromosomes and provide the basis for inheritance. He induced mutations in his flies using chemicals and radiation and began cross-breeding experiments to find mutations that were inherited. Despite the difficulty of spotting mutations in the tiny flies, he eventually noticed a white-eyed mutant male among the typical 'wild type' red-eyed flies. He bred white-eyed male flies with red-eyed females and all the offspring were red-eyed. The F_2 (second generation) cross produced white-eyed males so Morgan concluded that the white-eye mutation was a sex-linked recessive trait (Figure **10.10**). He also discovered a pink-eyed mutant which was not sex linked.

Morgan reported his discoveries of sex linkage and autosomal inheritance in the journal *Science* in 1911. Morgan's team discovered more mutants amongst thousands of flies they studied and also identified flies with multiple mutations. They studied more complex patterns of inheritance, finding more examples of crosses that did not fit the pattern of simple Mendelian ratios. To explain these discrepancies, Morgan went on to suggest that genes could be linked and inherited together. Morgan proposed the hypothesis of crossing over (Subtopic **3.3**) and the relationship between crossing over and linked genes. He suggested that cross-over frequency gave a measure of the distance separating genes on a chromosome.

His book *The Mechanism of Mendelian Heredity*, published in 1915, was a foundation for modern genetics.

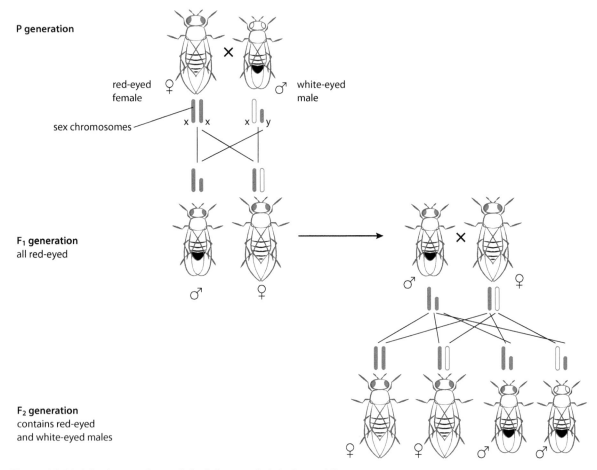

Figure 10.10 Inheritance of a sex-linked characteristic in *Drosophila*.

Test yourself

4 What are the gametes produced by the genotypes **Ttrr** and **HhGg**?
5 What is the name of the grid used to determine offspring genotypes and phenotypes?
6 State the generation from which individuals are crossed in order to produce an F_2 generation.

10.3 Gene pools and speciation

Allele frequency and evolution

Evolution is defined as the cumulative change in the heritable characteristics of a population. The 'heritable characteristics' referred to in this definition are all the alleles in the gene pool of a population. So, if the frequencies of these alleles in the gene pool do not change, then the population is not evolving. Allele frequencies in a population are always fluctuating, in fact, because they depend on the reproductive success of individuals. But if just a single allele shows a change in frequency over a prolonged period of time, then we can say that the population has evolved. Any process that allows the passing on of favourable alleles or prevents the transmission of unfavourable alleles can contribute to evolution.

Natural selection and change

In any population, an individual can have any combination of the alleles present in the gene pool. This gives rise to variation in the population. In most cases, a population will be well adapted to its environment and so the same alleles will be selected, maintaining a stable population. This is known as stabilising selection (Figure 10.11).

If the environment changes the population may also change. Some individuals may have alleles that are more favourable in the new conditions and these alleles will provide an advantage to those individuals, making them more likely to survive and reproduce successfully. This will

Learning objectives

You should understand that:
- A gene pool consists of all the genes and their different alleles that are present in an interbreeding population.
- As species evolve with time, allele frequencies in a population change.
- Reproductive isolation of populations may be temporal, spatial or behavioural.
- Speciation due to the divergence of isolated populations can be gradual.
- Speciation may occur abruptly.

Gene pool all the different genes in an interbreeding population at a given time

Allele frequency the frequency of a particular allele as a proportion of all the alleles of that gene in a population

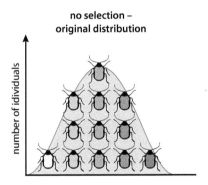

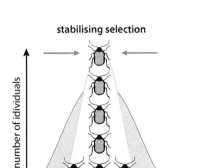
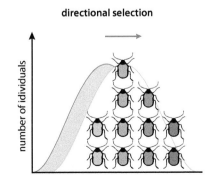
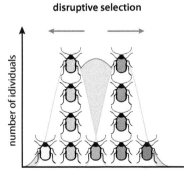

Figure 10.11 In these examples three kinds of selection change the colour of beetles — a feature that is controlled by several genes. Stabilising selection eliminates extreme forms, directional selection favours dark coloured beetles and disruptive selection eliminates the 'average' form and leads to two distinct forms.

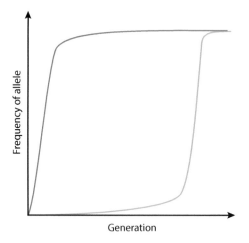

Figure 10.12 Graph to show how allele frequencies change under directional selection that favours a dominant advantageous allele (blue curve) and that favours a recessive advantageous allele (green curve).

result in a change in the population, known as **directional selection** and lead to the prevalence of new forms.

A third possibility is that natural selection results in the formation of two new forms from a single existing population. This is known as **disruptive selection**.

Allele frequencies in populations

Over a period of time, directional selection will lead to an increase the frequency of a favoured allele. For example, if three genotypes – **BB**, **Bb** and **bb** – vary so that **BB** individuals produce more offspring than the other genotypes, the **B** allele would become more common with each generation. The rate at which an advantageous allele approaches the point where it is fixed in the population can be shown in a graph (Figure 10.12). The initial increase in frequency of an advantageous, dominant allele that is rare at first is more rapid than that of a rare, but advantageous, recessive allele. A recessive allele cannot become fixed until it is frequent enough to occur in homozygous organisms but a new dominant allele has an immediate effect on heterozygous individuals.

Another important factor in determining allele frequency is **genetic drift**. Genetic drift is defined as a change in allele frequency in a gene pool due to chance events. It can cause gene pools of two isolated populations to become dissimilar, as some alleles are lost and others become fixed. Genetic drift can occur when a small number of 'founders' (new colonisers) separate from a larger group and establish a new population. Founding individuals carry only a small part of the total diversity of original gene pool, and the alleles that they have is determined by chance alone, which means that rare alleles, or rare combinations of alleles, may occur in higher frequencies in the new isolated population than in the general population. This is called the 'founder effect'.

Many island populations of species (such as Darwin's finches on the Galapagos Islands) display founder effects and may have allele frequencies that are very different from those of the original population. One human example is that of the German immigrants who first arrived in present-day Pennsylvania, USA, from Europe in the 16th century, and formed the Amish community. This group of new colonisers did not have all the genetic variation of the human species, or even of the European population, so the allele frequencies in their gene pool were different from those in the wider human population. For instance, a condition called Ellis-van Creveld syndrome (a form of dwarfism with polydactyly – additional fingers and toes) can be traced back to Samuel King and his wife who arrived in Pennsylvania in 1744. Today, the gene causing the syndrome is many times more common among the Amish people, who tend to marry within their own community, and demonstrates how the genes of the 'founders' of a new community are disproportionately frequent in the population from that point on.

Speciation

Speciation is the formation of new species from an existing population. Members of the same species and, therefore, the same gene pool can fail to reproduce as a result of a barrier that separates them. New species appear as a result of the population of a single species splitting into two or more new ones each with its own gene pool.

Speciation can only occur if there is a barrier dividing the population (Figure **10.13**). The barrier may take different forms, such as geographical separation, or temporal or behavioural differences. When one part of the divided population is isolated from the other, mutation and selection can occur independently in the two populations so that each has the potential to become a new species.

Speciation is said to be **sympatric** or **allopatric**.
- Allopatric speciation occurs in *different* geographical areas.
- Sympatric speciation occurs in the *same* geographical area.

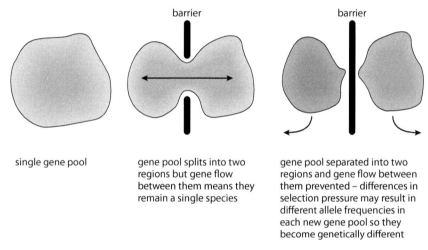

Figure 10.13 Speciation may occur when some kind of barrier divides the gene pool.

Geographic (allopatric) speciation

Allopatric speciation occurs when a physical barrier separates a species into two geographically isolated populations, which then develop independently under the different conditions in the two separated areas, and eventually become unable to interbreed. The barrier might be a natural feature such as a mountain range or a body of water, or it could be a result of human intervention in an environment, such as a major road system or a large conurbation. Many examples of this type of speciation occur on islands such as the Galapagos Islands or the islands of Hawaii.

One example of allopatric speciation resulting from a natural geographical barrier involves salamanders of the genus *Ensatina*, found on the west coast of the USA. Members of this genus all descended from a founder species that spread southward down each side of the San Joaquin Valley (Figure **10.14**). Conditions were slightly different on the east and west sides of the valley and the separated populations adapted to their own

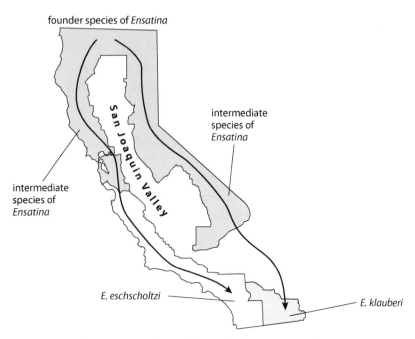

Figure 10.14 Speciation in salamanders, in the San Joaquin Valley, western USA.

Figure 10.15 Goldenrod gall fly and the gall it produces.

particular environments. The valley in between isolated them from one another. Eventually the two populations met at the southern end of the valley in Southern California but they could no longer interbreed. Two new species, *E. klauberi* and *E. eschscholtzi*, had formed and are said to be reproductively isolated from one another.

Sympatric speciation

A gene pool may become divided without the population being geographically split. Other factors may lead to groups within the population becoming reproductively separated within the same physical environment.

Temporal and behavioural isolation

Temporal isolation occurs when the time of reproduction or behaviour of members of one population of a species is incompatible with that of another. An example is the goldenrod gall fly (*Eurosta solidaginis*) found in eastern and midwestern North America. This fly causes the formation of galls on goldenrod plants (*Solidago* sp.) (Figure **10.15**). Over most of its range, the fly lays its eggs in *S. altissima* but in some places it uses *S. gigantea* as its host. Flies associated with *S. gigantea* emerge earlier in the season than flies associated with *S. altissima* so the males and females of the two forms of fly are ready to mate at slightly different times. This suggests the beginning of a temporal barrier to interbreeding between the two populations.

The two emerging groups of flies seem to actively prefer their own species of goldenrod plant and lay their eggs much more frequently in those plants. In addition, females prefer to mate with males of their own group and both of these factors indicate the development of additional behavioural barriers to interbreeding.

At the moment, the two populations can interbreed to produce fertile offspring and so are still a single species, but over time they could become separate species as the temporal and behavioural differences are reinforced and increase their separation.

Polyploidy

One of the most frequent causes of sympatric speciation is polyploidy. Polyploidy can act as a barrier between two gene pools because a polyploid organism has more than two sets of chromosomes. Polyploid organisms may have cells containing three or more sets of chromosomes and are said to be triploid ($3n$), tetraploid ($4n$) and so on.

Polyploidy is widespread in plants but rare in animals and it happens when sets of chromosomes are not completely separated during cell division so that one cell ends up with additional chromosomes. If the mistake occurs during mitosis and the cell fails to divide after telophase then the cell will become a tetraploid. Each chromosome will have a matching pair and will be able to undergo meiosis to form fertile gametes. A tetraploid can cross with another tetraploid to form fertile offspring in just the same way as normal plants. If a tetraploid crosses with a diploid plant they would produce triploid plants that would be sterile and unable to form gametes. In this case, polyploidy acts as a barrier between the diploid and tetraploid species. The two populations may become so different that they develop into new species. In many cases extra sets of chromosomes can produce plants that have improved 'vigour', such as a greater resistance to disease or which produce larger fruits.

Plants in the genus *Tragopogon*, a member of the sunflower family, demonstrate how speciation can occur. Three diploid species, *T. dubius*, *T. porrifolius*, and *T. pratensis*, were accidentally introduced into North America early in the 20th century. In 1950, two new species were discovered, both of which were tetraploid. Chromosome studies showed that *T. miscellus* ($4n$) was a **hybrid** produced by the interbreeding of *T. dubius* ($2n$) and *T. pratensis* ($2n$) whereas *T. mirus* ($4n$) was a hybrid of *T. dubius* ($2n$) and *T. porrifolius* ($2n$) (Figure **10.16**). In both these cases, the interbreeding between the two original species produced an infertile hybrid, in which chromosome doubling then occurred (because of

Polyploidy in alliums

The field garlic (*Allium oleraceum*) is a bulbous plant that occurs throughout Europe (Figure **10.17**). The species is a 'polyploidy complex' which has four different levels of polyploidy with either 3, 4, 5 or 6 sets of chromosomes. Tetraploids ($4n$) and pentaploids ($5n$) are the most common. Although the different plants do overlap in their preferred habitats, different ploidy levels are usually found in different ecological areas. It seems that polyploidy has led to the development of new characteristics, which have allowed the plants to occupy slightly different habitats (niches). *A. oleraceum* is extensively distributed over most parts of Europe, but the species that is thought to be its diploid predecessor has only a narrow range in southern Europe. This suggests that the polyploidy plants are more ecologically tolerant and therefore are able to extend their range and colonise harsher environments than diploid plants.

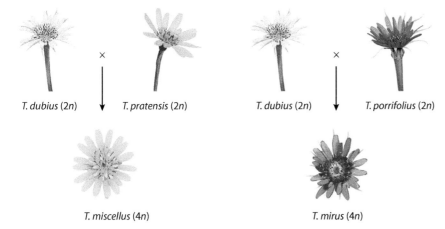

Figure 10.16 Tetraploidy can occur through hybridisation of diploid species.

Figure 10.17 Field garlic, *Allium oleraceum*.

mistakes during mitosis), giving rise to a tetraploid hybrid that could produce viable gametes, and was therefore fertile.

A summary of the two types of speciation is shown in Table **10.3**.

Sympatric speciation	Allopatric speciation
A new species arises from an existing species that is living in the same area.	A new species arises because a geographic barrier separates it from other members of an existing species.
Temporal or behavioural isolation can produce significant changes in the genetic make-up within a species so that a new species is formed.	Geographic barriers may include mountain ranges, valleys or bodies of water, or human-made features such as roads, canals or built-up areas.

Table 10.3 Comparison of sympatric and allopatric speciation.

The pace of evolution: gradualism and punctuated equilibrium

Darwin viewed evolution as a slow, steady process called gradualism, whereby changes slowly accumulate over many generations and lead to speciation. For many species, this seems to be true. A good example of gradualism is the evolution of the horse's limbs, which fossils indicate took around 43 million years to change from the ancestral form to the modern one (Figure **5.3**, Subtopic **5.1**).

In some cases, the fossil record does not contain any intermediate stages between one species and another. A suggested explanation is that fossilisation is such a rare event that fossils of intermediate forms simply did not form, or else have not been discovered. In 1972, Stephen J. Gould (1941–2002) and Niles Eldredge (b. 1943) suggested that such fossils have not been found because these intermediate forms never existed – that evolution had allowed an ancestral form to develop into a quite different descendent without any intermediate. To explain how this might happen, they proposed an additional mechanism for evolution called punctuated equilibrium. The driving force for evolution is selection pressure, so if the selection pressure is very mild or non-existent then species will tend to remain the same – that is, in equilibrium. When there is a sudden, dramatic change in the environment, there will also be new, intense selection pressures and therefore rapid development of new species (Figure **10.18**).

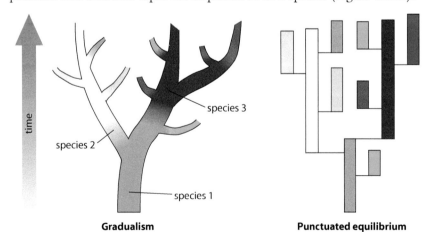

Figure 10.18 A gradualism view and a punctuated equilibrium view of evolution.

A good example of intense selection pressure in modern times is the use of antibiotics, which has resulted in the appearance of resistant species of bacteria in very short periods of time. Methicillin-resistant *Staphylococcus aureus* (MRSA) is an example of a bacterial pathogen resistant to several antibiotics. It has become adapted to new conditions in which there are many antibiotics in use and has developed new strategies to resist them (Subtopic **5.2**).

Paradigm shift

A paradigm shift is a change in the core beliefs or assumptions of an accepted scientific theory. It occurs when scientists are faced with anomalies that cannot be explained by the accepted paradigm. In modern science, a number of paradigm shifts have taken place in recent times. These include the acceptance of plate tectonics to explain large-scale changes in the continents and the replacement of Newtonian mechanics with quantum mechanics. In biology 'pangenesis' (Darwin's provisional theory that a reproductive cell contained 'gemmules' from every part of an organism in order to produce a new individual) was replaced with an acceptance of Mendelian genetics.

Two competing theories – gradualism and punctuated equilibrium – attempt to explain the appearance of new species and the absence of intermediate forms in the fossil record. The Darwinian view assumes that species gradually change over long periods of time while the theory of punctuated equilibrium proposes short periods of rapid evolution interspersed with long periods of equilibrium.

Some scientists reject punctuated equilibrium as being counter-Darwinian, but as it could be explained in terms of natural selection it is possible that both processes may have occurred. On the other hand, new evidence and analysis of gene sequences has lent support to the gradualism viewpoint. It was long thought that it was the mass extinction of the dinosaurs 65 million years ago that led to the rise of the mammals. This view was apparently supported by the fossil evidence. But recent genetic analysis using the *Genbank* database has indicated that early mammals were present at least 100 million years ago, 35 million years before the extinction of the dinosaurs and furthermore that their evolution followed a gradualism path, not a pattern of punctuated equilibrium.

Questions to consider

1 How does a paradigm shift take place in science?
2 What factors are involved in the acceptance of a paradigm shift?

Nature of science

Looking for trends and discrepancies – patterns of chromosome numbers and polyploidy

Doubling of chromosome number happens naturally in plants and has resulted in the formation of several new species. Bread wheat (*Triticum aestivum*), macaroni wheat (*T. durum*) and hempnettle (*Gaelopsis tetrahit*) are all examples of new polyploid species. Swedes are a polyploid species formed by the hybridisation of a type of cabbage and a turnip. Looking for patterns in chromosome numbers has led scientists to conclude that a new species of cord grass, Townsend's cord grass (*Spartina townsendii*), arose in the 1870s when a native British species hybridised with an introduced American species (Figure 10.19). Analysis of the chromosome number in the original species, the sterile hybrid between the two species and the new, fertile species enabled scientists to deduce the pattern of hybridisation and polyploidy.

Polyploidy can be induced in a plant by treating cells with colchicine during mitosis. This chemical prevents a spindle from forming. Chromosomes replicate but do not separate at anaphase. Many of the new cells are polyploid and contain twice the normal number of chromosomes.

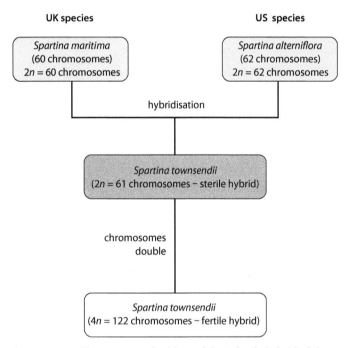

Figure 10.19 Chromosome doubling of the infertile hybrid of the two original species has given rise to a completely new tetraploid species, which is fertile.

Test yourself

7 State **one** difference between sympatric and allopatric speciation.
8 Outline the effect of directional selection on the features of a population.
9 Define 'gene pool'.

Exam-style questions

1. The genotypes **AABb** and **AaBb** were crossed. What would be the ratio of genotypes in the offspring?

   A 1 **AABB** : 1 **AaBB** : 1 **AAbb** : 1 **Aabb** : 2 **AABb** : 2 **AaBb**
   B 1 **AABB** : 1 **AaBb** : 1 **Aabb** : 1 **aaBb**
   C 2 **AABB** : 2 **aaBB** : 2 **AAbb** : 2 **Aabb**
   D 1 **AABb** : 1 **AaBb** : 1 **AABB** : 1 **AaBB** : 2 **AAbb** : 2 **Aabb** [1]

2. A maize plant homozygous for the genes for green leaves and round stem was crossed with a plant with yellow leaves and square stem. All the offspring had green leaves and round stem. The F_1 plants were crossed among themselves and the 480 F_2 generation individuals were of four different phenotypes.

   How many of the F_2 would be expected to have yellow leaves and round stem?

   A 60
   B 30
   C 120
   D 90 [1]

3. A fruit fly of genotype **GgTt** was test crossed. The offspring genotype ratio was 1 : 1. The reason for this was:

   A The genes assorted independently.
   B The genes were linked and crossing over occurred.
   C The genes were linked and no crossing over occurred.
   D Non-disjunction had taken place. [1]

4. A fruit fly of the genotype **RrBb** was crossed with another of the same genotype. The genes are linked. The offspring genotype ratio could be:

   A 1 : 1
   B 9 : 3 : 3 : 1
   C 1 : 1 : 1 : 1
   D 7 : 7 : 1 : 1 [1]

5. Outline how meiotic division results in almost infinite genetic variation in the gametes produced. [2]

6. Using a specific example, explain a cross between two autosomal linked genes, including the way in which recombinants are produced. You are advised to use a test cross. [9]

7 In the fruit fly, *Drosophila melanogaster*, the allele for dark body (**D**) is dominant over the allele for ebony body (**d**). The allele for straight bristles (**T**) is dominant over the allele for dichaete bristles (**t**). Pure-breeding flies with dark body and straight bristles were crossed with pure-breeding flies with ebony body and dichaete bristles.

   a State the genotype and the phenotype of the F_1 individuals produced as a result of this cross. [2]

   b The F_1 flies were crossed with flies that had the genotype **ddtt**. Determine the expected ratio of phenotypes in the F_2 generation, assuming that there is independent assortment. [3]

   The observed percentages of phenotypes in the F_2 generation are shown below.

dark body, straight bristles	37%
ebony body, straight bristles	14%
dark body, dichaete bristles	16%
ebony body, dichaete bristles	33%

   The observed results differ significantly from the results expected on the basis of independent assortment.

   c Explain the reasons for the observed results of the cross differing significantly from the expected results. [2]

8 Pure-breeding pea plants with round, yellow seeds were crossed with pure-breeding pea plants with wrinkled, green seeds.

   All the offspring had round, yellow seeds.

   These seeds were grown and their offspring were allowed to self-pollinate.

   The resulting offspring had the following characteristics.

Seed type	Number
Round yellow seeds	629
Round green seeds	202
Wrinkled yellow seeds	216
Wrinkled green seeds	63
Total number	1109

   a The expected ratio for this cross was 9 : 3 : 3 : 1

   The result of a chi-squared analysis produced a value for χ^2 of 0.47.

   Use Table **10.2** in Subtopic **10.2** to explain how the value calculated in the chi-squared test supports the hypothesis that these are two separate pairs of alleles. [2]

   b Explain, giving your reasons, how these two characteristics are inherited. [7]

Animal physiology (HL) 11

Introduction

Physiology is the study of the functioning of a living system. It examines how organ systems, organs, cells and molecules are integrated to carry out all the physical and biochemical functions in a living body. Much of our knowledge of human physiology has been derived from studies of animals. Whereas anatomy is the study of the structure of an organ, in this topic we will examine the function of both organs and the organ systems involved in immunity, movement, osmoregulation and reproduction.

11.1 Antibody production and vaccination

Immunity is based on the fact that an animal's body can recognise cells and proteins that are of its 'self' and distinguish those that are foreign or 'non-self'. Foreign material can be identified and destroyed.

Immunity: challenge and response

Resistance to an infection is known as immunity. Immunity is acquired from infancy onwards as the body is exposed to, and learns to recognise, many different types of pathogen that have the potential to cause disease. We become able to distinguish between cells that are our own 'self' and those that are 'non-self' and are therefore likely to be pathogens or cause harm. Cells are recognised by the proteins on their plasma membranes (Subtopic 1.3).

Certain leucocytes (a type of white blood cell) are able to recognise 'non-self' proteins, or antigens. Antigens may be on the surface of a pathogen, or may have been secreted by a pathogen in a toxin. Antigens are also likely to be present on the cell surfaces of transplanted tissues or organs.

If a pathogen enters the body, the immune system is stimulated to respond. As it is 'challenged' by the pathogen, it 'responds' by setting in motion processes that will destroy it. The first line of defence is phagocytic leucocytes. These are non-specific and will consume bacteria, viruses and other pathogens, as well as dead cells and cell fragments that might accumulate, for example, in a wound. The second line of defence is a specific response to antigens, involving antibodies. These proteins are the key to the body's immune response, and producing them effectively requires interaction between three types of cell: macrophages, B-lymphocytes and helper T-lymphocytes (also called B-cells and T-cells).

The immune response takes several days to become fully active and in the meantime we may become ill. Sometimes symptoms are mild, such as with the common cold, but sometimes they are severe, leading to permanent disability or even death.

Learning objectives

You should understand that:
- Every organism has unique molecules on its cell surfaces.
- Some pathogens are specific to one species while others can cross species barriers.
- In mammals, B lymphocytes are activated by T lymphocytes.
- When B lymphocytes are activated, they multiply to form a clone of plasma cells and memory cells.
- Antibodies are secreted by plasma cells.
- Antibodies help to destroy pathogens.
- White blood cells secrete histamine in response to allergens.
- Histamines cause allergic reactions.
- Immunity is dependent upon the persistence of memory cells in the body.
- Vaccines contain antigens that can trigger immunity but do not cause disease.
- Hybridoma cells are created by the fusion of a tumour cell with an antibody-producing plasma cell.
- Hybridoma cells are used to produce monoclonal antibodies.

Pathogens and species specificity

Pathogens that cause disease in one species do not always affect other species. For example, the pathogens responsible for syphilis, gonorrhoea, measles and polio infect humans, whereas canine distemper virus does not. *Shigella*, a bacterium that causes dysentery in humans and baboons, does not affect chimpanzees. The exact reasons for these differences are not fully understood, but it may be that cells in non-susceptible species do not have suitable receptors on their plasma membranes for the pathogens to bind to them. The temperature of the host organism may also be important: birds cannot be infected with mammalian tuberculosis because the bacteria that cause the disease cannot survive at the higher core temperatures of birds' bodies. Similarly, frogs are unaffected by anthrax-causing bacteria because their body temperatures are too low.

Occasionally, however, a disease does cross the species barrier. Most newly emerging diseases crossing the barrier from animal to human are caused by viruses. Crossing to a new species is a rare occurrence, but viruses that do so can cause severe outbreaks of disease, especially if they develop the ability to pass from human to human, rather than just from animal to human. The spread of the virus through human populations can then be rapid, exacerbated in modern times by increased international travel and trade.

For a virus to infect a new species, genetic adaptations must occur within the virus. Avian flu, commonly called 'bird flu', arose in this way. The expansion of both human and farm populations has made close interaction between birds and humans more common and enabled the virus to transfer from infected birds to humans via the birds' saliva, feces or nasal secretions. Many strains of bird flu have emerged but one of the most widely publicised is the H5N1 virus, a highly pathogenic Asian strain that caused a pandemic in birds in 2003. This virus can cause severe illness in humans who are infected by direct contact with infected birds. It does not appear that H5N1 can be spread by human-to-human contact at present, but because viruses can adapt and change quickly, it may evolve this ability at some point and would then have the potential to cause a human pandemic. Health agencies have instigated the preparation of vaccines in case this should happen.

Clonal selection

B-cells are antibody-producing lymphocytes, but each B-cell can only produce one particular type of antibody. Since the antibody–antigen response is highly specific, there must be a great many types of B-cell in order to be able to respond to all the possible types of antigen. At any time, there can only be a very few of each type of B-cell in the bloodstream because most of the blood volume is taken up with red cells. Look back at the micrograph of a stained blood smear in Figure **6.13** (Subtopic **6.2**) – this shows clearly how red cells heavily outnumber white cells in the blood.

In simple terms, when a pathogen enters the bloodstream, its surface antigen molecules are exposed to the antibodies attached to different B-cells in the blood. If there is a match between an antigen and an antibody, the B-cell with the matching antibody becomes 'selected' while all the other B-cells are rejected. The selected B-cell is stimulated to divide and produces a clone of antibody-secreting cells, in a process known as clonal selection.

It is likely that any pathogen will have many different antigenic molecules on its surface so several different types of B-cell will probably be selected. Each of these will result in clone of antibody-secreting B-cells. This is therefore called a polyclonal response and it will result in a more efficient destruction of the pathogen.

Antibody production

In reality, the response to pathogens is more complex than simple clonal selection and it involves two types of lymphocyte: B-cells and T-cells.

1. When a pathogen enters the bloodstream it is consumed by a macrophage, partly digested and antigen proteins from it are placed on the outer surface of the macrophage. This is called antigen presentation because the proteins are being 'presented' to other cells.
2. Helper T-cells with receptors matching the presented antigens bind to the macrophages and are activated.
3. Activated helper T-cells then start dividing into two clones of cells. One clone is of active helper T-cells, which are required for the next step in the process, and the other clone is of memory cells, which will be used if the same pathogen ever invades the body again.
4. B-cells with the matching antibody also take in and process antigen proteins from the pathogen and place them on their outer surface.
5. Active helper T-cells bind to these B-cells and, in turn, activate them.
6. Just like the T-cells, the B-cells now divide into two clones of cells. One is made up of active B-cells, or plasma cells, which secrete huge quantities of antibodies into the bloodstream. The second clone is made up of memory cells.
7. Antibodies in the bloodstream destroy pathogens and also help the macrophages to detect and consume more pathogens.
8. Memory cells remain and allow the body to make a large and rapid response should the same pathogen invade again. It is the persistence of memory cells that gives the organism immunity to that pathogen in the future.

Figure 11.1 summarises the process of antibody production.

Active and passive immunity

As we have seen, immunity develops as a result of exposure to a pathogen. This in turn causes the production of antibodies, but developing immunity to a disease can occur either actively or passively.

- Active immunity develops when an individual is exposed to an antigen and produces antibodies after their immune system has been

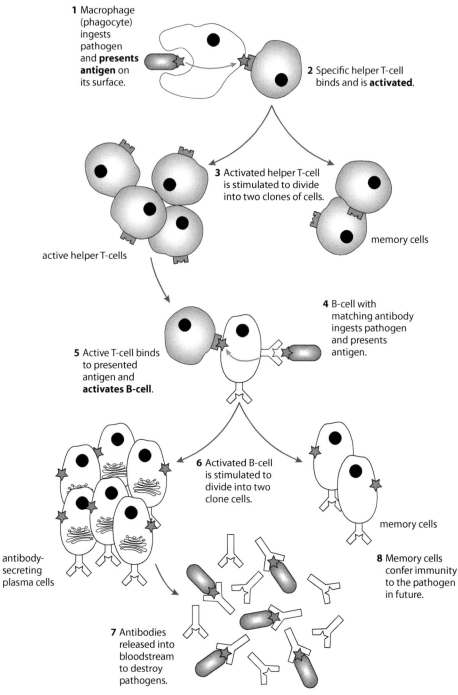

Figure 11.1 The production of antibodies, involving macrophages, helper T-cells and B-cells. The plasma cells contain large amounts of rER to synthesise the protein antibodies.

stimulated. The antigens may be present in the body as a result of infection or be intentionally introduced during vaccinations. In both cases, the body produces antibodies and specialised lymphocytes.
- **Passive immunity** is acquired when antibodies are transferred from one person (or other organism) to another. The antibodies will have been produced as a result of active immunity. Antibodies pass from mother to baby across the placenta and are also transferred in colostrum

(a special form of milk) in the first few days of suckling after birth. Pre-formed antibodies can be injected into a person in the form of a serum – for example, to treat a snake bite. Passive immunity is relatively short-lived. It lasts as long as the antibodies are present in the blood and the recipient does not produce any antibodies of their own.

Active immunity resistance to the onset of a disease due to the production of antibodies in the body after stimulation of the immune system by disease antigens; active immunity may occur as a result either of exposure to a disease or vaccination

Passive immunity resistance to the onset of a disease due to antibodies acquired from another organism in which active immunity has been stimulated

Allergies

Sometimes when microorganisms enter the body, or if the skin is injured, the immune system may trigger an inflammatory response. Inflammation is either a general response to an injury or a reaction that occurs in an area where phagocytes are destroying pathogens. The inflammatory response is brought about by two types of cell – these are **basophils**, which are a type of white blood cell, and **mast cells**, found beneath the skin and around blood vessels. Both types of cell can be stimulated to release a substance known as histamine into the affected area. Histamine relaxes the muscle in the walls of arterioles so that blood flow to the affected area is increased and it also loosens the cells in the capillary walls so that they become 'leaky'. Plasma can then escape from capillaries into the surrounding tissue causing swelling (known as oedema) as well as a slight increase in temperature. Histamine also stimulates sensory neurons leading to pain or itching, Figure **11.2**.

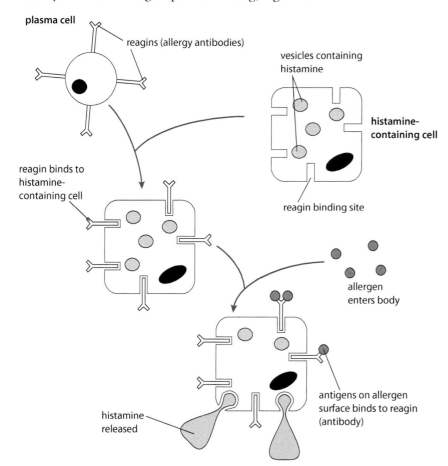

Figure 11.2 An allergic reaction occurs when plasma cells release antibodies against an allergen, which leads to the release of histamine.

In some cases, histamine release can lead to an excessive immune response known as an **allergy**. An allergy is an immune response to an antigen (known as an allergen) to which most people show no reaction. Asthma, eczema and hayfever are common allergic disorders. Allergens include substances such as pollen grains, animal fur, house dust and certain foods. These substances have proteins on their surfaces which act as antigens and stimulate plasma cells to produce antibodies called reagins. Unlike normal antibodies, reagins circulate in the blood and bind to cells that contain histamine, especially the mast cells in the skin and mucus membranes in the respiratory system. (These tissues are said to be hypersensitive.) Reagins cause the mast cells to release histamine, which binds to receptors on cells nearby and leads to inflammation and other symptoms of an allergy (Figure **11.2**). In the bronchi, inflammation can lead to constriction and breathing difficulties as well as the secretion of excess mucus.

Monoclonal antibodies

The body normally produces a polyclonal response to an invasion by a pathogen. A **monoclonal antibody** is artificially produced to target one specific antigen. This type of antibody is used both in commercial applications and for laboratory research. Figure **11.3** shows how monoclonal antibodies are obtained.

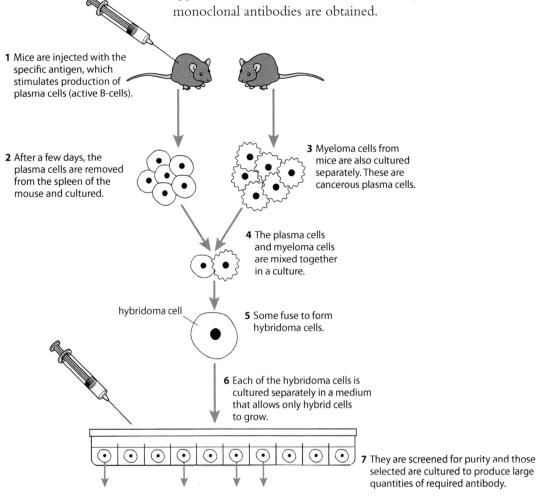

Figure 11.3 Formation of monoclonal antibodies.

B-cells are short lived and therefore of little value for commercial antibody production. **Hybridoma** cells, however, have characteristics of both B-cells and cancer cells: they produce antibodies *and* they are almost immortal if they are kept in culture medium. Monoclonal antibodies are produced by clones so they are all identical, they are highly specific and they can be produced in very large quantities.

Monoclonal antibodies are used in diagnosis to detect the presence of pathogens such as streptococcal bacteria and herpes virus. They are also used in pregnancy testing. Monoclonal antibodies can recognise HCG (human chorionic gonadotrophin), which is secreted by the early embryo to maintain the corpus luteum and progesterone production, and is therefore present in the bloodstream and the urine of a pregnant woman. In the pregnancy test kit, the glycoprotein HCG is detected using a simple 'dipstick' detector (Figure **11.4**). HCG in the urine first binds to mobile antibodies attached to coloured latex beads near the base of the stick. The coloured antibodies then travel up the stick, carried by the capillary action of the liquid, where they encounter a band of immobilised anti-HCG antibodies. If HCG is bound to the coloured antibodies, they combine with the immobilised anti-HCG antibodies here and a coloured (often blue) band shows in the test window to indicate a positive test. If the coloured antibodies do not have HCG bound to them (that is, there is no HCG in the urine) they are not immobilised at the band of anti-HCG antibodies, on a control band appears in the window and the test is negative.

A hybridoma cell is created from the fusion of a cancer cell with an antibody-producing plasma cell. These cells grow well in tissue culture and produce antibodies.

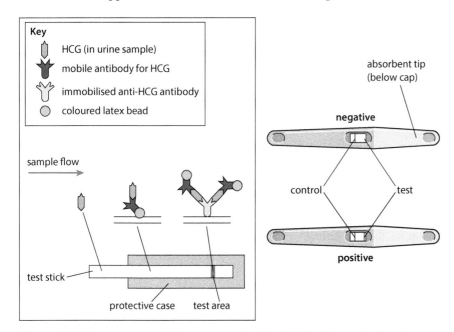

Figure 11.4 Pregnancy test sticks use monoclonal antibodies. The control band show that the test is working correctly. A second blue band in the window indicates a positive result.

Monoclonal antibodies may also prove to be invaluable in the treatment of cancer. Cancer cells carry specific antigens on their cell surfaces and if monoclonal antibodies can recognise these, they can be used to target these cancer cells and carry cytotoxic drugs to them. It is hoped that

Matching blood groups

When blood transfusions are given, blood types must be carefully matched, because antigens on the surface of red blood cells stimulate antibody production in a person with a different blood group. Blood cells from people with blood group A or B have antigen A and antigen B, respectively, and people with group AB have both antigens. People with blood group O have no antigens on their red cells – these people are said to be universal donors because their blood can be given to anyone. The distribution of the four blood groups O, A, B and AB varies throughout the world. It is due to the presence of three alleles (Subtopic **3.4**). Type O is the most common and AB is the rarest.

these treatments, which have been called 'magic bullets', could reduce the amounts of drugs that need to be taken during chemotherapy treatment. One disease, called mantle cell lymphoma, has already been treated in this way. Mantle cell lymphoma is a cancer of the B-cells (B-lymphocytes) and accounts for about 1 in 20 of all cases of the group of cancers known as non-Hodgkin lymphomas. This disease has been treated with manufactured monoclonal antibodies, with the generic name of rituximab. The antibodies are used in conjunction with chemotherapy and stick to particular surface proteins on the cancer cells, which they stimulate the body's immune system to destroy.

Vaccination

Immunity develops when a person has been exposed to a pathogen. For most mild illnesses, such as the common cold or tonsillitis, this happens naturally as a person comes into contact with the viruses or bacteria that cause them. But some pathogens cause diseases that have dangerous or life-threatening symptoms. For these diseases, which include tetanus, tuberculosis, cholera, poliomyelitis and measles, **vaccines** have been developed to provide a safe first exposure so that a vaccinated person will develop immunity but not the disease.

Vaccines are modified forms of the disease-causing pathogens. A vaccine may contain either weakened (attenuated) or dead pathogens, or their toxins. Vaccines are often produced by treating pathogens with heat or chemicals.

Most vaccines are injected into a person's body, although some, such as polio vaccine, can be taken orally. Antigens in the vaccine stimulate the immune response and the formation of sufficient memory cells to produce antibodies very quickly if the person is infected with the real pathogen later on.

A first **vaccination** produces a primary response but many vaccinations are followed up with another some time later. The second or 'booster' dose of vaccine causes a greater and faster production of antibodies and memory cells, known as a **secondary response** (Figure **11.5**), and provides long-term protection. The time that antibodies and memory cells persist depends on the disease. Rubella vaccination can provide protection for up to 20 years, while vaccinations for tetanus should be repeated every 10 years. Vaccines do not prevent infection by pathogens but they do enable the body to respond quickly to them and prevent serious illness.

Nature of science

Assessing ethics in science – the case of smallpox

Smallpox was a serious disease caused by the *Variola* virus, which killed thousands of people every year. It was transmitted by droplet infection. Edward Jenner (1749–1823) was a British scientist who developed a vaccine to protect people against the disease. He knew that dairymaids who had suffered from a similar but mild disease, called cowpox, were

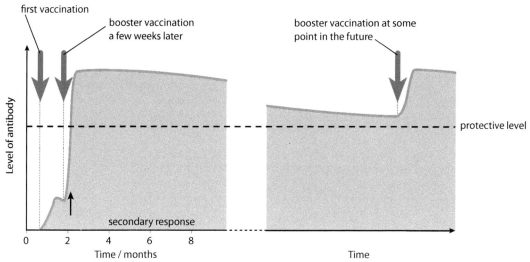

Figure 11.5 Antibody levels after vaccination. The persistence of antibodies varies and depends on the vaccine used.

protected against smallpox. In 1796, Jenner isolated pus from the lesions of a dairymaid with cowpox, and applied it to a cut in the skin of an 8-year-old boy, James Phipps. The boy caught cowpox and recovered. Next, Jenner inoculated the boy with smallpox viruses and discovered that James had developed immunity to the disease after his exposure to cowpox. Jenner also tested his ideas on himself and his family (Figure **11.6**). He named the procedure 'vaccination' from the Latin word *vacca* (cow).

Jenner was criticised and ridiculed for his ideas but Pasteur (1822–1895) later supported him and went on to investigate the use of vaccines for other diseases. Even today, many human vaccines are produced using the immune responses and antibodies produced by other animals. For example, polio, Japanese encephalitis and rotavirus vaccines have all been developed using cell lines from African green monkeys.

In 1967, the World Health Organization began a programme to eradicate smallpox from the world through systematic vaccination. By 1977 they had succeeded. The last recorded case of naturally occurring smallpox was in was detected in 1975 in Bangladesh. Smallpox is the first infectious disease of humans to have been eradicated by a campaign of vaccination.

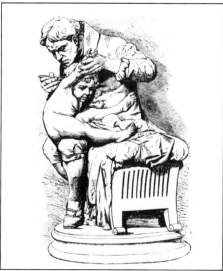

Figure 11.6 Jenner was so confident in his ideas about vaccination that he experimented on his own son.

Questions to consider

- Jenner carried out his research on himself, his family and a young boy, James Phipps. If he were carrying out such experiments today, how might the scientific community react to Jenner's work?
- Jenner's discovery has prevented suffering and saved the lives of millions of people since the 18th century. Were the risks he took with the lives of his subjects worth it?

Test yourself

1. Suggest **two** reasons why pathogens do not often cross from one species to another.
2. Outline the difference between an antibody and an antigen.
3. Outline the properties of a hybridoma cell.
4. Explain briefly how a vaccination can provide immunity to disease.

Learning objectives

You should understand that:
- Bones and exoskeletons provide anchorage for muscles and act as levers.
- Synovial joints allow a certain range of movements.
- Muscles work in antagonistic pairs to move the body.
- Skeletal muscle fibres contain specialised endoplasmic reticulum and are multinucleate.
- Muscles fibres contain many myofibrils.
- Each myofibril is composed of contractile sarcomeres.
- Skeletal muscle contracts as actin and myosin filaments slide past one another.
- ATP hydrolysis and the formation of cross-bridges are needed for the filaments to slide.
- Calcium ions and two proteins, tropomyosin and troponin, control muscle contraction.

11.2 Movement

Joints

A **joint** is a place where two or more bones meet. Joints between bones in the human body, together with the muscles that are attached to them, enable us to move and also support the body. Most joints involve bones, muscles, cartilage, tendons, ligaments and nerves.

- **Bones** provide a framework that supports the body. They protect vital organs such as the brain and the lungs. Blood cells are formed within bones, which contain bone marrow. Bones also act as a site for the storage of calcium and phosphate. At moveable joints such as the knee and elbow, bones act as levers so that forces generated in the muscles are able to cause movement.
- **Ligaments** attach bones to one another at a joint. Some strap joints together while others form a protective capsule around a joint. They are tough and fibrous and provide strength and support so that joints are not dislocated.
- **Tendons** attach muscles to bones. They are formed of tough bands of connective tissue made of collagen fibres and are capable of withstanding tension as muscles contract.
- **Muscles** provide the force needed for movement. They are able to contract in length and as they do so they move the joint into new positions. Muscles only cause movement by contraction, so they occur in antagonistic pairs – one muscle of the pair causes a movement in one direction while the other returns it to its original position.
- **Motor neurons** stimulate muscle contraction. Sensory neurons transmit information from proprioceptors (position sensors) in the muscles so that movements can be coordinated and monitored.

The elbow joint

The elbow is a hinge joint, so-called because it moves in a manner resembling the opening and closing of a door hinge (Figure 11.7). It is an example of a synovial joint. The capsule that seals the joint is lined by a membrane that secretes lubricating synovial fluid so that the bones move smoothly against one another and friction is reduced. Smooth cartilage covers the ends of the bones at the joint and also helps to reduce friction and absorb pressure as the joint moves.

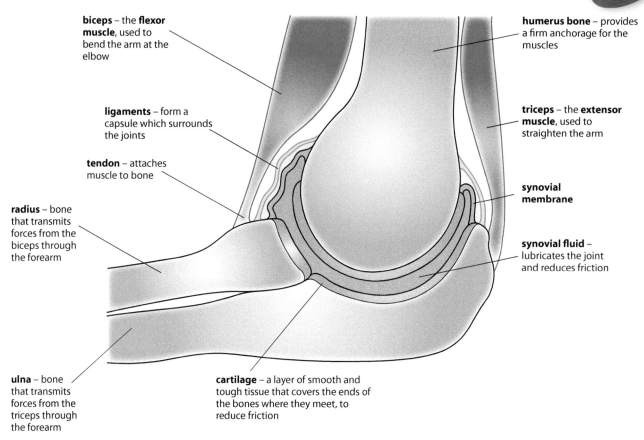

Figure 11.7 The hinge joint of the elbow.

The elbow joint is formed of three bones – the radius and ulna in the lower arm and the humerus in the upper arm. Tendons attach the biceps and triceps muscles to these bones. The biceps is attached to the radius and the shoulder blade. When it contracts the arm bends. The triceps is attached to the ulna, humerus and shoulder blade and it contracts to straighten the arm. The biceps and triceps are an example of an antagonistic pair of muscles.

The knee and hip joints

The knee joint is another example of a hinge joint and it moves in a similar way to the elbow, allowing movement in only one direction. The hip joint is a **ball-and-socket joint** with the ball-shaped head of the thigh bone (the femur) fitting into a socket in the hip. Ball-and-socket joints allow movement in more than one direction and also permit rotational movements (Figure **11.8**).

Exoskeletons and muscles

Like vertebrates, insects and other arthropods also have a complex system of muscles and joints that allow them to flex their bodies and move. However, their muscles are not attached to bones but instead to the inside of the hard exoskeleton that encloses the whole body. Arthropod muscles work in particular segments of their bodies. The muscles of an insect's jointed legs are arranged in antagonistic pairs, just as in vertebrate bodies (Figure **11.9**).

The human body contains three types of muscle: **skeletal** or **striated muscle** shown in Figure 11.10 is attached to the skeleton and enables us to move, **cardiac muscle** found only in the heart, and **smooth** or **non-striated muscle**, makes up the walls of hollow organs such as the intestines and bladder. Striated muscle can be controlled voluntarily and enables us to move.

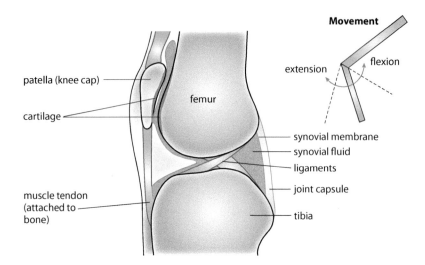

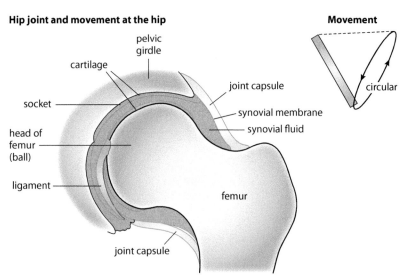

Figure 11.8 Longitudinal sections of the knee and hip joints, and the degree of movement they allow.

> **Exam tip**
> Make sure you can label all the parts of a muscle fibre during the various stages of contraction.

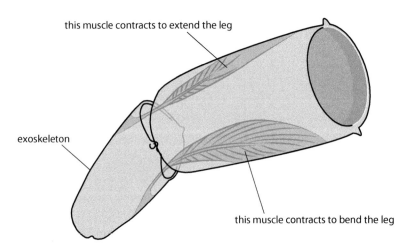

Figure 11.9 Insect muscles are attached to the exoskeleton, rather than to bones, but work in antagonistic pairs just as in vertebrates.

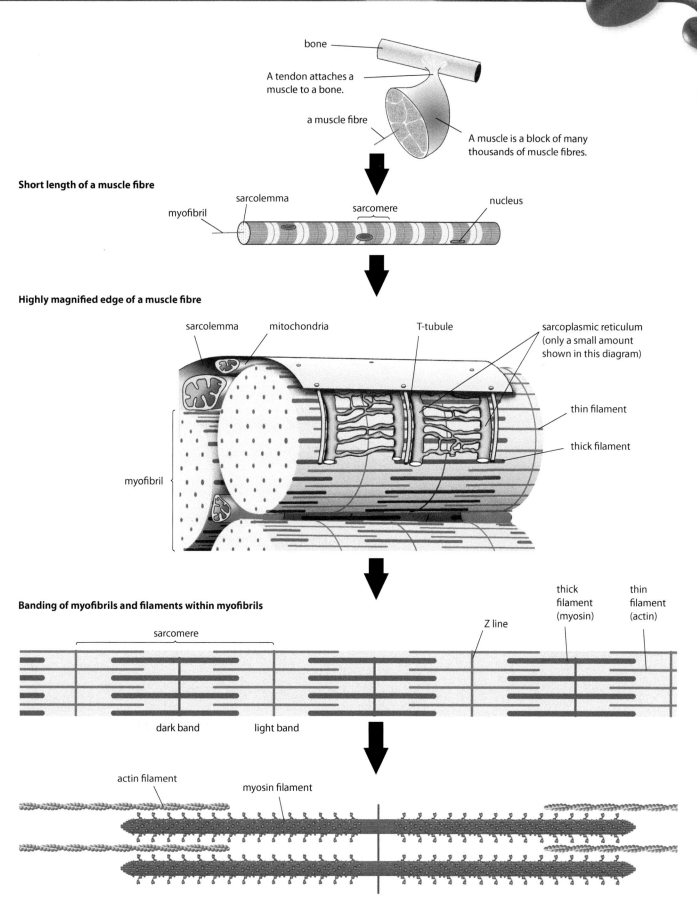

Figure 11.10 The structure of skeletal muscle.

11 ANIMAL PHYSIOLOGY (HL)

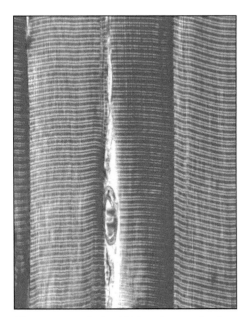

Figure 11.11 Stained light micrograph of striated muscle, stained to show the banding in muscle fibres.

Muscle tone

Contraction of a muscle causes shortening, and this in turn moves bones into a new position. If only a few fibres in a muscle contract, the muscle tightens but does not cause movement. Partial contraction produces muscle tone, which is important in maintaining posture and body shape.

Muscles

Skeletal or striated muscle is the muscle that causes the movement of our joints. Viewed under the light microscope (Figure 11.11) it has a striped appearance made up of multinucleate cells known as muscle fibres. Surrounding the muscle fibre is a plasma membrane called the sarcolemma. Each fibre is made up of many myofibrils running parallel to one another (Figure 11.10).

If skeletal muscle is examined with an electron microscope, it is possible to see that surrounding each myofibril is a system of membranes called the sarcoplasmic reticulum (which resembles smooth endoplasmic reticulum) and between the closely packed myofibrils are many mitochondria (Figure 11.10). Myofibrils are made up of repeating subunits called sarcomeres, which produce the striped appearance of a muscle fibre and are responsible for muscle contraction. The ends of a sarcomere are called the Z lines.

There are two types of filament that form the striped pattern of a muscle. These filaments are formed from the contractile proteins actin and myosin. The narrow filaments of actin are attached to the Z lines and extend into the sarcomere. Thicker filaments of myosin run between them. Where myosin is present, the myofibril has a dark appearance and a light band is seen where only actin is present. Myosin filaments have 'heads' which protrude from their molecules and are able to bind to special sites on the actin filaments.

Muscle contraction

Muscle contraction is explained by the 'sliding filament theory', which describes how actin and myosin filaments slide over one another to shorten the muscle. Contraction is initiated by the arrival of a nerve impulse from a motor neuron, which stimulates the sarcolemma of the muscle fibre. This, in turn, causes the release of calcium ions (Ca^{2+}) from the sarcoplasmic reticulum and begins the process that causes actin filaments to slide inward towards the centre of the sarcomere. The series of events is shown in Figure 11.12.

1. Nerve impulses (action potentials) travel along the muscle fibre membrane, or sarcolemma, and are carried down into the fibre through infoldings called T-tubules. The impulses then spread along the membrane of the sarcoplasmic reticulum, causing Ca^{2+} ions to be released.
2. Before contraction, binding sites for myosin heads on the actin filaments are covered by two molecules, troponin and tropomyosin. The myosin heads are prepared in an erect position as ATP binds to them.
3. Now Ca^{2+} ions bind to the actin filaments, causing the troponin and tropomyosin to change shape and expose the myosin binding sites. The myosin heads bind to the actin filaments at the exposed binding sites, forming cross-bridges.

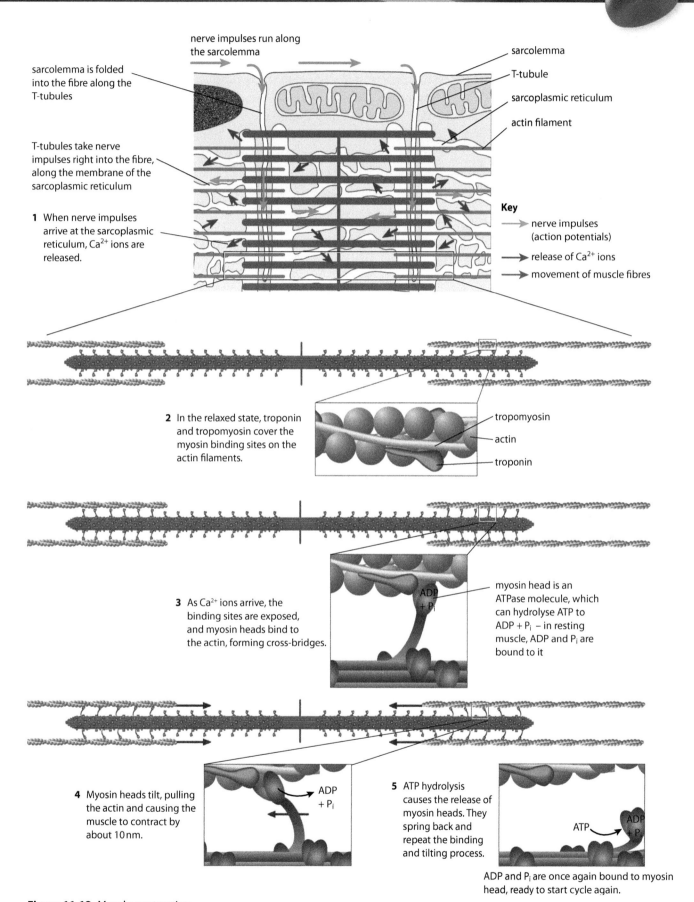

Figure 11.12 Muscle contraction.

Rigor mortis

Rigor mortis is a partial contraction of muscles that occurs after a person has died. The muscles become stiff and locked in position. It is caused because ATP production stops and myosin heads are unable to detach from actin filaments. It lasts for about 24 hours, at which time decomposition begins to break down the muscle tissues.

4 This causes inorganic phosphate (P_i) to be released and, as each cross-bridge forms, ADP is also released. The myosin heads bend towards the centre of the sarcomere, pulling the actin filaments inward past the myosin filaments, by about 10 nm. This produces a 'power stroke'.

5 New ATP molecules bind to the myosin heads, breaking the cross-bridges and detaching them from the actin filaments. ATP is used and the myosin heads return to the start position. If the muscle receives further stimulation, the process is repeated and the myosin heads attach further along the actin filaments.

Although the actin and myosin filaments do not change in length when a muscle contracts, the appearance of the banding patterns in the sarcomere is changed. The light bands become reduced, and as the overall length of the sarcomere decreases the dark bands take up a greater proportion of the length (Figure **11.13**).

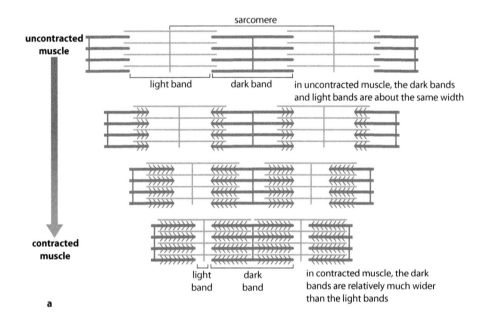

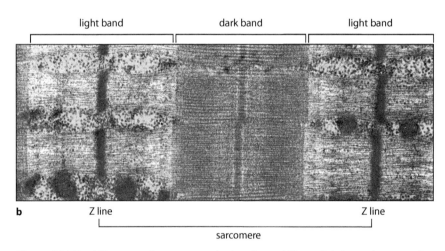

Figure 11.13 a When muscle contracts, the interleaved fibres slide inward, past each other. This makes the light bands appear narrower, but the dark bands remain the same width. **b** Coloured electron micrograph of a longitudinal section through striated muscle. (× 16 300)

Nature of science

Scientific advance follows technical innovation – using radioactive isotopes to track ion movements

In the 1940s, Alan Hodgkin, Andrew Huxley and Bernard Katz first explained action potentials and resting potentials in terms of the movement of potassium, sodium and chloride ions across nerve cell membranes.

It was not until radioactive ions were used to track movements of both potassium ions across nerve axons and calcium ions in muscles in the mid-1950s that Hodgkin and Huxley were able to explain the existence of ion channels. In the years that followed, the role of Ca^{2+} ions in muscle contraction has been extensively studied and has been shown to regulate the initiation of contraction. Calcium-dependent mechanisms in the membrane, such as ion channels, pumps and enzymes, have been studied, as well as the role of calcium ions in activating the enzymes that enable myosin to interact with actin and induce contractions. Our knowledge of these processes is largely due to the use of radioactively labelled calcium ions.

Test yourself

5 Explain why muscles occur in antagonistic pairs.
6 Outline the functions of cartilage and synovial fluid in the elbow joint.
7 Compare the movement of a hinge joint and a ball-and-socket joint.
8 Explain how actin and myosin filaments produce the striped appearance of skeletal muscle.
9 Describe the role of ATP in muscle contraction.

11.3 The kidney and osmoregulation

Learning objectives

You should understand that:
- Animals are either osmoregulators or osmoconformers.
- The kidney and the Malpighian tubule system in insects remove nitrogenous waste and carry out osmoregulation.
- Blood composition in the renal artery is different from that in the renal vein.
- The structure of the glomerulus and Bowman's capsule enable ultrafiltration to take place.
- Selective reabsorption of useful substances occurs by active transport in the proximal convoluted tubule.
- The loop of Henle maintains hypertonic conditions in the medulla of the kidney.
- Antidiuretic hormone (ADH) controls the reabsorption of water in the collecting duct.
- There is a positive correlation between the length of the loop of Henle and the need for water conservation in animals.
- The type of nitrogenous waste an animal species produces depends on its habitat and its evolutionary history.

Osmoregulation

Osmoregulation is the control of the water potential of body fluids to maintain a constant internal environment in the blood, tissue fluid and cytoplasm. This is essential to ensure that all cell processes occur effectively. Osmoregulation is achieved by regulating the water and salt balance of body fluids.

Animals may either be 'osmoregulators', which means they are able to osmoregulate in this way, or 'osmoconformers', which means they cannot. Most marine invertebrates are osmoconformers. For example, jellyfish and echinoderms have body fluids with the same solute concentration as sea water. Fish, on the other hand, are osmosregulators – although marine and freshwater fish have different problems in controlling the water potential of their body fluids, and different strategies to overcome them.

A freshwater fish placed in sea water will die because the salt concentration of sea water is approximately twice as high as that of its blood. This means that the fish will lose water by osmosis via its gills and mouth, which are permeable to water. The same fish living in freshwater, has the opposite problem – water will enter its body by osmosis. To overcome this, freshwater fish have very efficient kidneys that remove water and reabsorb vital salts. Their gills are also able to absorb sodium and chloride ions from the water against their concentration gradient.

In the sea, a marine fish has an internal solute concentration lower than that of sea water. This means that water will tend to leave its body by osmosis and so it must conserve water and get rid of salts. Marine fish have secretory cells in their gills that expel chloride ions, and they excrete nitrogenous waste in the form of urea, which uses only a little water as it is removed.

On land, animals tend to lose water by evaporation from skin and any other surfaces, such as the mouth and nose, which are open to the atmosphere. Different animals use different methods to conserve water. Some, such as reptiles, have a waterproof skin, insects have a waxy, waterproof cuticle and many terrestrial animals, including humans, have complex, efficient kidneys. Many animals behave in a way that ensures they avoid the hottest part of the day by burrowing, or by being active at night, so that evaporation losses are reduced to a minimum.

The kidney

Many animals carry out osmoregulation – that is, they control the solute concentration of their blood and tissue fluid – by filtering out unwanted substances at the kidneys, thereby regulating the composition and volume of urine produced. Mammalian kidneys are among the most complex, but fish and birds also have an efficient filtering system in their kidneys. Insects, too, remove excess water and metabolic waste in simple structures known as Malpighian tubules.

The many metabolic processes occurring in mammalian cells result in waste substances that must be removed. For example, **urea** is produced as a waste product from the metabolism of amino acids. Waste products

Marine birds and marine reptiles, such as the turtle, have to get rid of excess salt. Turtles have salt-secreting glands near their eyes and are sometimes said to cry salt tears. Marine iguanas on the Galapagos Islands snort out the salty spray from their salt glands, which runs into their nose. Marine birds such as penguins and cormorants also lose salty liquid through their nasal passages making them look as though they have a runny nose.

such as this are carried away from cells by the bloodstream, but they must be continuously removed from the blood so that they do not reach toxic levels. One of the main functions of the kidneys is to act as filters, removing waste molecules from the blood passing through them. This process is called excretion.

Structure of the kidney

In humans, the kidneys are situated in the lower back, one on either side of the spine, as shown in Figure **11.14**. Each receives a blood supply from a renal artery, which is a branch of the main aorta. After filtration, blood leaves the kidney via a renal vein that joins the vena cava. Because of the processes occurring in the kidney, the composition of the blood in the renal vein is quite different from that in the renal artery. Urea, water content and salt levels are adjusted by the kidney so that they are at the correct levels as blood leaves the kidney, but glucose, protein and the cellular content of blood remain unchanged. Figure **11.15** shows a kidney in longitudinal section. Three regions are visible – the outer cortex, the central medulla and the inner renal pelvis. Urine produced by the kidney collects in the renal pelvis and is carried down to the bladder in the ureter.

Each kidney is made up of more than one million tiny structures called nephrons. These are the functional units of the kidney, selectively filtering and reabsorbing substances from the blood. Figure **11.16** shows the structure of a nephron, which consists of a filtering unit (a complex of capillaries called a glomerulus surrounded by a Bowman's capsule) together with a tube that extends from the filtering unit to the renal pelvis. This tube is divided into four regions – the proximal convoluted tubule, the loop of Henle, the distal convoluted tubule and finally a collecting duct. Each of these regions has a specific role to play in urine formation.

Excretion the removal from the body of the waste products of metabolic pathways

Osmoregulation control of the water potential of body fluids by the regulation of water and salt content

Kidney functions at a glance

- excretion of waste products, water-soluble toxic substances and drugs
- regulation of the water and salt content of the body
- retention of substances vital to the body such as protein and glucose
- maintenance of pH balance
- endocrine functions

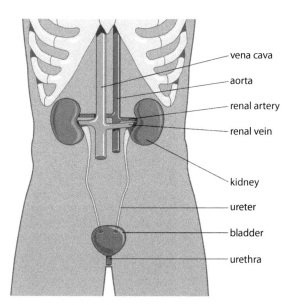

Figure 11.14 Location of the kidneys in the human body.

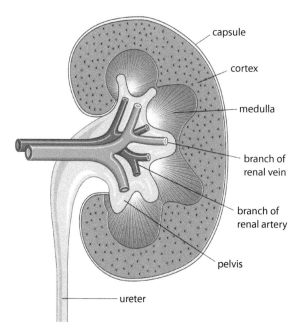

Figure 11.15 Longitudinal section of a human kidney.

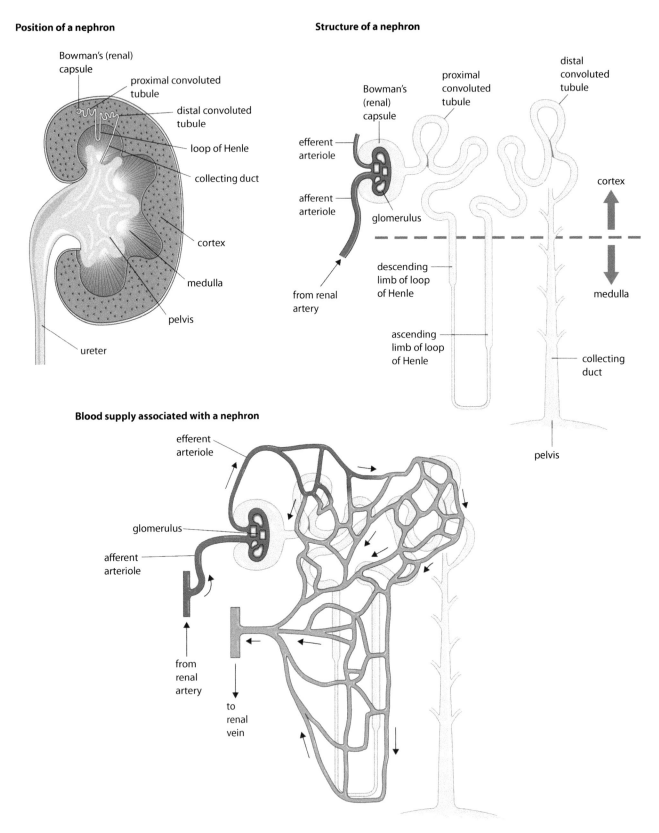

Figure 11.16 The location, structure and blood supply of a nephron.

How the kidney works

The kidney's complex structure allows it to carry out its functions with amazing efficiency. The process begins as blood from the renal artery reaches each glomerulus.

Ultrafiltration and the glomerulus

Ultrafiltration occurs in the glomerulus, as various small molecules leave the blood.

Because the incoming **afferent** arteriole (a branch of the renal artery) has a wider diameter than the outgoing **efferent** arteriole, blood pressure in the glomerulus capillaries is very high – so high, in fact, that about 20% of the blood plasma leaves the capillaries in the glomerulus and passes into the Bowman's capsule.

The blood plasma passes through three layers – the wall of the capillary in the glomerulus, the **basement membrane** (which acts as a molecular filter) and finally the epithelium of the Bowman's capsule. It leaves the capillaries through small pores or **fenestrations**, which the high blood pressure causes to open in the capillary walls. The fenestrations allow all molecules to pass through easily, but filtration occurs at the basement membrane, which is made of a glycoprotein. Only molecules with a molecular mass smaller than 68 000 are able to pass through. So water, salts, glucose, amino acids and small proteins can all pass through the basement membrane, but it effectively prevents blood cells and large molecules such as plasma proteins from leaving the blood (Figure **11.17**).

Blood supply to the kidney is normally about 20% of the heart's output. Approximately 99% of this blood flow goes to the cortex.

Exam tip
Remember the **A**fferent vessel **A**pproaches the glomerulus but the **E**fferent vessel **E**xits it.

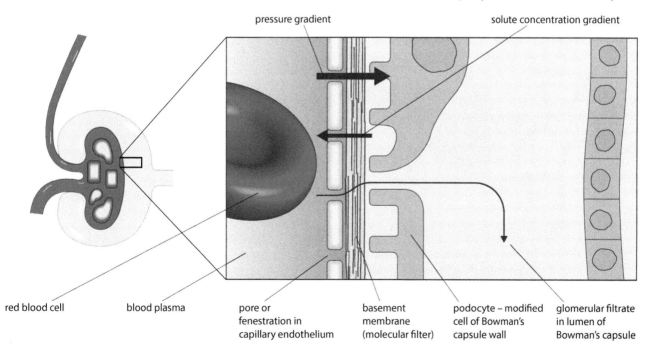

Figure 11.17 Detail of the Bowman's capsule showing the basement membrane, fenestrated capillary and podocytes. Podocytes are supportive, foot-shaped cells of the capsule wall that form a network of slits. Filtrate passes through these slits, and into the capsule.

The blood plasma that has passed through the basement membrane is now known as **filtrate**. It now passes through the epithelium of the Bowman's capsule into the nephron, and enters the proximal convoluted tubule. Blood cells and large molecules remain in the blood in the glomerulus capillaries and flow on into the efferent arteriole.

Reabsorption in the proximal convoluted tubule

Along with the unwanted molecules, many useful substances (water, glucose and ions that the body needs) enter the Bowman's capsule during ultrafiltration. These must be **reabsorbed** into the bloodstream. Between 80% and 90% of the filtrate is reabsorbed in the proximal convoluted tubule of the nephron (Figure **11.18**). The wall of the tubule is a single layer of cells and each one has a border of microvilli to increase its surface area. The cells have many mitochondria fuelling active transport through membrane pumps that selectively reabsorb ions and glucose from the tubular fluid. All the glucose in the filtrate is actively reabsorbed together with almost 80% of sodium (Na^+), potassium (K^+), magnesium (Mg^{2+}) and calcium (Ca^{2+}) ions. Chloride ions (Cl^-) are absorbed passively and water follows by osmosis as the solute concentration of the cells rises due to the active uptake of ions and glucose.

The remaining filtrate now moves into the loop of Henle.

The loop of Henle

The filtrate that enters the **loop of Henle** still contains a good deal of the water that was filtered from the blood. The wall of the descending limb of the loop is permeable to water but relatively impermeable to salts, whereas the ascending limb is impermeable to water, but allows salt to be passed

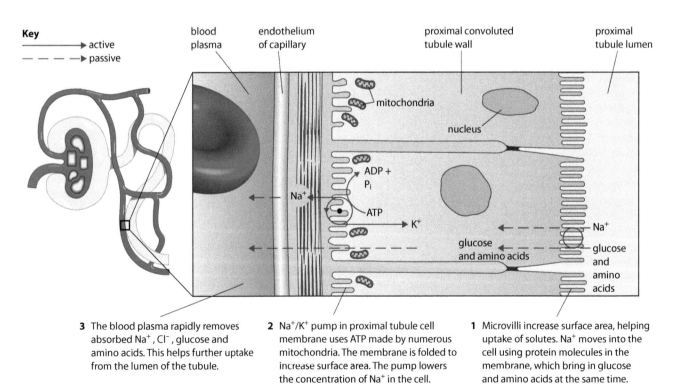

Figure 11.18 Reabsorption in the proximal convoluted tubule.

through its walls – Na⁺ and Cl⁻ ions move by active transport out of the ascending limb into the tissue fluid of the medulla, creating **hypertonic** conditions there (a high salt concentration). This means that as the descending limb of the loop of Henle passes down into the medulla, water leaves passively by osmosis and enters the surrounding blood capillaries (Figure **11.19**). The hypertonic environment in the medulla of the kidney produced by the loop of Henle is also essential for the fine-tuning of the water content of the blood by the collecting duct at a slightly later stage.

Despite the loss of water from the loop of Henle, the filtrate that enters the next section of the tubule still has relatively high water content. The length of the loop of Henle is different in different species and its length is related to an animal's need to conserve water. Terrestrial animals, such as camels, that live in dry environments and need to conserve water produce small volumes of very concentrated urine. Relatively speaking, these animals have a longer loop of Henle than species such as otters and beavers, which live in places where dehydration is not a problem. Animals that live in wet environments tend to have very short loops of Henle and excrete dilute urine.

> The kidneys secrete several hormones including erythropoietin which stimulates red blood cell production.

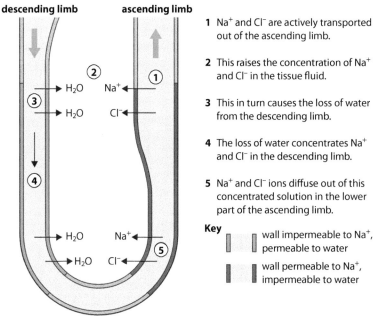

Figure 11.19 The counter-current mechanism in the loop of Henle builds up a high Na⁺ ion and Cl⁻ ion concentration in the tissue fluid of the medulla.

The distal convoluted tubule and the collecting duct

Ions are exchanged between the filtrate and the blood in the distal convoluted tubule. Na⁺, Cl⁻ and Ca²⁺ ions are reabsorbed into the blood while H⁺ and K⁺ ions may be actively pumped into the tubule.

The last portion of the nephron is the collecting duct where the final adjustment of water is made (Figure **11.20**). The permeability of the duct depends on the presence or absence of **antidiuretic hormone** (**ADH**). If ADH is present, the duct develops membrane channels called **aquaporins**

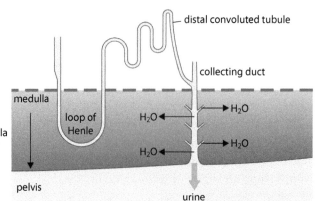

Figure 11.20 Water can be drawn out of the collecting duct by the high salt concentration in the surrounding tissue fluid of the medulla.

ADH and control of water loss

ADH is secreted by the posterior lobe of the pituitary gland, in the brain. If blood volume is low and more water is needed in the body, ADH causes the production of more concentrated urine. Osmoreceptors in the hypothalamus monitor water levels in the blood and control the release of ADH.

Caffeine, alcohol and cold conditions suppress ADH production and can lead to dehydration if too much water is lost in urine. Stress and nicotine increase ADH production producing the opposite effect.

Although the kidney can conserve water already present in the body, only intake of water by drinking or in foods can replace water that has already been lost.

so that it becomes more permeable and water is taken back into the blood. If the water content of the blood is high, ADH is not produced so the duct becomes impermeable and water remains inside the nephron, producing more dilute urine.

The urine now flows from the collecting ducts into the renal pelvis and down the ureter to the bladder.

A summary of kidney function

Table 11.1 compares the concentration of glucose, urea and protein in the blood plasma that enters the glomerulus, inside the glomerular filtrate and in urine.

Protein should not be present in the urine of a healthy person whose kidneys are working properly because protein molecules are too large to fit through the membrane filters in the glomerulus.

Urea is toxic in high concentrations. Its content (mg 100 cm^3) increases in urine as water is absorbed from the filtrate.

Glucose, filtered from the blood, forms part of the glomerular filtrate but is reabsorbed by active transport and should not be present in the urine of a healthy person. Glucose in urine is frequently a sign of untreated diabetes. Glucose concentration in a diabetic person's blood rises to a high level because their blood sugar level is not regulated properly by insulin. High blood glucose levels mean the pumps in the proximal convoluted tubule cannot remove it all from the filtrate and return it to the bloodstream. As a result, some glucose remains in the

Substance	Content in blood plasma / mg 100 cm^3	Content in glomerular filtrate / mg 100 cm^3	Content in urine / mg 100 cm^3
urea	30	30	2000
glucose	90	90	0
proteins	750	0	0

Table 11.1 The concentrations of some key substances in blood plasma, glomerular filtrate and urine.

nephron and is lost in the urine because it cannot be reabsorbed in any other regions of the tubule.

Osmoregulation in insects

Insects and some other arthropods have an excretory and osmoregulatory system made up of Malpighian tubules. The tubules are branches extending from the posterior region of the gut of the animal (Figure 11.21). They absorb water, waste and solutes from the insect's hemolymph (the fluid that fills the body cavity). Each of the slender tubules is blind ending and its walls are made up of a single layer of cells. The cells lining the tubules contain the protein actin to give them structural support, and have microvilli that move material along the tubules.

Urine formation begins as salts and nitrogenous waste move into the tubules either by diffusion (in the case of urea and amino acids) or by active transport (for Na^+ and K^+). Water follows by osmosis. The mixture that is produced is known as 'pre-urine' and it enters the hind gut of the insect and mixes with digested food. Here uric acid precipitates out and ions and water are reabsorbed. Uric acid is mixed with feces and is excreted in a solid form.

Types of nitrogenous waste

The type of nitrogenous waste an organism produces correlates with the habitat in which it lives and also with its evolutionary history (Table 11.2).

More energy is needed to produce uric acid than the other forms of nitrogenous waste but uric acid has the lowest toxicity and little water is lost as it is excreted.

As animals evolved and moved from an aquatic to a terrestrial environment, the need to conserve water became more and more important. Vertebrates produce either uric acid or urea as these substances require less water to excrete from the body.

Urine tests

Doctors regularly carry out urine tests to check for the presence of substances that may indicate illness or infection. A urine test can indicate the presence of protein, glucose, nitrites, ketones, bilirubin and blood cells, as well as measuring pH. High protein levels can indicate that the kidneys are inflamed, ketone and glucose in urine are signs of uncontrolled diabetes and leucocytes or nitrite in urine can suggest a bacterial infection. Urine may also contain residues of hormones (for example, HCG used in pregnancy testing) and drugs, such as steroids or alcohol. For this reason, urine tests are used by athletics authorities and the police checking for illegal use of certain substances.

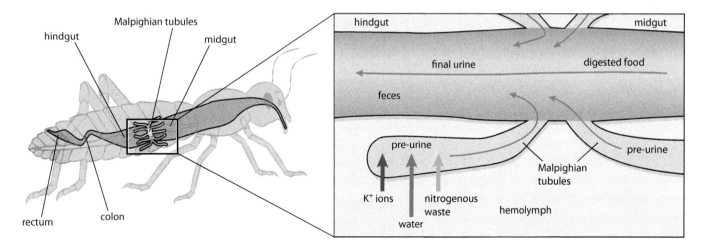

Figure 11.21 Diagram to show the Malpighian tubules of a typical insect.

Organism	Habitat	Type of nitrogenous waste	Reasons
aquatic invertebrates, most fish	aquatic	ammonia	ammonia is very toxic but is quickly diluted in water
birds, insects, land snails, some reptiles	terrestrial	uric acid	uric acid has low solubility and is non-toxic
mammals, some fish, amphibians	terrestrial and some aquatic	urea	urea has low toxicity and is soluble; it can be concentrated for excretion

Table 11.2 Excretory products of different organisms.

Figure 11.22 When water has been lost from the body, it is vital to replace it to maintain the correct balance of water and salts in body fluids.

Dehydration

Dehydration occurs if the body loses more fluid than it takes in. If this happens, the balance of salts and sugar in the blood will be upset and metabolic processes will not take place efficiently. Early signs of dehydration are extreme thirst, light-headedness and concentrated, dark urine. Dehydration may occur as a result of excessive sweating – for example, during vigorous exercise on a hot day – or due to the loss of water in cases of diarrhoea and vomiting. Children and infants are particularly vulnerable to this type of dehydration, which may leave them with an imbalance of salt and water in their blood.

During sporting events, participants are advised to drink water or drinks containing glucose and salts (isotonic drinks), which ensure that salt and water balance are maintained (Figure **11.22**). In cases of diarrhoea in young children, oral rehydration therapy may be used to replace lost water and salts (sometimes called electrolytes in this context). A typical recipe for oral rehydration therapy is six teaspoons of sugar + half teaspoon of salt + one dm^3 of water. This simple remedy can save the life of a child who has suffered dehydration due to diarrhoea.

Overhydration

In normal circumstances, drinking too much water is unlikely to happen. Most cases in which it has occurred have happened during endurance events such as a marathon, where runners have drunk too much simple water to try to avoid dehydration and upset the natural balance of salts in the blood. Overhydration can be potentially fatal and cause a dangerous disruption of brain function. Long distance runners are usually advised to drink no more than $0.5\,dm^3$ of water per hour and to replenish their levels of salts and glucose by taking isotonic drinks, which contain a similar balance of ingredients to oral rehydration mixtures.

Treating kidney failure

If a person's kidneys fail, they will be unable to remove toxic substances from their blood or regulate the water potential of their body fluids, which is ultimately a fatal condition. However, they can be kept alive if their blood is filtered by hemodialysis, also known as renal dialysis. The patient is connected to a dialysis machine, which acts as an artificial

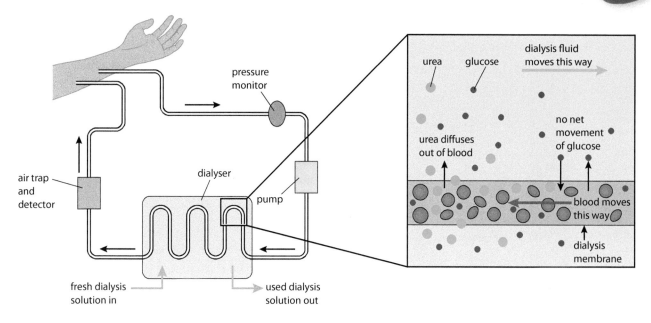

Figure 11.23 Dialysis machine.

kidney. Blood is withdrawn from a vein in the patient's arm and passed through narrow tubes made of a partially permeable dialysis membrane (Figure **11.23**). The tubes are surrounded in dialysis fluid, which contains the components of blood plasma – including sodium ions and glucose – at the correct, healthy levels, but it contains no urea. As blood passes through the tubes, large proteins and blood cells are held inside but small molecules of unwanted substances such as urea and excess salts diffuse through the membrane into the dialysis fluid. Desirable substances are present in the same concentrations on each side of the membrane so there is no net movement. The dialysis fluid is renewed at regular intervals so that a steep concentration gradient of the unwanted substances is maintained across the dialysis membrane. In addition to salts and urea, water can also be drawn out of the blood if necessary by increasing the solute concentration of the dialysis fluid – water will then pass into the fluid by osmosis. Dextran is a solute that is ideally suited to this purpose because it increases the concentration of the fluid but will not pass through the membrane. After blood has passed through the machine, it is returned to the patient's vein. Dialysis patients may have to be connected to a dialysis machine for up to 8 hours at a time and as often as three times a week. This indicates just how efficient a properly working human kidney is and how important our kidneys are.

Although a person can receive dialysis treatment for many years, a kidney transplant using a donated kidney from another person is the only permanent solution to kidney failure.

Ethical issues and kidney donation

Today many organs can be transplanted between well-matched human donors and recipients. Kidneys, corneas, bone marrow and skin are all transplanted regularly for certain medical conditions. Donors usually carry a card or express the wish that they will donate their organs should they die – for example, in an accident. Successful kidney transplants and the drugs needed to prevent rejection of a donated kidney cost less than keeping a person alive using renal dialysis and the quality of life of the recipient is also better. Nevertheless, there is still a shortage of people willing to donate a kidney after their death.

Questions to consider

- Should governments adopt policies that make organ donation compulsory rather than voluntary unless an individual has strong moral objection?

Nature of science

Using theories to explain natural phenomena – investigating desert animals

Scientists have been interested in the adaptations of desert animals for a long time and have investigated the structure of the kidney and urine content of many species. Different animals adopt different strategies to conserve water. Some adaptations are physiological and others are structural or behavioural. Unexpected adaptations have been found in some species. Some of these are summarised in the Table 11.3.

Animal	Adaptations
camel	- extracts water from food - produces very concentrated urine (long loop of Henle) - metabolises fat stores from the hump to provide metabolic water - tissues are adapted to tolerate dehydration
American kangaroo rat	- does not sweat or pant - produces very dry feces and concentrated urine - burrows during the day to avoid water loss from the lungs
Australian desert frogs	- glomerulus is slow to filter - urine is retained in a bladder for use in the dry season (frogs may swell up like a ball and have been used as a source of water by aboriginal Australians) - burrowing behaviour reduces water loss
Andean chinchilla	- does not sweat or pant - produces very dry feces and concentrated urine - obtains metabolic water from seeds - hairs in the nostrils cause water from respiration to condense so that it can be reabsorbed
desert scorpions and insects	- waxy, waterproof covering - some can absorb water from the atmosphere

Table 11.3 Adaptations of some desert species to conserve water.

Test yourself

10. Define the terms 'excretion' and 'osmoregulation'.
11. Draw and label the functional sections that make up the nephron.
12. Outline the role of the loop of Henle in regulating the content of urine.
13. Outline the role of the efferent and afferent arterioles in ultrafiltration.
14. Explain the role of microvilli in the cells of the proximal convoluted tubule.
15. List the layers through which a glucose molecule passes during ultrafiltration.
16. Outline the role of ADH in controlling the water content of urine.
17. Suggest why a diabetic person may produce urine containing glucose.

11.4 Sexual reproduction

Spermatogenesis

Spermatogenesis is the production of mature sperm cells (spermatozoa) in the testis. More than 100 million sperm cells are produced each day in a process that takes place in the narrow seminiferous tubules making up each testis (Figures 11.24 and 11.25).

Sperm production and development takes place from the outer part of the seminiferous tubules towards the central lumen, where sperm cells are eventually released. Each tubule is enclosed in a basement membrane beneath which is an outer layer of germinal epithelium cells. These diploid cells (2n) divide regularly by mitosis to produce more diploid cells, which enlarge and are known as **primary spermatocytes**.

Primary spermatocytes divide by meiosis and their first division produces two haploid (n) cells called **secondary spermatocytes**. The second division of these two cells results in four **spermatids** (n).

Developing sperm are attached to Sertoli cells (Figure 11.25), which are also called nurse cells. These large cells assist the differentiation of immature spermatids into **spermatozoa** and provide nourishment for them.

Spermatozoa that have developed their tails (Figure 11.26) detach from the Sertoli cells and are carried down the lumen of the tubule to the epididymis of the testis.

Hormones and sperm production

Sperm production is controlled by three hormones – follicle-stimulating hormone (FSH) and luteinising hormone (LH) from the pituitary gland, and testosterone produced by the testes.
- **FSH** stimulates meiosis in spermatocytes, to produce haploid cells.
- Testosterone stimulates the maturation of secondary spermatocytes into mature sperm cells.
- **LH** stimulates the secretion of testosterone by the testis.

Learning objectives

You should understand that:
- Spermatogenesis and oogenesis involve mitosis, cell growth, the two divisions of meiosis and differentiation.
- Spermatogenesis and oogenesis result in the production of different numbers of gametes with different amounts of cytoplasm.
- Fertilisation in animals may be external or internal.
- Mechanisms during fertilisation prevent polyspermy.
- For pregnancy to occur, it is essential that the blastocyst implants in the endometrium.
- HCG stimulates the ovary to secrete progesterone in the early stages of pregnancy.
- The placenta enables mother and fetus to exchange materials. It secretes estrogen and progesterone once it is fully formed.
- Birth is mediated by positive feedback mechanisms involving estrogen and oxytocin.

Epididymis, seminal vesicles and semen production

Sperm cells are stored and mature in the **epididymis**, where they also develop the ability to swim. Sperm cells are released at ejaculation in a nutrient-rich fluid known as **semen**. Semen is produced by two **seminal vesicles** and the **prostate gland**. It is mixed with the sperm cells as they leave the epididymis and move along the **vas deferens** (sperm duct). Fluid from the seminal vesicles makes up about 70% of semen. It is rich in fructose, which provides energy for the sperm cells to swim, and it also contains protective mucus. The prostate gland produces an alkaline fluid that helps the sperm cells to survive in the acidic conditions of the vagina.

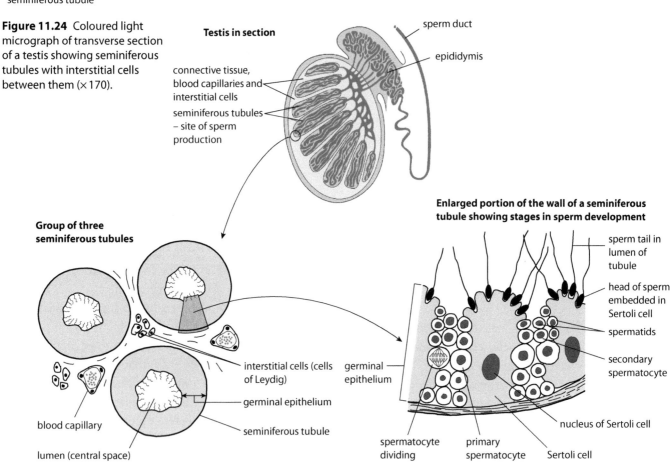

Figure 11.24 Coloured light micrograph of transverse section of a testis showing seminiferous tubules with interstitial cells between them (×170).

Figure 11.25 Structure of the testis.

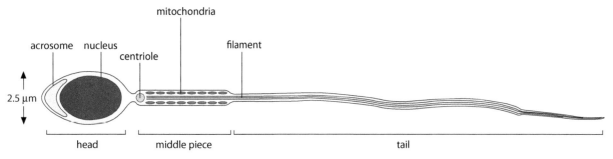

Figure 11.26 Structure of a human sperm cell. Total length is 60 μm.

Oogenesis

Oogenesis produces female gametes, the **ova**. Unlike spermatogenesis, which takes place in an adult male, oogenesis begins in the ovaries of a female when she is still a fetus. **Oogonia**, the germinal epithelial cells within the ovaries of the female fetus, divide by mitosis to produce more diploid (2*n*) cells. These enlarge to form **primary oocytes**, which are also diploid.

Primary oocytes undergo the first stages of meiosis but this stops during prophase I, leaving the primary oocyte surrounded by a layer of follicle cells in a structure known as the **primary follicle**. Development now ceases but the ovaries of a baby girl contain around 300 000 primary follicles at birth. The remaining stages of oogenesis are shown in Figure **11.27**.

At puberty, development of the primary follicles continues. During each menstrual cycle, a few follicles proceed to complete the first division of meiosis, although usually just one will complete its development. Two haploid cells (*n*) are produced but the cytoplasm divides unequally so that one cell is much larger than the other. The larger cell is known as the **secondary oocyte** (*n*) and the smaller cell is the **polar body** (*n*). The polar body degenerates and does not develop further.

The secondary oocyte, protected within its follicle, begins meiosis II but stops in prophase II. At the same time, the follicle cells divide and produce a fluid that causes the follicle to swell. At the point of **ovulation**, the follicle bursts, releasing the secondary oocyte, which floats towards the **oviduct** (Fallopian tube). Although ovulation is often described as the release of the

To remind yourself of the structure of the male reproductive system, look back at Figure **6.39** in Subtopic **6.6**.

Exam tip
Make sure you can draw and label diagrams of the male and female reproductive systems.

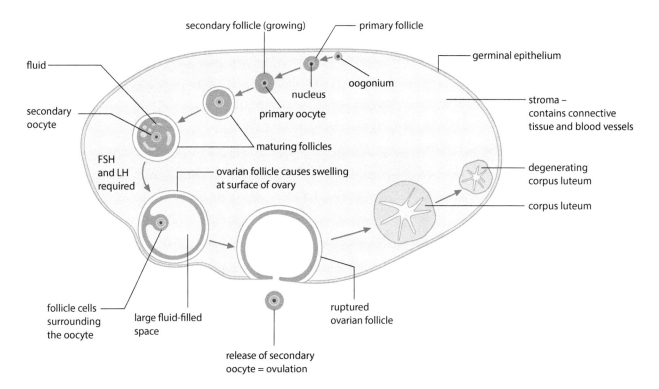

Figure 11.27 Stages in the development of one follicle in a human ovary. The arrows show the sequence of events.

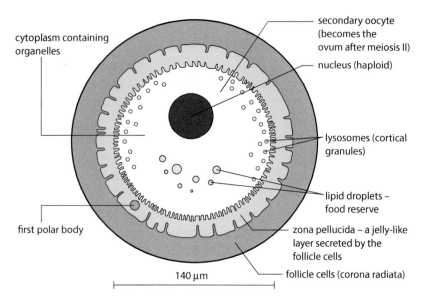

Figure 11.28 Structure of the secondary oocyte and surrounding structures at ovulation.

ovum, the cell that is released is in fact still a secondary oocyte. The detailed structure of a secondary oocyte is shown in Figure **11.28**, and Figure **11.29** shows secondary oocytes in a rabbit ovary in section.

After fertilisation, the secondary oocyte completes meiosis II, becoming a mature ovum, and expels a second polar body, which degenerates. The empty follicle in the ovary develops to become the **corpus luteum**, or 'yellow body', which produces the hormone progesterone.

To remind yourself about the production of progesterone by the corpus luteum, look back at Subtopic **6.6**.

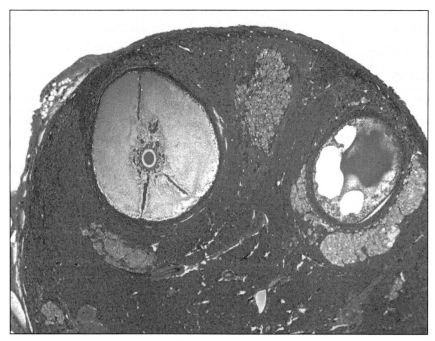

Figure 11.29 Longitudinal section of the ovary of a rabbit showing a mature follicle (×22.5).

Comparing spermatogenesis and oogenesis

There are a number of similarities and also several differences between the processes of spermatogenesis and oogenesis, as shown in Figure **11.30**. Both involve the division of cells in the germinal epithelium by mitosis, and the growth of cells before they undergo meiosis and differentiation. In both cases, meiosis produces haploid gametes. Table **11.4** summarises the differences and similarities in the two processes.

	Oogenesis	**Spermatogenesis**
Similarities	both begin with production of cells by mitosis	
	in both, cells grow before meiosis	
	in both, two divisions of meiosis produce the haploid gamete	
Differences	usually only one secondary oocyte is produced per menstrual cycle	millions of sperm cells are produced continuously
	only one large gamete is produced per meiosis	four small gametes are produced per meiosis
	occurs in ovaries, which tend to alternate oocyte production	occurs in testes, which both produce sperm cells
	early stages occur during fetal development	process begins at puberty
	ova released at ovulation during the menstrual cycle	sperm cells released at ejaculation
	ceases at menopause	continues throughout an adult male's life

Table 11.4 Oogenesis and spermatogenesis compared.

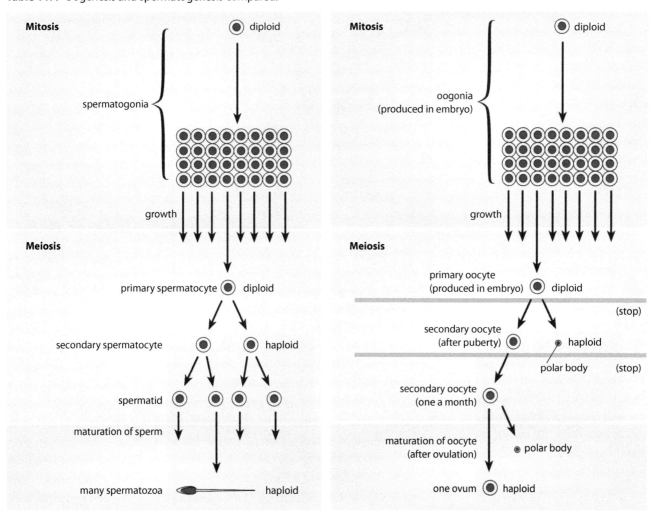

Figure 11.30 A comparison of spermatogenesis and oogenesis.

Fertilisation

Gametes produced during oogenesis and spermatogenesis come together and fuse in the process of fertilisation. Fertilisation may take place either inside or outside the body of the female.

External fertilisation occurs in many aquatic species including invertebrates and fish, and also frogs and toads, which return to the water to breed (Figure **11.31**). These animals release both male and female gametes into the water, where fertilisation takes place. To ensure the best chance for gametes to meet, species with external fertilisation often take part in courtship rituals so that spawning (the release of gametes) occurs at the same time. They produce large numbers of gametes because many will be wasted in the water.

Animals including reptiles, birds and mammals have internal fertilisation. Sperm cells are transferred into the female body during copulation (sexual intercourse in humans). Internal fertilisation has two main advantages: it increases the chance that gametes will meet so species with internal fertilisation produce fewer gametes and it means that the fertilised ovum can be enclosed in a protective covering such as a shell. In mammals, the embryo remains inside the female's body where it develops until birth.

Figure 11.31 Female frogs lay their ova in water where they are fertilised by sperm cells from the male.

Human fertilisation

Fertilisation usually occurs in one of the oviducts and is the moment when one sperm cell fuses with the secondary oocyte to form a zygote. The sequence of events is summarised in Figure **11.32**.

During sexual intercourse, millions of sperm cells are ejaculated into the vagina and some of them make their way through the cervix and uterus towards the oviducts. Only a very small number of the ejaculated sperm will complete the journey, which is a considerable distance for the tiny cells. As sperm cells approach the zona pellucida, they go through a process known as the **acrosome reaction**. The contents of their acrosomes, which include many enzymes, are released to penetrate the outer layers of follicle cells covering the secondary oocyte and allow the sperm cells through. For fertilisation to be successful, many sperm cells must be present to release the contents of their acrosomes, but only one sperm cell will eventually break through and reach the plasma membrane, where the membrane of its empty acrosome fuses with it.

After fusion has occurred, changes known as the cortical reaction take place in the membrane of the oocyte, which modify the zona pellucida and also prevent **polyspermy** (the entry of more than one sperm cell nucleus). Cortical granules, the enzyme-containing vesicles found just inside the oocyte, fuse with the plasma membrane in wavelike fashion, away from the point of fusion of the sperm, and release their contents. Some of the enzymes digest away the sperm cell receptor proteins on the oocyte plasma membrane so that sperm cells are no longer able to attach and fuse.

To remind yourself of the structure of the female reproductive system, look back at Figure **6.40** in Subtopic **6.6**.

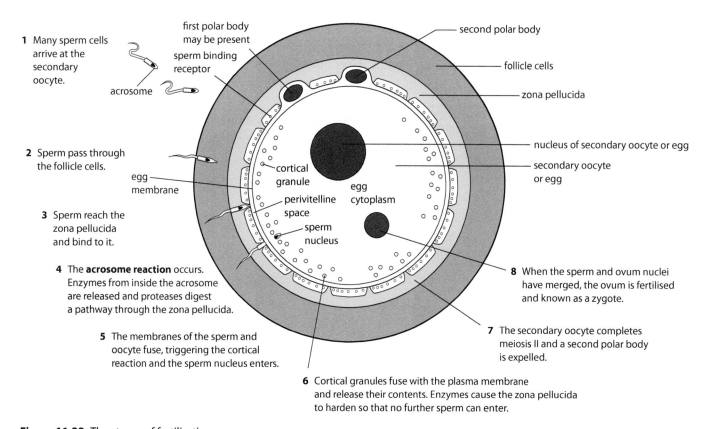

Figure 11.32 The stages of fertilisation.

Pregnancy

Approximately 24 hours after fertilisation, the zygote begins to divide by mitosis. Mitosis continues and, after about five days of division, produces ball of around 100 cells known as a **blastocyst**, as shown in Figure **11.33**. As these divisions are occurring, the ball of cells is moved down the oviduct towards the uterus. After about seven days, it reaches the uterus and settles in the **endometrium** lining, where it implants itself and continues to divide and develop into an **embryo**.

Once the blastocyst has become established in the endometrium, it begins to secrete the hormone **human chorionic gonadotrophin (HCG)**. HCG travels in the bloodstream to the ovary, where its role is to maintain the corpus luteum, the mass of cells that developed from the empty follicle. The corpus luteum produces progesterone and estrogen, which – in a non-pregnant woman – maintain the endometrium until the end of the menstrual cycle, when the corpus luteum degenerates. During pregnancy, it is important that the lining remains in place. HCG stimulates the corpus luteum so that it grows and continues to produce its hormones for the first three months (**trimester**) of pregnancy. Thereafter, the placenta is fully formed and produces placental progesterone and estrogen, so the corpus luteum degenerates.

HCG is excreted in the urine of a pregnant woman and it is this hormone that is detected in a pregnancy test with the use of monoclonal antibodies (Subtopic **11.1**).

The embryo grows and develops. After about one month, it is only 5 mm long but has a beating heart and the beginnings of a nervous system. From two months onwards, it is known as a **fetus**. The fetus at this stage is 30–40 mm long and has recognisable limbs with developing bones. The uterus lining provides nourishment for the early embryo but the placenta soon forms from the endometrium and fetal membranes and by about 12 weeks it is fully functioning. The fetus is connected to the placenta by the **umbilical cord** and is surrounded by a fluid-filled

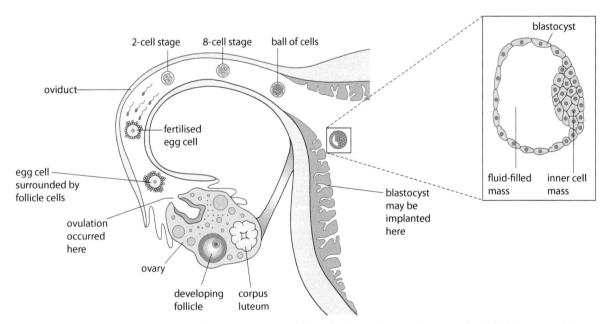

Figure 11.33 The blastocyst consists of an outer layer of cells enclosing an inner cell mass and a fluid-filled space. The outer layer forms part of the placenta and the inner cell mass develops to become the body of the embryo.

sac called the amnion, which contains amniotic fluid. The fetus is supported in this fluid throughout its development and is protected by it from bumps and knocks, as the fluid is an effective shock absorber. Amniotic fluid also enables the growing fetus to move and develop its muscles and skeleton.

The placenta

The developing fetus depends on its mother for all its nutrients and oxygen and for the disposal of its waste carbon dioxide and urea. The placenta allows these materials to be exchanged between the mother and the fetus and also acts as an endocrine gland, producing estrogen, progesterone and other hormones that maintain the pregnancy.

The placenta is a disc-shaped structure, about 180 mm in diameter and weighing about 1 kg when it is fully developed. It is made up of the maternal endometrium and small projections, or villi, from the outer layers of the chorion, which surrounds the embryo. These **chorionic villi**, which are rich in capillaries, grow out into the endometrium to produce a very large surface area for the exchange of gases and other materials. Fetal blood remains inside these capillaries, which penetrate the endometrium tissue until they are surrounded by maternal blood flowing into blood sinuses (spaces) around them. In this way, the mother's blood is brought as close as possible to the fetal blood to allow for efficient diffusion without the two ever mixing. These features are shown in Figure **11.34**.

Exchange of materials

Fetal blood is carried to the placenta in two umbilical arteries, which divide to form capillaries in the villi. Nutrients and oxygen from the mother's blood diffuse into the fetal capillaries and are carried back to the fetus in a single umbilical vein. Waste products and carbon dioxide are carried to the placenta in the two umbilical arteries and diffuse into the mother's blood.

Many materials pass to the fetus from its mother. Some of these – such as drugs (both prescription and illegal), nicotine and alcohol – have the potential to seriously harm the fetus, which is why pregnant women are encouraged not to smoke or drink alcohol during pregnancy and to be careful with any medicines they may take.

Table **11.5** lists some of the main substances that are exchanged between mother and fetus.

Substances that pass from mother to fetus include	Substances that pass from fetus to mother include
• oxygen • nutrients including glucose, amino acids, vitamins, minerals • water • hormones • alcohol, nicotine and other drugs • some viruses, including rubella and sometimes HIV	• carbon dioxide • urea • water • some hormones, such as HCG

Table 11.5 Substances exchanged between mother and fetus.

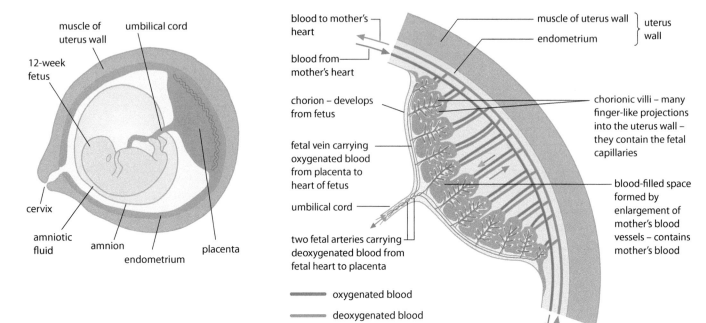

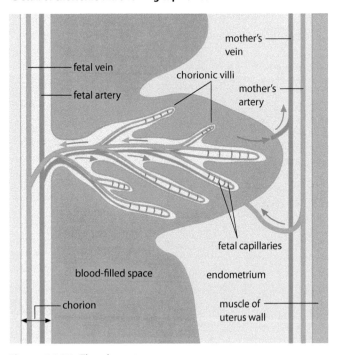

Figure 11.34 The placenta.

Prenatal testing – amniocentesis and CVS

Amniocentesis is one of a number of techniques used in prenatal testing to check human fetuses for abnormalities. A fine needle is inserted through the mother's abdomen into the amniotic sac and a small sample of amniotic fluid is taken. The fluid contains fetal cells, which can be cultured for 3–4 weeks until the cells divide and chromosomes become visible. The chromosomes are stained to produce a karyogram (Subtopic **3.3**), which can be checked for mutations.

An alternative procedure, which can be done earlier in the pregnancy, is chorionic villus sampling (CVS). In this case, a sample of cells is taken for examination from the chorionic villi, via the cervix. More fetal cells are obtained in this way and the results are produced more quickly. However, CVS does have a greater risk of inducing a miscarriage than amniocentesis.

Hormonal changes and childbirth

For the first 12 weeks of pregnancy, the corpus luteum produces progesterone to maintain the uterus lining (endometrium). After this, the placenta takes over and produces progesterone and estrogen, which suppress the menstrual cycle and promote the growth of breast tissue for lactation (milk production).

After about 9 months, as the end of pregnancy approaches, the levels of progesterone and estrogen produced by the placenta fall (Figure **11.35**) and this signals the onset of the uterine contractions known as **labour**. At this time, the endometrium secretes a group of hormones known as prostaglandins, which initiate the contraction of the uterus. Stretch receptors in the cervix then stimulate the hypothalamus, which triggers the release of the hormone oxytocin, secreted by the posterior lobe of the pituitary gland in the brain. Oxytocin stimulates the uterus muscles to continue their contractions. At first, the contractions are mild and infrequent but oxytocin is a hormone that is controlled by positive feedback. A small contraction of the uterus muscle stimulates the release of further oxytocin, which in turn stimulates more and stronger contractions. As the uterus contracts, the cervix widens and the amniotic sac breaks, releasing the amniotic fluid. Contractions continue for several hours and the baby is pushed through the cervix and out of the mother's body down the vagina. Gentle contractions continue until the placenta, now known as the afterbirth, is also expelled from the uterus.

After birth, blood levels of the hormone prolactin, from the anterior pituitary gland, increase. This hormone stimulates milk production by the mammary glands. As a baby suckles, prolactin secretion is maintained and oxytocin is also released from the posterior pituitary gland. Oxytocin causes milk to be released from milk ducts.

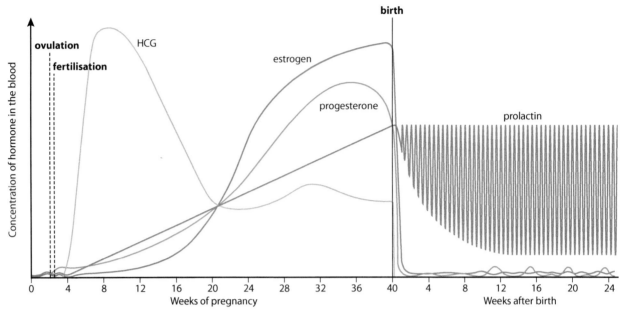

Figure 11.35 Changes in the levels of hormones during pregnancy and after the birth.

Nature of science

Assessing risk in science – effects of oral contraceptive residues

In recent years, some scientists have raised concerns about the apparent decline in human male fertility. This has been attributed in part to the excretion of residues from oral contraceptives into drinking water supplies. There is some evidence that synthetic estrogens such as estradiol (or EE2), used in contraceptive pills and patches, can affect the 'maleness' of fish in rivers and streams – intuitively, some people fear its presence in drinking water may have similar effects on humans.

Understanding the possible effects of such compounds in the water supply is clearly important and further study into their effects is ongoing. But it is also important to note that the source of these estrogenic compounds is not only oral contraceptives but also veterinary contraceptives and residues from other industrial processes. EE2 is part of a larger group of chemicals known as endocrine disrupters, which alter hormonal balance in humans and other animals. When these chemicals were first discovered, risks to human fertility were unknown and could not be adequately assessed. It is important now to weigh such risks against the benefits that the use of oral contraceptives has brought to the human population as a whole.

Is intuition a reliable way of knowing?

Some people intuitively feel that if synthetic estogens are harming fish, they will also be harmful to humans. Others are sceptical of intuition because it means drawing a conclusion without facts to back it up.

Questions to consider

- Why is it difficult to trust intuition to provide reliable knowledge?
- Can the reasoning behind intuitive feelings ever be rationalised?

Test yourself

18 State the role of testosterone in spermatogenesis.
19 Explain why unequal divisions of cytoplasm are necessary in oogenesis.
20 List **three** differences between spermatogenesis and oogenesis.
21 Describe the role of the amnion in the development of the fetus.
22 Describe the role of oxytocin at birth.

Exam-style questions

1 Which of the following statements about defence is correct?

    A Passive immunity occurs as a result of the body being challenged by antigens.
    B Active immunity occurs as a result of the body being challenged by antigens.
    C Monoclonal antibodies are produced by fusing a hybridoma cell with a myeloma cell.
    D As a result of vaccination, the primary response of the body will be greater when it is invaded by an antigen. [1]

2 Which of the following statements about muscles and movement is correct?

    A During muscle contraction, the distance between Z lines decreases, and the dark bands get wider but the light bands get narrower.
    B Bones are held together at a joint by tendons, which also protect the joint.
    C Synovial fluid, secreted by the cartilage, lubricates the joint to prevent friction.
    D The elbow is an example of a ball-and-socket joint. [1]

3 Which of the following statements about the kidney is correct?

    A Osmoregulation is the maintenance of the correct concentration of water in the urine.
    B Ultrafiltration occurs in the Bowman's capsules situated in the medulla region of the kidney.
    C Glucose is sometimes present in the urine of people with diabetes because the lack of insulin prevents its reabsorption in the loop of Henle.
    D Human urine can become more concentrated than blood plasma as a result of water reabsorption due to the release of ADH from the pituitary gland. [1]

4 Which of the following statements about reproduction is correct?

    A The hormones LH and FSH from the pituitary gland and testosterone from the Sertoli cells are involved in spermatogenesis.
    B During oogenesis, polar bodies are formed by meiosis in the follicle cells of the germinal epithelial layer of the ovary.
    C The role of the hormone HCG early in pregnancy is to prevent further ovulation by blocking the production of estrogen.
    D During the process of fertilisation, substances released from the cortical granules result in the formation of the fertilisation membrane. [1]

5 Explain antibody production. [5]

6 Outline how the fluid in the proximal convoluted tubule is produced by the process of ultrafiltration. [4]

7 Explain how the cells of the proximal convoluted tubule are adapted to carry out selective reabsorption. [3]

8 Outline how muscle contraction is brought about, starting with the arrival of a nerve impulse at the muscle. [5]

9 Outline the process of birth, including hormonal control. [5]

10 The graph shown below shows data on a kidney patient who received dialysis as a result of kidney failure. Dialysis began on 18th October and was needed for two weeks.

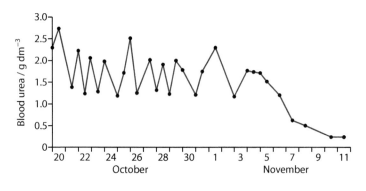

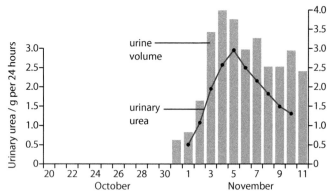

a State how many times the patient received dialysis. [1]

b Explain your reasoning. [1]

c Identify the date when the patient began to produce urine again. [1]

d The body mass of the patient fell by 6 kg between 30th October and 11th November. Outline two reasons for this change. [2]

11 An investigation was carried out into the relationship between the birth weight of babies and the concentrations of the elements zinc and lead in the mother's placenta. Results are shown in the histogram below. The standard deviation of each value is given at the top of each bar.

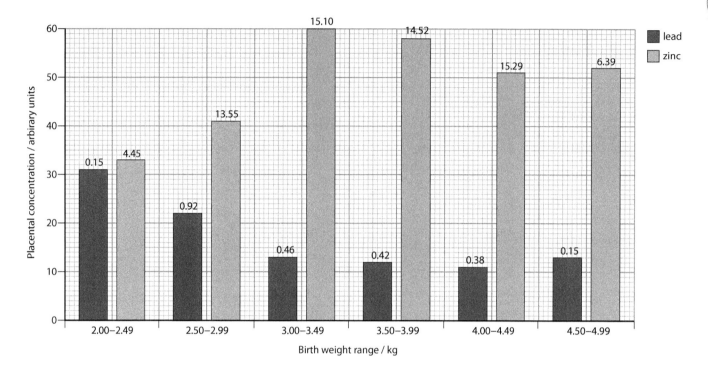

a i Describe the relationships between levels of lead and zinc and birth weight, as shown in the histograms. [2]
   ii Suggest reasons for these relationships. [2]
b State the **two** ranges of birth weight that appear to show the lowest overall variability in placental metal concentration. [2]
c Explain how the efficiency of the human placenta is increased by the presence of villi. [2]

Answers to test yourself questions

1 Cell biology

1. As surface area increases, ratio of surface area to volume decreases.
2. 10 cells
3. For example:
   nerve cells work with muscles to coordinate flight
   lungs and vocal cords produce song
   petals and nectaries attract insects
   petals and anthers work together to achieve pollination.
4. any of: treatment of leukemia; treatment of Alzheimer's disease; treatment of diabetes
5. Each specialised cell differentiates to carry out its function, e.g. producing insulin, absorbing food, photosynthesis. Genes not relating to this function are switched off.
6. 1000 µm or 1 mm
7. Eukaryotic cells have a nuclear envelope, organelles such as mitochondria, and DNA that is associated with protein. Prokaryotes do not.
8. a The cell wall is made of cellulose and surrounds a plant cell, whereas a plasma membrane surrounds both plant and animals cells.
   b A lysosome is a vesicle found in the cytoplasm, containing enzymes that can digest the contents of phagocytic vesicles, while a ribosome is a small organelle that is the site of protein synthesis.
9. Different chemical processes can occur at the same time in different parts of the cell.
10. The endoplasmic reticulum is the site of synthesis for proteins that will be exported from a cell.
11. 'Fluid' because phospholipids and proteins can move and occupy any position in the membrane, which can break and reform. 'Mosaic' because the surface view of a membrane resembles a mosaic made of many small separate parts.
12. Fatty acids are hydrophobic (water-hating) and always orientate away from water.
13. Integral proteins are partially or completely embedded in the plasma membrane. Peripheral proteins are attached to its surface.
14. Simple diffusion is the movement of molecules across a permeable membrane from a higher to a lower concentration. Facilitated diffusion involves the movement of larger molecules or ions through specific integral protein channels in a membrane. Both are passive processes.
15. three from: diffusion; facilitated diffusion; protein pumps; osmosis; phagocytosis; exocytosis
16. Requires energy from ATP: phagocytosis or active transport. Does not require energy from ATP: diffusion or facilitated diffusion.
17. Exocytosis exports material from the cell, while endocytosis takes materials in. Both involve invagination of the membrane.
18. It showed that bacteria could not grow in a sealed, sterilised container of chicken broth.
19. The theory of endosymbiosis suggests that some organelles found inside eukaryotes were once free-living prokaryotes.
20. three from: UV light; X rays; environmental chemicals such as tobacco smoke
21. formation of a tumour
22. prophase, metaphase, anaphase, telophase
23. death of specific cells in a growing organism

2 Molecular biology

1. four
2. building up new molecules from simpler components
3. a small molecule that can bind chemically to other similar molecules to form a polymer
4. Water has a high specifc heat capacity and energy is needed to break bonds between its molecules. It evaporates below its boiling point as energy is used to break hydrogen bonds.
5. 'water loving'
6. Many hydrogen bonds in a matrix of water molecules produce a strong force between all the molecules. This is known as cohesion.
7. maltose, sucrose
8. A saturated fatty acid has no double bonds between the carbon atoms in its molecule, whereas an unsaturated fatty acid has one or more double bonds.
9. condensation
10. Hydrolysis is the separation of two molecules by the addition of water, whereas condensation is the joining of two molecules with the loss of water.
11. It is determined by the sequence of codons in DNA which are transcribed and translated in the cell.
12. all the proteins expressed by a genome at a certain time
13. the area in an enzyme where substrate binds and which catalyses a reaction involving the substrate
14. If the temperature is below the optimum temperature, increasing it will increase the rate of enzyme action. If the temperature is increased above the optimum, the rate of activity will decrease until the enzyme is denatured and activity ceases.
15. production of lactose-free milk

16 two from:

DNA	RNA
double stranded	single stranded
contains bases A, C, G and T	contains bases A, C, G and U
sugar = deoxyribose	sugar = ribose

17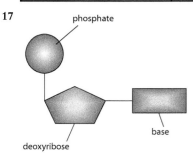

18 No DNA molecule is ever completely new. Every double helix contains one 'original' and one 'new' strand.

19 the enzyme involved in linking the new nucleotides into place during replication

20 production of a strand of mRNA from a DNA template

21 specific pairing of base A with base T or base U, and base C with base G

22 in the cytoplasm

23 ATP and lactic acid (lactate)

24 cytoplasm in mitochondria

25 cytoplasm

26 Yeast ferment sugar in flour to produce alcohol, which evaporates, and carbon dioxide, which causes the bread dough to rise.

27 Red light was used for photosynthesis by *Spirogyra*, which produced oxygen. The bacteria moved towards the oxygen.

28 Light of this wavelength was not used for photosynthesis, so no oxygen was produced.

29 blue or red

30 Growth would be inhibited; it might become thin and weak.

31 All except green light, but mainly red and blue.

32 directly by measuring oxygen production or carbon dioxide release and indirectly by measuring the increase in biomass

3 Genetics

1 438 nucleotides

2
  a a heritable factor that controls a specific characteristic
  b one specific form of a gene, differing from other alleles by one or a few bases only and occupying the same gene locus as other alleles of the gene
  c the whole of the genetic information of an organism
  d a change in the sequence of bases in a gene

3 the replacement of a base in a gene with a different base

4 DNA in a eukaryote is enclosed in a nuclear membrane. Prokaryotes do not have protein associated with their DNA.

5 a diagram or photograph of the chromosomes from an organism

6 a chromosome that does not carry genes determining the sex of an organism

7
  a anaphase I
  b prophase I and II
  c anaphase II
  d metaphase II
  e prophase
  f anaphase I

8 chromosomes with the same genes but not necessarily the same alleles

9 The number of chromosomes is halved as a result of meiosis.

10 four cells

11 a nucleus that contains three copies of one chromosome instead of two

12 Down syndrome

13 any two of: banding pattern; position of centromere; length

14 diploid number

15
  a the alleles possessed by an organism
  b the characteristics of an organism
  c an allele that has the same effect on the phenotype whether in homozygous or heterozygous state
  d an allele that only has an effect on the phenotype in the homozygous state
  e having two identical alleles at a gene locus
  f having two different alleles at a gene locus

16
  a red
  b red
  c yellow

17
  a R
  b r
  c R and r

18

	gametes from green parent	
	G	g
gametes from green parent — G	GG green	Gg green
gametes from green parent — g	Gg green	gg purple

19
  a 2
  b 2
  c 1

20 blood group O

21 colour blindness and hemophilia

22 ●

23 codominant

24 carrier

25 to amplify small quantities of DNA

26 their size and charge

27 any two of: paternity testing, forensic examination of crime, scenes, animal pedigree testing

28 the fact that the code is universal

29 plasmids

30 restriction enzyme and DNA ligase

31 mammary gland cell and egg cell

4 Ecology

1. **a** a group of organisms that can interbreed and produce fertile offspring
   b the study of relationships between living organisms and between organisms and their environment
   c a group of organisms of the same species which live in the same area at the same time
   d a group of populations of different species living and interacting with each other in an area
   e a community and its abiotic environment

2. An autotroph makes its own food/organic compounds; a heterotroph takes in organic materials as food.

3. A consumer ingests other organic matter that is living or recently dead; a detritivore is an organism that ingests non-living organic matter; a saprotroph lives on or in non-living organic matter and absorbs digested food.

4. **a** detritivore
   b carnivore
   c herbivore

5. **a** trophic level 3/secondary consumer
   b trophic level 4/tertiary consumer

6. the transfer of nutrients and energy

7. a series of interlinking food chains

8. the Sun

9. the position of an organism in a food chain

10. not consumed, heat, excretion

11. $2000 \text{ J m}^{-2} \text{ y}^{-1}$

12. Energy enters and leaves the ecosystem but nutrients cycle within the ecosystem

13. Peat is produced in wetlands from accumulation of partly decayed vegetation in the absence of oxygen. Coal is produced as a result of compression and fossilisation of organic material over millions of years.

14. any three of: plants; animals; bacteria; fungi

15. photosynthesis

16. any three of: carbon dioxide; water vapour; methane; oxides of nitrogen; fluorocarbons

17. burning fossil fuels in factories and vehicles; rice farming; cattle farming; rainforest destruction; HFCs from refrigerants

18. changing the temperature of the oceans, acidification of the ocean

5 Evolution and biodiversity

1. a cumulative change in the heritable characteristics of a population

2. anatomical features showing similarities in shape (but not necessarily function) in different organisms

3. Individuals with specific characteristics are selected to breed, so that more of the next generation have these useful features than if the parents had not been artificially selected.

4. It ensures that offspring are varied and have different combinations of characteristics from one another and their parents.

5. competition

6. survival of the fittest

7. for example: resistance to antibiotics in bacteria; change in phenotype of the peppered moth

8. domain; kingdom; phylum; class; order; family; genus; species

9. genus and species

10. conifers/Coniferophyta

11. Bryophyta

12. Arthropoda

13. Annelida

14. 'Large' is a subjective term and not suitable for use in a key.

15. group of organisms that have evolved from a common ancestor

16. Homologous structures have evolved from a common ancestral form and may have different function, while analogous structures are structurally very different even though they perform a similar function.

17. Cladistics is an objective method of classification; using it, existing classifications can be tested to determine if relationships are correct or whether they should be abandoned in favour of a competing hypothesis.

6 Human physiology

1. Contraction of the bands of circular muscle squeezes the intestine behind food, while longitudinal muscles relax and extend to receive the food.

2. Absorption is the uptake of nutrients into blood or cells, while assimilation is the use of absorbed molecules by the body.

3. Enzymes speed up chemical reactions so that food can be digested as it passes through the intestine.

4. Its microvilli provide a large surface area, it has a rich blood supply for absorption, and the diffusion distance between the intestine and the blood is very small.

5. vena cava, right atrium, atrioventricular valve, right ventricle, semilunar valve, pulmonary artery, lung capillaries, pulmonary vein, left atrium, atrioventricular valve, left ventricle, semilunar valve, aorta

6 It sets the pace of the heart beat.
7 Arteries must resist higher blood pressure than veins.
8 The left ventricle pumps blood all round the body, whereas the right ventricle only needs to drive blood the short distance to the lungs, so the left ventricle must be stronger and thicker.
9 They prevent blood flowing back from the ventricles to the atria as the heart relaxes.
10 to supply additional oxygen to working muscles and to carry away the carbon dioxide they produce
11 it stimulates the sinoatrial node to increase heart rate
12 an organism or a virus that causes disease
13 a protein that provokes an immune response
14 Antibiotics interfere with bacterial metabolism. Viruses do not have metabolism of their own.
15 Phagocytes engulf bacteria and cell fragments, whereas lymphocytes secrete antibodies.
16 atmosphere, nasal passages or mouth, trachea, bronchi, bronchioles, alveolus
17 rich blood supply; large surface area; short diffusion distance
18 diaphragm and intercostal muscles (external)
19 Gas exchange is the exchange of oxygen and carbon dioxide between the alveoli and the blood; ventilation is the movement of air into and out of the lungs.
20 Sodium ions are pumped out of the neuron and potassium ions are pumped in.
21 the depolarisation and repolarisation of a nerve cell membrane
22 calcium ions enter the pre-synaptic neuron, vesicles fuse with the pre-synaptic membrane, neurotransmitter is released into synaptic cleft, neurotransmitter binds to receptor on post-synaptic membrane, post-synaptic membrane is depolarised
23 An action potential must jump from node to node along a myelinated axon.
24 neurotransmitter
25 insulin and glucagon
26 leptin acts on receptor cells in the hypothalamus
27 Y chromosome
28 It stimulates the thickening of the endometrium and also inhibits the production of FSH and LH.

7 Nucleic acids

1 the two strands of DNA in the double helix run in opposite directions
2 consists of two twists of DNA wound around eight histone proteins, which are held in place by an additional histone
3 5' to 3'
4 because DNA replication can only occur in a 5' to 3' direction
5 to protect the ends of chromosomes from damage during cell division
6 a portion of the primary RNA transcript that codes for part of a polypeptide in eukaryotes (as compared with introns, which are part of the transcript but are not translated to form part of a polypeptide)
7 The sense strand has the same base sequence as mRNA (with thymine instead of uracil). The antisense strand is the one that is transcribed to produce mRNA.
8 The promoter region contains specific DNA sequences that are the initial binding site for RNA polymerase during transcription.
9 External environmental factors such as drugs, chemicals, temperature and light can cause some genes to be turned on or off and influence development, e.g. temperature affects the fur colour of Himalayan rabbits.
10 Transcription is the production of mRNA from DNA, while translation is the production of a polypeptide from mRNA.
11 It is used in the cell; it is not exported for use elsewhere.

8 Metabolism, cell respiration and photosynthesis

1 the amount of energy required to destabilise chemical bonds in a substrate so that a chemical reaction can occur
2 If the end product begins to accumulate because it is not being used, it inhibits an enzyme earlier in the pathway to switch off the assembly line. Usually, the first enzyme in a process is inhibited.
3 gain of electrons; loss of oxygen; gain of hydrogen
4 phosphate from ATP
5 lysis
6 in the matrix of mitochondria
7 on the membranes (cristae) of mitochondria
8 acetyl CoA
9 to provide a large surface area for the enzymes involved in respiration
10 the pH decreases
11 ATP synthase
12 in the stroma of the chloroplast
13 green
14 to boost electrons to a higher energy level
15 oxygen
16 ATP and $NADPH^+$
17 ribulose bisphosphate

9 Plant biology

1. stomata are found deep in grooves; rolled leaf; thick cuticles
2. mass flow of water carrying the ions, or by facilitated diffusion of ions
3.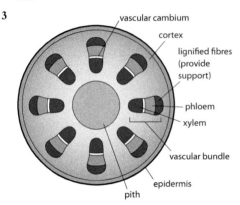
4. Xylem is composed of dead cells whose walls are thickened with lignin, whereas phloem is made of living cells and adjacent companion cells. Both xylem and phloem are found in vascular bundles in the stem and leaves. The xylem carries water and salts up the plant, whereas phloem carries sugars, vitamins and other nutrients both up and down.
5. water and mineral salts
6. Transpiration is increased by an increase in temperature, wind or light intensity and by a decrease in humidity.
7. Sugar enters the phloem in leaves and is followed by water. The pressure gradient created pushes water to other parts of the plant.
8. The presence of auxin on the shaded side of a shoot causes the shoot to bend towards a source of light as cells on the shaded side elongate.
9. Apical meristems occur at the end of a shoot and growth here results in a taller plant.
10. These pumps are special transport proteins in plasma membranes, which allow auxin to move in one direction only.
11. Germination requires activation of enzymes to digest food reserves and initiate respiration. All enzyme-controlled reactions are temperature sensitive and so a suitable temperature (which varies from species to species) is needed for germination.
12. Pollination is the transfer of pollen from anthers to stigma, whereas fertilisation is the fusion of the male and female gamete in the ovary.
13. anthers and filaments (together known as the stamen)
14. Seeds that germinate at some distance from the parent plant do not have to compete with it for water, light or nutrients and so are more likely to survive.

10 Genetics and evolution

1. During meiosis, non-sister chromatids may touch and break, and the two segments re-join at the corresponding position on the other chromatid.
2. anaphase 2
3. During metaphase 1, bivalents line up on the equator, but the way they line up is random for each chromosome.
4. Ttrr: Tr and tr in a ratio of 1 : 1
   HhGg: HG, Hg, hG and hg in a ratio of 1 : 1 : 1 : 1
5. Punnett grid
6. members of the F_1 generation
7. In sympatric speciation, a gene pool becomes divided without the population being geographically split. In allopatric speciation, a physical barrier separates a species into two geographically isolated populations.
8. leads to the prevalence of one form from a varied population
9. all the different genes in an interbreeding population at a given time

11 Animal physiology

1. Natural immunity or resistance to a particular pathogen is related to the genetics of the host organism, which differ between species. For a virus to infect a new species, genetic adaptations must occur within the pathogen.
2. Antibody is a substance produced by lymphocytes to counteract an antigen. Antigen is a protein that stimulates an immune response.
3. It has characteristics of both B-cells and cancer cells: it produces antibodies *and* cells are immortal if kept in culture medium.
4. It stimulates the immune response and the formation of memory cells to produce antibodies very quickly if the person is infected with the real pathogen later on.
5. One member of the pair causes flexing of a joint and the other causes extension of it.
6. Cartilage reduces friction between two bones in a joint and synovial fluid acts as a lubricant.
7. A hinge joint (e.g. the elbow joint) moves in one direction, whereas a ball-and-socket joint can allow movement in a circle.
8. Actin is made up of thin fibres, whereas myosin fibres are thicker. Each type of filament is arranged in bands that alternate in skeletal muscle. The lighter bands of thin myosin alternate with darker bands of thicker myosin, which produces the striped appearance.
9. ATP provides the energy for muscle contraction. ATP molecules bind to myosin heads; the energy from the ATP breaks cross-bridges and detaches the myosin heads from the actin filaments.
10. Excretion is the removal from the body of waste products produced as a result of metabolic activity. Osmoregulation is the maintenance of the correct level of water in the body.

11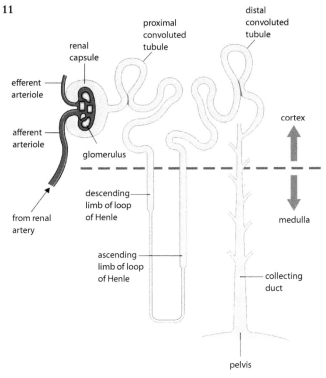

12 The descending limb of the loop of Henle is permeable to sodium and chloride ions, which leave the tubule and are concentrated in the medulla region of the kidney. The high level of salt in the medulla causes water to enter the capillaries in the medulla from the descending limb, increasing the concentration of the solution in the loop of Henle.

13 The afferent arteriole is wider in diameter than the efferent arteriole. This causes a build-up of pressure in the glomerulus that enables ultrafiltration of molecules.

14 Microvilli increase the surface area of the cells in the proximal convoluted tubule, which is the region where useful molecules are reabsorbed into the bloodstream from the kidney tubule. The large surface area ensures this process occurs efficiently.

15 capillary endothelium, basement membrane, cell of Bowman's capsule wall

16 ADH influences the permeability to water of the cell wall in the collecting duct. If ADH is released, the collecting duct is more permeable and water is taken back into the blood so that more concentrated urine is produced.

17 A diabetic person cannot regulate the amount of glucose in their blood. If blood glucose is very high, not all glucose can be reabsorbed from the kidney tubule and some passes out of the body in the urine.

18 Testosterone stimulates secondary spermatocytes to mature into sperm cells.

19 An ovum must contain sufficient nutrients to provide fuel for the first few divisions after fertilisation. Unequal cell division during oogenesis produces one large cell that contains these nutrients and three small polar bodies that do not become gametes.

20 Spermatogenesis produces four gametes, begins at puberty and continues throughout a man's life and produces sperm continuously, whereas oogenesis produces one gamete, occurs from puberty to menopause and produces one gamete each month.

21 The amnion forms a protective, fluid-filled sac around the fetus that protects it from pressure and allows it to move.

22 Oxytocin stimulates the contraction of the muscles in the wall of the uterus. It works by positive feedback so that more hormone is produced as the muscles contract more strongly.

Glossary

Terms in **_bold italic_** refer to keywords from online material.

abiotic factor   aspect of the environment that is not living – for example, humidity, temperature, salinity, wind, soil particles

absorption   the taking up of a substance by a tissue; for example, the process by which nutrients pass from the lumen of the intestine into the lymph or blood, or by which the intensity of a beam of light energy is reduced as it passes through a leaf

absorption spectrum   the range of wavelengths of light that a pigment is able to absorb

acinus   berry-shaped termination of an exocrine gland where its secretion is produced

actin   globular protein that polymerises to form the thin filaments in muscle myofibrils

action potential   a rapid wave of depolarisation at a cell surface causing an impulse in a neuron

activation energy   the energy a substrate molecule must achieve before it can change chemically

active immunity   immunity due to the production of antibodies by an organism in response to an antigen

active site   the region on the surface of an enzyme to which the substrate molecule binds

active transport   transport of a substance across a membrane against the concentration gradient, involving a carrier protein and energy expenditure

addiction   regular or poorly controlled use of a psychoactive substance, despite adverse physical, psychological or social consequences, often with the development of physiological tolerance and withdrawal symptoms

adhesion   a force that attracts water molecules to a surface by hydrogen bonding

aerobic respiration   respiration that requires oxygen, and produces carbon dioxide and water from the oxidation of glucose

aggregate   a group of distinct organisms, such as bacteria, living together to form a colony that has characteristics not displayed by the individual organisms

aleurone layer   layer of specialised cells between the seed coat and endosperm of grass seeds, which produces enzymes that help to digest the food reserves for the developing embryo

algal bloom   rapid increase or accumulation in the population of algae in an aquatic ecosystem often due to the run-off of agricultural fertiliser

alimentary canal   the gut – a tube that runs from the mouth to the anus of a vertebrate in which food is digested and absorbed

allele   an alternative form of a gene found at a specific locus on a chromosome

allele frequency   the commonness of occurrence of any particular allele in a population

allergy   a disorder of the immune system in which the immune system overreacts to a substance (known as an allergen) which is usually harmless

allopatric speciation   separating a species into two geographically isolated groups by a physical barrier so eventually they can no longer interbreed

allosteric   regulation of an enzyme by a molecule that does not have the same structure as the substrate, binding at a site away from the active site

allosteric site   the region on the surface of an enzyme to which an allosteric effector binds

altruist   an organism that performs actions benefitting other individuals rather than itself

alveolus (alveoli)   an air sac in the lungs

amino acid   one of the building blocks of protein; each amino acid has an amine group $-NH_2$ and an acid carboxyl group $-COOH$, and the general formula NH_2-CHR-COOH, where R is one of 20 side groups

amnion   a membrane which surrounds the embryo of reptiles, birds and mammals during their development

amniotic fluid   the fluid contained within the amnion in which the embryo is suspended.

amphipathic   an unsymmetrical molecule which has one hydrophilic end and one hydrophobic end

amylase   an enzyme that digests starch

anabolism   an anabolic reaction is one in which large molecules are built up from small ones

anaerobic respiration   respiration that occurs in the absence of oxygen in which glucose is broken down to lactic acid or to ethanol and carbon dioxide

analgesic   a drug or other treatment that relieves or reduces pain

analogous structures   structures with similar functions but which have different evolutionary origins

anaphase   the stage in cell division in which homologous chromosomes (in meiosis I) or chromatids (meiosis II and mitosis) separate and go to opposite poles

anesthetic   a drug or other treatment that produces a loss of feeling or sensation

anorexia nervosa   severe malnutrition brought on by a psychological condition that causes an aversion to food

antagonistic pair   a pair of muscles that work in opposition to each other at a joint

anther   pollen-bearing structures on the stamens of flowers

antibiotic   organic compound produced by microorganisms to kill or inhibit other microorganisms

antibody   one of millions of blood proteins produced by plasma cells in response to specific antigens, which are then neutralised or destroyed

anticodon   a triplet of bases in tRNA that pair with a complementary triplet (codon) in mRNA

antidiuretic hormone (ADH)   hormone secreted by the posterior pituitary gland to control the water permeability of the collecting ducts in the kidney (also called vasopressin)

antigen a substance that stimulates the production of antibody

antigen presentation the process by which a cell takes in a pathogen, or molecules from it, and holds them in its plasma membrane where they may be encountered by a lymphocyte

antiparallel running in opposite direction; the two polynucleotide strands in a DNA molecule are antiparallel

antisense strand the strand of DNA that is transcribed during transcription

apical meristem the actively growing tissue at the tip of a stem or root

apoptosis the natural destruction of certain cells in a growing organism

aquaporins proteins embedded in a plasma membrane that regulate the movement of water through the membrane

arteriole a small artery

artery a muscular blood vessel that carries blood away from the heart under high pressure

asexual reproduction reproduction that does not involve gametes or fertilisation

assimilation uptake and use of nutrients by cells and tissues

association areas parts of the cerebrum that integrate information from the primary sensory areas and other parts of the brain

atherosclerosis hardening and loss of elasticity in the arteries

ATP (adenosine triphosphate) a universal energy storage nucleotide formed in photosynthesis and respiration from ADP and P_i; when it is formed, energy is stored; when it is broken down, energy is released

ATP synthase/synthetase an enzyme that catalyses the production of ATP

atrioventricular node (AVN) small patch of tissue in the right atrium which acts as a pacemaker in the heart

atrioventricular valve valve between the atrium and the ventricle

atrium (atria) one of the two upper chambers of the heart that receive blood from veins

autonomic nervous system (ANS) the part of the nervous system comprised of the sympathetic and parasympathetic nervous systems, consisting only of motor neurons; it controls involuntary functions (such as activities of the glands and digestive system, and blood flow) without our awareness

autosome any chromosome other than a sex chromosome

autotroph organism able to make its own food from simple inorganic materials and an energy source

auxin plant growth substance (also known as indoleacetic acid, IAA)

axon a cytoplasmic process that transmits action potentials away from the cell body of a neuron

B-cell a type of lymphocyte involved in the immune response; B-cells develop into plasma cells, which secrete antibody

ball-and-socket joint a joint such as the hip, which permits movement in more than one plane

basal metabolic rate the amount of energy used by the body per day when at rest

basement membrane thin layer of cells which separates epithelium from underlying tissue

batch culture large-scale culture of microorganisms in a closed fermenter, to which nothing is added during the fermentation; cells initially grow at a maximum rate, but as nutrients are used up and waste products accumulate, growth slows and eventually stops, at which time the product is harvested, the fermenter cleaned, and a fermentation is set up

behaviour the manner in which an organisms acts under specified conditions or circumstances, or in relation to other things

binary fission cell division producing two genetically identical daughter cells, as in prokaryote reproduction

binomial the two-part name that gives the genus and the species of an organism in Latin – for example, *Homo sapiens*

biochemical oxygen demand (BOD) the uptake rate of dissolved oxygen by the biological organisms in a body of water

biofilm colony of bacteria living in a layer on a surface, in a matrix of extracellular polysaccharides (EPS), which is resistant to removal by physical or antimicrobial agents, and which has emergent properties including quorum sensing

biofuel any fuel derived from biomass e.g. bioethanol, biodiesel

biogas fuel gas obtained typically by the bacterial decomposition of animal or plant waste in the absence of oxygen, which produces mainly methane

biogeography the spatial distribution of organisms, species and ecosystems, or the study of these

bioinformatics the use of databases – searchable, up-to-date electronic records of information from experiments, scientific papers and computer analyses of data, such as information about gene function, structure and location

biomagnification the process by which chemical substances accumulate at progressively more concentrated levels at each trophic level in a food chain

biomarker any measurable characteristic that may be used as an indicator of some biological state or condition

biomass the total dry mass of organisms in a given area

biopharming the use of genetic engineering to add genes to animals or plants so that they produce useful pharmaceuticals

bioreactor a vessel in which microorganisms can be cultured in order to harvest useful products

bioremediation the process by which microorganisms are used to treat areas of land or sea that have been contaminated by pesticides, oil or solvents, in order to remove the contaminants and restore the environment

biotechnology commercial application of biology, particularly the industrial use of microorganisms, enzymes and genetic engineering

biotic index a measure of the biodiversity, or level of pollution, of a given ecosystem or habitat obtained by surveying the types of organisms present in it

bipolar cells cells in the retina that transmit signals between the photoreceptors (rod and cone cells) and ganglion cells

blastocyst early stage in embryo development – a hollow ball of cells, which implants into the endometrium

Bohr shift the reduction in hemoglobin's affinity for oxygen caused by an increase in the partial pressure of carbon dioxide, which means the dissociation curve shifts to the right

Bowman's capsule the cup-shaped part at the beginning of a nephron, into which ultrafiltration takes place (also known as renal capsule)

Broca's area a part of the left cerebral hemisphere involved in speaking and writing

bronchiole a small branch of a bronchus

bronchus (bronchi) one of the major tubes which branch from the trachea

buffering keeping pH at a nearly constant value

Calvin cycle cycle of light-independent reactions in the stroma of the chloroplast in which carbon dioxide reacts with RuBP, producing GP, TP and regenerating RuBP

cambium a cylinder of cells in plant stems and roots – the cells are able to divide and produce new cells so that the stem or root grows wider

capillary tiny, thin-walled blood vessels that form a network which links arteries and veins

captive-breeding programme the process of breeding threatened species in controlled *ex situ* conservation environments, such as wildlife reserves or zoos, often with a view to releasing organisms back into the wild when appropriate

carbohydrate organic compound usually containing only C, H and O – examples include sugars, starch, cellulose and glycogen

carbon fixation the conversion of carbon atoms into a combined organic form

carcinogen a substance that causes the transformation of cells to form malignant tumours (cancer)

cardiac arrest when the heart muscle stops contracting so that normal circulation of the blood is prevented

cardiac cycle sequence of events in the heart during one complete heart beat

cardiac muscle specialised muscle found only in the heart

carrier an individual who has one copy of a recessive allele that causes a genetic condition in individuals who are homozygous for the allele

carrying capacity the maximum population size that can be sustained within an area

catabolism the breaking down of complex molecules in the biochemistry of cells

catalyst a substance that changes the rate of a chemical reaction, by lowering the activation energy, but remains unchanged at the end; enzymes are biological catalysts

cell cycle the sequence of events that takes place from one cell division until the next; it is made up of interphase, mitosis and cytokinesis

cell respiration controlled release of energy from organic compounds in cells to form ATP

cell theory the theory that organisms consist of cells and that all cells come from pre-existing cells

cell wall firm structure that surrounds the plasma membrane of cells of plants, fungi and bacteria; it gives cells their shape and limits their expansion

cellulose a polymer of glucose; it is the primary constituent of most plant cell walls

central nervous system (CNS) the brain and spinal cord of a vertebrate

centriole an organelle in an animal cell that forms and organises the spindle microtubules in cell division

centromere the region where sister chromatids are joined and where the spindle microtubule attaches during cell division

cerebellum part of the hind brain concerned with posture, movement and muscle tone

cerebral cortex a highly folded layer of nerve cell bodies that forms the surface of the cerebrum

cerebral hemispheres (cerebrum) the main part of the human brain, the coordinating system of the nervous system consisting of two hemispheres made of neurons and nerve fibres

channel protein proteins found in a membrane which form a pathway through which specific ions or molecules can pass in and out of a cell

chemiosmosis the passive flow of protons down a concentration gradient from the intermembrane space of mitochondria or from the thylakoid interior of chloroplasts through a protein channel

chemoautotroph an organism that uses energy from chemical reactions to generate ATP and produce organic compounds from inorganic raw materials

chemoheterotroph an organism that uses energy from chemical reactions to generate ATP and obtains organic compounds from other organisms

chemoreceptor sensory cell that responds to pH or the concentration of a chemical such as carbon dioxide

chiasma (chiasmata) point of crossing over between homologous chromosomes in prophase 1 of meiosis

chloride shift process in which bicarbonate (HCO_3^-) and chloride (Cl^-) are exchanged across the membrane of red blood cells

cholesterol a lipid formed in the liver and carried in the blood as lipoprotein; it is a precursor of steroid hormones, and is also found in the plasma membrane of animal cells

chlorophyll the most important photosynthetic pigments of green plants, found in the grana of chloroplasts and responsible for trapping light energy (some bacteria have a chemically different form called bacteriochlorophyll)

chloroplast organelle found in some plant cells that is the site of photosynthesis

chorion outermost membrane that protects mammal embryos; it forms part of the placenta

chromatid one of the two copies of a chromosome after it has replicated and before the centromeres separate at anaphase

chromosome in eukaryotes, a structure consisting of a long thread of DNA and protein that carries the genetic information of the cell; in bacteria, the DNA molecule that contains the genetic information of the cell

circadian rhythm a cycle of activity or physiology which is repeated, usually every 24 hours

cirrhosis disease caused by excessive alcohol consumption, in which liver tissue is replaced by fibrous scar tissue leading to loss of liver function

cisternae space enclosed by the membranes of the endoplasmic reticulum

clade all the organisms both living and fossil descended from a particular common ancestor

cladistics a way of classifying organisms which uses lines of descent rather than physical (phenotypic) similarities

cladogram diagrammatic representation of cladistic relationships

classical conditioning a type of learning in which an animal learns to respond to a stimulus that is different from the one that normally elicits that response

classification the system of arranging species into categories based on physical and evolutionary relationships

climax community the stable stage at the end of a succession of communities in an ecosystem

clonal selection the way in which exposure to antigen results in activation of selected T-cell or B-cell clones producing an immune response

clone genetically identical cells or organisms produced from a common ancestor by asexual reproduction

cochlea spiral shaped tube in the inner ear, which contains sensory cells involved with hearing

codominant alleles pairs of alleles that both affect the phenotype when present in a heterozygous state

codon a triplet of three nucleotides in mRNA that specify the position of an amino acid in a polypeptide

cohesion forces of attraction between water molecules, which enable them to stick together in a lattice

community a group of populations of organisms living and interacting within a habitat

companion cell specialised cell found adjacent to a sieve tube cell in the stem of flowering plants

competition the requirement of a limited resource by two or more organisms in the same place at the same time

competitive exclusion the principle that two species competing for the same resources cannot coexist indefinitely – if one species has even a slight competitive advantage, then it will come to exclude the other in the long term

competitive inhibitor substance similar to the substrate of an enzyme, which binds to the active site and inhibits a reaction

complementary base pairing pairing of bases A–T and G–C in double stranded DNA, and of A–U and C–G between DNA and RNA during transcription, and between tRNA and mRNA during translation

concentration gradient a difference in concentration of a substance between one area and another

condensation reaction a reaction in which two molecules become bonded by a covalent bond and a molecule of water is released

cone cells photoreceptors in the retina, which are not very sensitive to light; however, three different types of cone are sensitive to three different wavelengths of light, which enables us to see in colour

conjugated protein protein combined with a non-protein group

consumer an organism that feeds on another organism

continuous culture large-scale culturing of microorganisms in a fermenter to which nutrients are supplied continuously at a steady rate, so that the cells are kept in an exponential growth phase; an uninterrupted supply of product is harvested and equipment can be used without breaks for cleaning and recharging

continuous variation variation in a species which is controlled by several genes producing features that cannot be divided into discrete categories e.g. human height

contralateral processing process in which the right brain processes information from the left visual field and vice versa

coronary thrombosis a blockage in one of the arteries that supplies the heart

corpus callosum a band of tissue containing neurons, which connects the two sides of the cerebrum

corpus luteum mass of cells which develops from an ovarian follicle after the release of the oocyte

cortical reaction release of cortical granules at fertilisation so that the plasma membrane of the oocyte is changed to prevent polyspermy

cotyledon a 'seed leaf' found in a plant embryo which stores food reserves and may become the first leaf when a seed germinates

covalent bond a bond between atoms in which electrons are shared

cristae folds of the inner membrane of mitochondria, where oxidative phosphorylation takes place

critical period the limited time period in which a young organism is able to learn a particular skill, given the appropriate stimulus

crossing over exchange of genetic material between homologous chromosomes during meiosis

cytokinesis division of cytoplasm after the nucleus has divided

cytoplasm contents of a cell enclosed by the plasma membrane, not including the nucleus

decarboxylation reaction removal of carbon dioxide during a chemical reaction

decomposer organism that feeds on dead plant and animal matter so that it can be recycled; most decomposers are microorganisms

deficiency when a person does not have enough of a particular nutrient and suffers health problems as a result

denaturation a change in the structure of a protein that results in a loss (usually permanent) of its function

dendrite a short cytoplasmic process of a neuron, which conducts action potentials towards the cell body

denitrification metabolic activity of some soil bacteria by which nitrogen containing ions are reduced to nitrogen gas

deoxynucleoside triphosphate (dNTP) a building block for DNA – deoxyribose, three phosphate groups and one of the four bases

depolarise to temporarily reverse the membrane potential of an axon as an impulse is transmitted

detoxification the removal of toxic substances from the body of an organism

detritivore an organism that feeds on dead organic matter

diabetes a condition in which the blood sugar level is not maintained within normal limits either because the body does not produce sufficient insulin or because cells do not respond to insulin properly

diaphragm sheet of fibrous and muscular tissue that separates the thorax from the abdomen in mammals

diastole relaxation of the heart muscle during the cardiac cycle

dichotomous key a key in which organisms are separated into pairs of smaller and smaller groups by observation of their characteristics

dietary fibre the indigestible parts of plant-based foods such as cereals, fruits and vegetables, which pass relatively unchanged through alimentary canal, and are important in a healthy diet: soluble fibre lowers LDL ('bad') cholesterol, while insoluble fibre helps to reduce the risk of colon cancer

differentiation a process by which originally similar cells follow different developmental pathways because particular genes are activated and others are switched off

diffusion passive, random movement of molecules (or other particles) from an area of high concentration to an area of lower concentration

digestion enzyme-catalysed process by which larger molecules are hydrolysed to smaller, soluble molecules

dipeptide molecule composed of two amino acids joined by a peptide bond

diploid organisms whose cells have two sets of chromosomes in their nuclei

directional selection natural selection that favours a single phenotype and causes a change in allele frequency in one direction

disaccharide molecule composed of two monosaccharide molecules linked in a condensation reaction

discrete (or discontinuous) variation variation in characteristics so that types can be grouped into specific categories with no intermediate phenotypes

disruptive selection natural selection that leads to the formation of two separate phenotypes

DNA (deoxyribonucleic acid) the fundamental heritable material, contained in the nucleus in eukaryotes; DNA consists of two strands of nucleotide subunits containing the bases adenine, thymine, guanine and cytosine

DNA gyrase an enzyme that relieves stress in a DNA molecule by adding negative supercoils as it is being unwound

DNA helicase an enzyme that unwinds and separates the two strands of a DNA molecule

DNA ligase enzyme that links Okazaki fragments (strands of DNA) during DNA replication; it is also used in biotechnology to join sticky ends of double-stranded DNA fragments

DNA microarray a glass or silicon chip with DNA probes attached to its surface, which allows scientists to carry out thousands of genetic tests at the same time – each spot is used to hybridise a different target sequence of DNA

DNA polymerases a group of enzymes that catalyse the formation of DNA strands from a DNA template

dogma a principle or belief that is not subject to scientific testing

domain the highest rank of taxonomy of organisms; there are three domains – Archaea, Eubacteria and Eukaryota

dominant allele an allele that has the same effect on a phenotype whether it is present in the homozygous or heterozygous condition

ear drum a thin membrane (also called the tympanic membrane) that separates the external ear from the middle ear; it transmits sound waves from the air to the ossicles inside the middle ear, helping to amplify the vibrations

ecology the study of the relationships between living organisms and their environment, including both the physical environment and the other organisms that live in it

ecosystem the organisms of a particular habitat together with the physical environment in which they live – for example, a tropical rain forest

ectoderm the outer of the three layers of cells in a very early animal embryo, which differentiates to form the nervous system, tooth enamel, the linings of the mouth, anus, nostrils and sweat glands, and the hair and nails

edema the build-up of tissue fluid in the tissues, caused by poor drainage through the lymphatic system

effector an organ or cell that responds to a stimulus – for example, a muscle that contracts or a gland that produces a secretion

egestion removal of waste from the body during defecation

electron transport chain (ETC) a series of carriers that transfer electrons along a redox pathway, enabling the synthesis of ATP during respiration

electrophoresis separating components in a mixture of chemicals – for example, lengths of DNA – by means of an electric field

embryo the earliest stage of development of a young animal or plant that is still contained in a protective structure such as a seed, egg or uterus

embryonic stem cells cells derived from an embryo that retain the potential to differentiate into any other cell of the organism

emergent property a property of a complex system that arises from the relatively simple interactions of its individual component parts

emigration deliberate departure of an organism from the environment in which it has been living

emphysema a disorder caused by the breakdown of the alveolar walls, making it difficult to obtain sufficient oxygen in the blood

endemic a species that is found naturally in only one geographic area

end-product inhibition control of a metabolic pathway in which a product within or at the end of the pathway inhibits an enzyme found earlier in the pathway

endocrine gland a ductless hormone-producing gland that secretes its products into the bloodstream

endocytosis the movement of bulk liquids or solids into a cell, by the indentation of the plasma membrane to form vesicles containing the substance; endocytosis is an active process requiring ATP

endoderm the inner of the three layers of cells in a very early animal embryo, which forms the epithelial lining of multiple systems

endometrium lining of the uterus of a mammal

endoparasite a parasite, such as a tapeworm, that lives inside another organism

endoplasmic reticulum a folded system of membranes within the cytoplasm of a eukaryotic cell; may be smooth (sER), or rough (rER) if ribosomes are attached

endosperm the food reserves found in a seed of a monocotyledonous flowering plant

endosymbiosis an organism that lives inside the cell or body of another organism; also, a theory which proposes that mitochondria and chloroplasts evolved from endosymbiotic bacteria

energy pyramid   diagram that shows the total energy content at different trophic levels in an ecosystem

enzyme   a protein that functions as a biological catalyst

epidemiology   the study of the occurrence, distribution and control of disease

epididymis   coiled tubules in the testes that store sperm cells and carry them from the seminiferous tubules to the vas deferens

epigenetics   changes in gene expression or phenotype not caused by changes in the basic DNA sequence

epinephrine (adrenalin)   the 'flight or fight' hormone produced by the medulla of the adrenal gland

erythrocyte   a red blood cell

estrogen   female sex hormone produced in the ovaries

ethology   the study of the behaviour of an animal in its environment

eukaryotic   an organism whose cells contain a membrane-bound nucleus

Eustachian tube   a tube about 35 mm long linking the middle ear to the back of the nose and throat

eutrophication   an increase in the concentration of nutrients such as nitrate in an aquatic environment so that primary production increases, which can lead to depletion of oxygen

evolution   the cumulative change in the heritable characteristics of a population

ex situ *conservation*   conservation of species away from their natural habitat, e.g. in a zoo breeding programme

excitatory synapse   synapse in which, when a neurotransmitter is released from the pre-synaptic membrane, the post-synaptic membrane is depolarised as positive ions enter the cell and stimulate an action potential

excretion   the removal of waste products of metabolism (such as urea or carbon dioxide) from the body

exhalation   breathing out

exocrine gland   a gland whose secretion is released via a duct

exocytosis   the movement of bulk liquids or solids out of a cell by the fusion of vesicles containing the substance with the plasma membrane; exocytosis is an active process requiring ATP

exon   portion of the primary RNA transcript that codes for part of a polypeptide in eukaryotes (compare with ***intron***)

exoparasite   organism that lives as a parasite on the outside of the body of another organism e.g. fleas, lice

exoskeleton   inextensible outer casing of arthropods, providing support for locomotion, as well as protection

expiratory reserve   the additional volume of air which can be expelled from the lungs after normal exhalation by additional effort

expressed sequence tag   a short nucleotide sequence from one or both ends of a known gene, which computers can use to scan databases to help make a match with unknown genes and to map their positions within a genome, assisting in the construction of genome maps

F_1 generation   the immediate offspring of a mating; the first filial generation

F_2 generation   the immediate offspring of a mating between members of the F_1 generation

facilitated diffusion   diffusion across a membrane through specific protein channels in the membrane, with no energy cost

fatty acid   molecule which has a long hydrocarbon tail and a carboxyl group

feces   waste that leaves the digestive system by egestion

fenestration   space between the cells of a capillary, which enable materials to pass through

fermenter   an enclosed vessel in which microorganisms are cultured; industrial fermenters are usually very large and made of stainless steel

fertilisation   the fusion of male and female gametes to form a zygote

fetus   a human embryo from 7 weeks after fertilisation

fibrin   a plasma protein that polymerises to form long threads, which provide the structure of a blood clot

fibrinogen   the soluble plasma protein, produced by the liver, which is converted to fibrin during the blood-clotting process

filament   a long protein molecule in a myofibril; thin filaments are made of actin and thick filaments are made of myosin

flagellum   a long, thin structure used to propel unicellular organisms; the structure of a flagellum in a prokaryote is different from that in a eukaryote

fluid mosaic model   the generally accepted model of the structure of a membrane that includes a phospholipid bilayer in which proteins are embedded or attached to the surface

fluxomics   the study of the flow of fluids and molecules within cells

follicle-stimulating hormone (FSH)   hormone produced by the anterior pituitary gland that stimulates the growth of follicles at the start of the menstrual cycle

food chain   a sequence of organisms in a habitat, beginning with a producer, in which each obtains nutrients by eating the organism preceding it

food web   a series of interconnected food chains

fovea   the point on the retina, directly behind the pupil, where vision is most distinct; contains only cone cells

fundamental niche   the potential mode of existence of a species, given its adaptations

G_1 and G_2   phases of the cell cycle

gametes   haploid sex cells – for example, sperm, ovum or pollen

ganglion cell   a type of neuron in the retina of the eye, which receives impulses from the photoreceptors via bipolar cells and transmits them to the brain via the optic nerve

gas exchange   the exchange of oxygen and carbon dioxide between an organism or its cells and the environment

gene   a heritable factor that controls a specific characteristic

gene cluster   two or more genes that code for the same or similar products

gene mutation   a change in the base sequence of a gene that may or may not result in a change in the characteristics of an organism or cell

gene pool   all the genes and their alleles present in a breeding population

genetic modification   introduction or alteration of genes, often using genes from a different species, in order to modify an organism's characteristics

genetic predisposition   a susceptibility to a particular disease due to genetic factors, which makes it more likely that the person will suffer the disease than a person without the predisposition, given the same environmental conditions

genome   the complete genetic information of an organism or an individual cell

genomics   the study of the genomes of organisms

genotype   the exact genetic constitution of an individual feature of an organism; the alleles of an organism

germination   the growth of an embryo plant using stored food in a seed

gibberellin   plant growth substance important in elongation of stems and in seed germination

global warming   the hypothesis that the average temperature of the Earth is increasing due to an increase in carbon dioxide and other greenhouse gases in the atmosphere

glomerulus   network of capillaries within the Bowman's (renal) capsule

glucagon   hormone released by the pancreas, which stimulates the breakdown of glycogen in the liver and brings about an increase in blood glucose

glycogen   polymer of glucose used as a storage carbohydrate in liver and muscle tissue

glycolysis   the first (anaerobic) stage of respiration during which glucose is converted to pyruvate

goitre   deficiency disease leading to enlargement of the thyroid gland and mental impairment; caused by a lack of iodine

Golgi apparatus   a series of flattened membranes in the cytoplasm important in the modification of proteins produced by the rER

gradualism   a theory of evolution which suggests that most speciation occurs in a slow and gradual way

granum (grana)   layers of discs of membranes (thylakoids) in the chloroplast, which contain photosynthetic pigments; the site of the light-dependent reactions of photosynthesis

gross primary production   the total amount of energy that flows through the producers in an ecosystem as they make carbohydrates during photosynthesis; it is measured in kilojoules of energy per square metre per year (kJ m^{-2} y^{-1})

guard cell   one of a pair of modified epidermis cells that control the opening of a stoma

habitat   the locality or surroundings in which an organism usually lives

halophile   an organism that lives in an environment with a high salt concentration

haploid   cells containing one set of chromosomes

helicase   (see DNA *helicase*)

helper T-cells   cells that activate B cells and other T cells in the immune response; target of the HIV virus

hemodialysis   a method of treating kidney failure by filtering blood through a machine which removes urea and other unwanted substances

hemoglobin   protein found in red blood cells that combines with and carries oxygen from areas of high partial pressure to areas of low partial pressure

hepatocyte   liver cell

herbivory   the consumption of plant tissue for the purposes of nutrition

heterotroph   an organism that feeds on organic molecules

heterozygous   having two different copies of an allele of a gene

hinge joint   a joint such as the knee, which permits movement in one plane

histamine   a molecule released during allergic reactions which causes smooth muscles to contract and capillaries to dilate

histone   one of a group of basic proteins that form nucleosomes and act as scaffolding for DNA

homeostasis   maintenance of a constant internal environment

homologous chromosomes   chromosomes in a diploid cell that contain the same sequences of genes but which are derived from different parents

homologous structures   similar structures due to common ancestry

homozygous   having two identical copies of an allele

hormone   a chemical substance produced by an endocrine gland, which is transported in the blood and which affects the physiology or biochemistry of specific target cells

host   an organism that harbours a parasite and provides it with nourishment

human chorionic gonadotrophin (HCG)   hormone released by an implanted blastocyst, which maintains the corpus luteum so that progesterone is produced during early pregnancy; used in pregnancy testing kits

hybrid   the offspring of a cross between genetically dissimilar parents

hybridoma   a cell formed by the fusion of a plasma cell and a cancer cell; it can both secrete antibodies and divide to form other cells like itself

hydrocarbon chain   a linear arrangement of carbon atoms combined with hydrogen atoms, forming a hydrophobic tail to many large organic molecules

hydrogen bond   weak bond found in biological macromolecules in large numbers, formed by the attraction of a small positive charge on a hydrogen atom and a slight negative charge on an oxygen or nitrogen atom

hydrolysis reactions   reaction in which hydrogen and hydroxyl ions (water) are added to a large molecule to cause it to split into smaller molecules

hydrophilic   water loving

hydrophobic   water hating

hydroponic culture   a way of growing plants with their roots submerged in nutrient solutions, which provide all the inorganic ions they need, rather than in soil

hyperglycemia   an excess of glucose in the blood

hypertension   high blood pressure

hypertonic   a more concentrated solution (one with a less negative water potential) than the cell solution

hypothalamus   control centre of the autonomic nervous system and site of release of releasing factors for the pituitary hormones; it controls water balance, temperature regulation and metabolism

hypotonic a less concentrated solution (one with a more negative water potential) than the cell solution

hypothesis a testable explanation of an observed event or phenomenon

imbibition the process by which a seed takes in water to bind to dry starch and protein

immigration the movement of organisms into an area, augmenting the population there, or beginning a new population

immune response the production of antibodies in response to the presence of antigen

immune system the organs and cells in the body that help to destroy pathogens

immunity resistance to the onset of a disease after infection by the agent causing the disease

imprinting a type of learning in which, during a particular critical period early in its development, an organism comes to treat a particular object or organism as something to which it should remain near, as though it were its mother

in situ conservation conservation of a species in its natural habitat

indicator species species that, by their presence, abundance or lack of abundance, demonstrate some aspect of a habitat

inflammatory response the series of events that take place when tissues are injured by bacteria, toxins or other trauma

ingestion taking in food, eating

inhalation taking in air, breathing in

inhibitor a molecule that slows or blocks enzyme action, either by competition for the active site (competitive inhibitor) or by binding to another part of the enzyme (non-competitive inhibitor)

inhibitory synapse synapse at which the pre-synaptic neuron releases neurotransmitters that cause the hyperpolarisation of the post-synaptic membrane, therefore making it harder to trigger an action potential

innate behaviour behaviour that is genetically determined and common to all members of a species

inorganic compounds of mineral, not biological, origin

inspiratory reserve the additional volume of air which can be inhaled after a normal inhalation

insulin hormone produced by the pancreas that lowers blood glucose levels

intercostal muscles muscles between the ribs capable of raising and lowering the rib cage during breathing

interphase the period between successive nuclear divisions when the chromosomes are extended; the period when the cell is actively transcribing and translating genetic material and carrying out other biochemical proc

intertidal zone the area of the seashore between the low tide and high tide marks

intraspecific competition competition between members of the same species (as opposed to interspecific competition, which occurs between different species)

intron a non-coding sequence of nucleotides in primary RNA, in eukaryotes (compare with *exon*)

invasive species an alien species that grows rapidly in its new, non-native habitat, becoming a threat to endemic (native) species, which it out-competes and may eventually eliminate

ionic bond a bond formed by the attraction between oppositely charged ions (for example, NH^{3+} and COO^-), important in tertiary and quaternary protein structure

isomers chemical compounds of the same chemical formula but different structural formulae

isotonic being of the same osmotic concentration and therefore of the same water potential

isotope different isotopes of an element contain the same number of protons in the nucleus of each atom but different numbers of neutrons – they therefore have a different atomic mass

jaundice yellow colouration of the skin and whites of the eyes caused by increased levels of bilirubin in the blood, which may be due to abnormal functioning of the liver

karyogram a diagram or photographic image showing the number, shape and types of chromosomes in a cell

karyotype the number, shape and types of chromosomes in a cell

keystone species a species that is important in maintaining community structure in an ecosystem because it reduces competition in other trophic levels, so that community diversity is sustained

kinesis orientation behaviour in which an organism moves at an increasing or decreasing rate, which is not directional, in response to a stimulus

Krebs cycle a cycle of biochemical changes that occur in the mitochondrial matrix during aerobic respiration

lactation secretion of milk in mammary glands

lagging strand the daughter strand that is synthesised discontinuously in DNA replication

lateral meristem tissues in the side shoots of plants which have the capacity to divide and produce new cells

leaching the loss of soluble nutrients from the soil as rain or irrigation water passes through it

leading strand the daughter strand that is synthesised continuously in DNA replication

learned behaviour behaviour patterns that are acquired during an animal's lifetime as a result of experience, which are not genetically determined

leucocyte white blood cell

light-dependent reactions series of stages in photosynthesis that occur on the grana of the chloroplasts in which light is used to split water, and ATP and NADPH + H^+ are produced

light-independent reactions series of stages in photosynthesis that take place in the stroma and use the products of the light-dependent reactions to produce carbohydrate

lignin chemical found in the cellulose walls of certain plant cells, such as xylem, which gives strength and support

limiting factor a resource that influences the rate of processes (such as photosynthesis) if it is in short supply; with respect to populations, any factor that prevents the growth of the population above a certain value

limits of tolerance upper and lower limits of the range of an environmental factor within which an organism can survive; outside this range are the 'zones of intolerance' in which conditions are too extreme for the organism to survive

linkage group in genetics, the genes carried on one chromosome that do not show random or independent assortment

lipid   a fat, oil, wax or steroid; organic compound that is insoluble in water but soluble in organic solvents such as ethanol

lipophilic   having an affinity for lipids

lipoprotein   a complex of lipid and protein that can be classified by density e.g. LDL (low density lipoprotein) and HDL (high density lipoprotein); the form in which cholesterol and other lipids are transported in the blood

liposome   tiny vesicles with a phospholipid bilayer around them, which contains other substances such as DNA or proteins

locus   the specific location on a chromosome of a gene

long-day plant   a plant that requires fewer than a certain number of hours of darkness in each 24-hour period to induce flowering

loop of Henle   the section of a nephron between the proximal and distal convoluted tubules that dips down into the medulla and then back up into the cortex of the kidney

luteinising hormone (LH)   a hormone produced by the anterior pituitary gland that stimulates the production of sex hormones by ovaries and testes

lymphatic system   network of fine capillaries throughout the body of vertebrates, which drain lymph and return it to the blood circulation

lymphocyte   a type of white blood cell that is involved in the immune response; unlike phagocytes they become active only in the presence of a particular antigen that 'matches' their specific receptors or antibodies

lysis   breakdown of cells

malnutrition   condition(s) caused by a diet which is not balanced and may be lacking in, or have an excess of, one or more nutrients

marker   (see *biomarker*)

marker genes   genes linked to sequences of DNA that are being transferred to new organisms so that researchers can check that the insertion has been successful

maximum sustainable yield   the largest proportion of fish that can be caught without endangering the population

medulla oblongata   the part of the brain stem that connects to the spinal cord and controls breathing and other reflex actions

meiosis   a nuclear division that produces cells containing half the number of chromosomes of the parent cell

memory   the process of encoding, storing and accessing information

memory cells   cloned lymphocytes which remain in the blood stream after an infection to give protection against the same infection (antigen) at a later date

menstrual cycle   the process of shedding the lining of the uterus at monthly intervals if fertilisation does not occur

menstruation   the shedding of the lining of the uterus during the menstrual cycle

meristem   area of plant tissue that divides to produce new cells and tissues

mesoderm   the middle of the three layers of cells in a very early animal embryo, which forms the connective tissue, muscles and part of the gonads

messenger RNA (mRNA)   a single-stranded transcript of one strand of DNA, which carries a sequence of codons for the production of protein

metabolic pathway   a series of chemical reactions that are catalysed by enzymes

metabolism   integrated network of all the biochemical reactions of life

metabolomics   the analysis of metabolites in a blood or urine sample

metaphase   stage in nuclear division at which chromosomes become arranged on the equator of the spindle

micropyle   opening in a plant ovule through which a pollen grain enters

microvilli   folded projections of epithelial cells, such as those lining the small intestine, that increase cell surface area

mitochondrion (mitochondria)   organelle in the cytoplasm of eukaryotic cells; the site of respiration reactions, the Krebs cycle and the electron transport chain

mitotic index   a measure of cell proliferation; it is the ratio between the number of cells undergoing mitosis and the total number of cells observed in the sample

mitosis   cell division that produces two daughter cells with the same chromosome compliment as the parent cell

model organism   a non-human species used for study and experimentation in order to understand particular pathways or processes

monoclonal antibody   antibody produced in the laboratory by a single clone of B cells, which gives rise to many identical antibody molecules

monohybrid cross   a breeding experiment that involves one pair of different characteristics from two homozygous parents

monomer   a small molecule which can link with other identical molecules to form a polymer

monosaccharide   a simple sugar (monomer) that cannot be hydrolysed

mortality   a measure of the number of deaths in a population in a certain amount of time; the death rate

motor areas   parts of the cerebrum generating nerve impulses that are sent to effectors, via motor neurons

motor neuron   nerve cell that carries impulses away from the brain

muscle fibre   a single muscle cell that is multinucleate in striated muscle

mutagen   an agent that causes mutation

mutation   a permanent change in the base sequence of DNA

mutualism   a type of symbiotic relationship in which both organisms benefit

myelin sheath   a fatty covering around the axons of nerve fibres, which provides insulation

myofibril   a unit or contractile filament making up muscle

myogenic   a contraction of heart muscle that originates in heart muscle cells

myosin   one of the two proteins found in muscle; it makes up thick filaments

natality   a measure of the number of births in a population in a certain mount of time; the birth rate

natural selection   the mechanism of evolution proposed by Charles Darwin in which various genetic types make different contributions to the next generation

negative feedback a regulating mechanism in which a change in a sensed variable results in a correction that opposes the change

nephron the functional unit of the kidney

net primary production the amount of energy in plants that is available to herbivores, after some energy has been lost by the plants through respiration; it is measured in kilojoules of energy per square metre per year (kJ m^{-2} y^{-1})

neural groove cells of the neural plate that gradually develop into upward folds, eventually meeting and closing over to form the neural tube

neural network series of interconnected neurons in the brain

neural patterning a process in embryonic development during which neurons take up their positions and start to establish connections

neural plate an area of embryonic ectoderm cells that develops and becomes the neural groove; these cells eventually become the brain and spinal cord

neural tube embryonic structure whose outer cells form the foundation of the nervous system – cells at the anterior (front) end become the brain while those in the posterior (back) region become the spinal cord

neuro-adaptation the brain's attempt to counteract the effect of a psychoactive drug when used frequently and in excess, which is the basis of addiction and drug tolerance

neurogenesis the origination and proliferation of new neurons in the brain

neuron a nerve cell that can carry action potentials and which makes connections to other neurons, muscles or glands by means of synapses

neurotransmitter a substances produced and released by a neuron, which passes across a synapse and affects a post-synaptic membrane

niche the habitat an organism occupies, its feeding activities and its interactions with other species

nitrification oxidation of ammonia to nitrites and nitrates by nitrifying bacteria

nitrogen fixation conversion of nitrogen gas into nitrogen compounds by nitrogen fixing bacteria

nodes of Ranvier gaps in the myelin sheath of an axon where the membrane can initiate action potentials

non-competitive inhibitor an inhibitor of an enzyme that binds at a site away from the active site

non-disjunction failure of sister chromatids to separate in mitosis or meiosis II, or of homologous chromosomes to separate in meiosis I

non-striated muscle (see *smooth muscle*)

normal distribution a frequency distribution represented by a symmetrical bell shaped curve

notochord a supporting dorsal rod found in the embryos of all chordates

nuclear envelope a double membrane that surrounds the nucleus in a eukaryotic cell

nucleolus a small body found in the nucleus of eukaryotic cells where ribosomal RNA is synthesised

nucleoside triphosphate (NTP) a building unit for RNA – a ribose nucleotide with two additional phosphates, which are chopped off during the synthesis process

nucleosome a part of a eukaryotic chromosome made up of DNA wrapped around histone molecules and held in place by another histone protein

nucleotide the basic chemical unit of a nucleic acid – an organic base combined with pentose sugar (either ribose or deoxyribose) and phosphate

nucleus organelle found in eukaryotic cells that controls and directs cell activities; it is bounded by a double membrane (envelope) and contains chromosomes

nucleus accumbens the 'pleasure centre' in each hemisphere of the cerebrum, which has a role in pleasure, addiction, fear, laughter and reinforcement learning

null hypothesis a hypothesis used in statistics that proposes there is no statistical significance between two measured sets of observations

nutrient a substance taken in by an organism that is required for its metabolism

Okazaki fragments newly formed DNA fragments that form part of the lagging strand during replication and which are linked by DNA ligase to produce a continuous strand

oncogene a cancer-initiating gene

oogenesis female gamete production

oogonia cells in which an egg is produced in some fungi and algae

open reading frame a significant length of DNA beginning at a start codon and ending with a stop codon; it has no codons that terminate transcription within it

operant conditioning a type of trial-and-error learning that develops as a result of the association of reinforcement (reward) with a particular response

optic chiasma region of the brain near the thalamus and hypothalamus at which portions of each optic nerve cross

organelle a cell structure that carries out a specific function – for example, ribosome, nucleus, chloroplast

organic compounds of carbon, excluding carbon dioxide and carbonates

osmoregulation control of the water balance of the blood, tissues or cytoplasm of a living organism

osmosis the diffusion of water molecules from an area where they are in high concentration (low solute concentration) to a area where they are in a lower concentration (high solute concentration) across a partially permeable membrane

osteomalacia softening or malformation of the bones due to a lack of calcium and phosphorus, often caused by deficiency of vitamin D; in children the condition it known as rickets

ovary organ in the female body in which female gametes are formed

oviduct (Fallopian tube) the tube with an opening close to the ovary and connects to the uterus

ovulation the release of ova from the ovary

oxidation gain of oxygen or loss of electrons in a chemical reaction, usually associated with release of energy

oxidative phosphorylation ATP formation in the mitochondria as electrons flow through the electron transfer chain

oxytocin hormone produced by the pituitary gland which stimulates the contraction of the uterus during the birth of a baby

ovum (ova) a female gamete

parasite an organism that lives on or in a host for most of its lifecycle and derives its nutrients from it, causing the host harm

parasympathetic nervous system (PNS) part of the autonomic (involuntary) nervous system which produces effects such as decreased blood pressure and heart rate

parental generation in a genetic cross, the first pair of animals or plants which are mated

parietal cells stomach epithelial cells that secrete gastric acid

partial pressure the proportion of the total pressure that is due to one component of a mixture of gases

passive immunity immunity due to the acquisition of antibodies from another organism in which active immunity has been stimulated

passive transport the diffusion of substances across a plasma membrane without the expenditure of energy

pathogen an organism or virus that causes disease

pathway engineering a way of manipulating the regulatory and genetic processes that occur in microorganisms so that particular metabolites of interest are produced in optimal quantities

peat an organic substance made up of partially decomposed plant material found in wetlands such as swamps and bogs

pentadactyl having limbs that end in five digits

pentose a five-carbon monosaccharide

peptide bond a covalently bonded linkage between two amino acids in which the α amino group of one links to the carboxyl group of the next

peripheral nerves parts of the nervous system not including the brain and spinal cord, which transmit information to and from the CNS

peripheral nervous system (PNS) all the nerves that do not form the central nervous system (brain and spinal cord) – that is, the sensory neurons and the autonomic nervous system (ANS)

peristalsis wave-like contractions that propel food along the alimentary canal

petal modified leaf, often brightly coloured, found in angiosperm flowers

phagocyte a type of white blood cell which removes harmful particles by engulfing them

phagocytosis the process of modifying the shape of a phagocytic cell so that it can engulf bacteria or other particles

phenotype the characteristic or appearance or an organism which may be physical or biochemical

phenylketonuria an autosomal recessive metabolic genetic disorder characterised by a lack of enzyme needed to utilise the amino acid phenylalanine

phloem tissue that carries food in the stem of a plant

phospholipid important constituent molecules of membranes, formed from a triglyceride in which one fatty acid is replaced by a phosphate group

phosphorylation the addition of a phosphate group to a molecule

photoautotroph an organism that uses light energy to produce ATP and produces organic compounds from inorganic molecules

photoheterotroph an organism that uses light energy to generate ATP and obtains organic compounds from other organisms

photolysis the splitting of water molecules in the light-dependent stage of photosynthesis

photoperiodism the control of flowering in plants in response to day length

photophosphorylation the formation of ATP using light energy in the grana of chloroplasts

phototropism the tropic response of plants to light

phylogenetics a classification system based on evolutionary relationships

physiomics the study of an organism's physiome, the interconnections of aspects of physiology that result from genes and proteins in the organism

phytochrome a pigment found in plants that regulates several processes including the flowering pattern in response to day length

pilus (*plural* pili) an extension of the surface of some bacteria, used to attach to another bacterial cell during conjugation

pinna the external part of the ear; a sound-collecting device that in many animals can be rotated by muscles to pick up sounds from all directions

pinocytosis 'cell drinking', a form of endocytosis, taking extra cellular fluids into a cell by means of vesicles

pituitary gland so-called 'master gland' whose hormones control the activities of other glands

placenta a structure of maternal and fetal tissues on the lining of the uterus; the site of exchange of materials between maternal and fetal blood systems

plaque the build up of fatty deposits in the wall of an artery

plasma cells antibody-secreting cells that develop from a B cell

plasma membrane lipoprotein bilayer that surrounds and encloses a cell

plasmid an independent chromosome; a small circle of DNA found in bacteria

plasticity the ability of the nervous system to change in both structure and function over an organism's life, as it reacts to the changes in its environment

platelets cell fragments found in the blood that are concerned with blood clotting

plumule part of an embryo plant that will become the shoot

pluripotent a cell that is able to differentiate into many different types of cell in the correct conditions

pneumocyte Either of two types of cell (type I and type II) found in the alveoli responsible for gas exchange or secretion of surfactant

polar body a non-functioning nucleus produced during meiosis; three are produced during human oogenesis

pollen the fertilising element of flowering plants; a microspore containing the male gamete, formed in the anthers

pollen tube grows out of a pollen grain attached to a stigma, and down through the style tissue to the embryo sac

pollination transfer of pollen from the anther to the stigma of the ovary in plants

polyclonal response activation of many different B-cells in response to an infection

polygene   multiple loci whose alleles affect a continuously variable phenotypic characteristic, such as height in humans

polygenic inheritance   inheritance of phenotypic characteristics that are determined by the collective effects of several different genes

polymerase chain reaction (PCR)   process by which small quantities of DNA are multiplied for forensic or other examination

polymer   a substance built up from a series of monomers

polypeptide   a chain of amino acids linked by peptide bonds

polysaccharide   large carbohydrates formed by condensation reactions between large numbers of monosaccharides – for example, cellulose, glycogen and starch

polysome   an arrangement of many ribosomes along a molecule of mRNA

population   a group of organisms of the same species that live in the same area at the same time

positive feedback   a control mechanism in which a deviation from the normal level stimulates an increase in the deviation

precautionary principle   principle that those responsible for change must prove that it will not do harm before they proceed

precursor   an inactive form of a molecule, such as an enzyme, which is converted to its active form at the site when required

predation   the killing and eating of animals for the purposes of nutrition

prey value   the ratio of the energy content of a prey item to the energy a particular predator expends in catching and eating it; prey value varies both with the size of the prey item and the size of the predator

primary consumer   an organism that feeds on a primary producer

primer   short strand of nucleic acid that forms a starting point for DNA synthesis.

probe   a specific DNA sequence attached to the surface of a DNA microarray, which is used to hybridise samples of target DNA

producer   an autotrophic organism

progesterone   a female sex hormone that maintains pregnancy

prokaryote   an organism whose genetic material is not contained in a nucleus

prolactin   a hormone released by the pituitary gland that stimulates lactation

promoter region   a sequence of DNA that initiated transcription of a particular gene and is needed to turn a gene on or off

prophase   first stage in cell division by meiosis or mitosis

prostate gland   male gland that produces a white alkaline secretion making up to 30% of semen volume

prosthetic group   a non-protein part of an enzyme that often forms part of the active site

proteome   the complete set of proteins expressed by a genome

proteomics   the study of the structure and functions of proteins

prothrombin   a glycoprotein found in blood plasma essential for blood clotting

punctuated equilibrium   a form of evolutionary change whereby species remain stable for long periods of time followed by brief periods of rapid change in response to significant environmental change; leading to the formation of new species

pupil reflex   a constriction of the pupils, caused by contraction of the circular muscles in the iris, which occurs when bright light shines into the eye; it is used to test the functioning of the brain stem and its loss is a key indicator of brain death

pure breeding   an organism that is homozygous for a specified gene or genes

purine   a nitrogenous base such as adenine or guanine found in nucleic acids

pyramid of energy   diagram which shows the total energy content at different trophic levels in an ecosystem

pyrimidine   a nitrogenous base such as cytosine, thymine or uracil found in nucleic acids

quadrat   a frame which encloses a sampling area

quorum sensing   method by which microorganisms within a biofilm communicate through the release of signalling molecules into the environment, allowing the cells to coordinate their behaviour according to their local density

R group   any of a number of side chains that may be attached to an amino acid

radicle   part of an embryo plant that will become the root

realised niche   the actual niche occupied by an organism, differing from the fundamental niche because of competition with other species or predation

receptor   specialised cell or neuron ending which receives a stimulus

recessive allele   an allele that has an effect on the phenotype only when present in the homozygous state

recombinant   a chromosome (cell, or organism) in which DNA has been rearranged so that genes originally present in two individuals end up in the same haploid cell

recombinant DNA   DNA that has been artificially changed, often involving the joining of genes from different species

recombination   the rearrangement of genetic material, or of the corresponding heritable characteristics, by crossing over during meiosis between pairs of homologous chromosomes derived from each parent, so as to produce phenotypes which are different from both parents

redox reaction   a reaction in which reduction and oxidation occur simultaneously

reduction   gain of electrons; the opposite of oxidation

reduction division   meiosis, when the chromosome number of a diploid cell is halved

reflex action   a rapid automatic response

reflex arc   a response which automatically follows a stimulus and usually involves a small number of neurons

refractory period   the time after an action potential during which another action potential cannot occur

regulator gene   a gene determining the production of a protein that regulates or suppresses the activity of one or more structural genes

relay neuron   neuron through which a connection between a sensory and a motor neuron can be made

replication fork   the point at which DNA is replicating

repolarisation   reestablishment of the resting potential following depolarisation of a neuron

residual volume   volume of air in the lungs after a maximum exhalation

resolving power the capacity of an instrument, such as a microscope, to allow two points that are close together to be distinguished as separate

respiration (see *cell respiration, aerobic respiration, anaerobic respiration*)

response a reaction to a specific stimulus

resting potential the potential difference across the membrane of a neuron when it is not being stimulated

restriction enzyme (endonuclease) one of several enzymes that cut nucleic acids at specific sequences of bases

retina the light-sensitive layer at the back of the eye

ribosome a small organelle that is the site of protein synthesis

rickets (see osteomalacia)

RNA (ribonucleic acid) a nucleic acid that contains the pentose sugar ribose and bases adenine, guanine, cytosine and uracil

RNA polymerase an enzyme that catalyses the formation of RNA from a DNA template

RNA primase an enzyme that catalyses the synthesis of RNA primers as the starting point for DNA synthesis

RNA splicing modification of RNA to remove introns and rejoin the remaining exons

rod cells photoreceptors in the retina, which are more sensitive to light than cones, absorbing all wavelengths and functioning well at low light intensities; however, all rod cells are of the same type, so in dim light we cannot differentiate colour

RuBP carboxylase the five-carbon acceptor molecule for carbon dioxide found in the light-independent reactions of photosynthesis

S phase in the cell cycle, the period during interphase when DNA is replicated

saltatory conduction impulse conduction 'in jumps', between nodes of Ranvier

sample a small part of the whole, taken to be representative of the whole in investigations

saprotroph organism that feeds on dead organic matter

sarcolemma membranous covering of a muscle fibre

sarcomere contractile unit of skeletal muscle between two Z-lines

sarcoplasmic reticulum network of membranes surrounding myofibrils of a muscle fibre

saturated hydrocarbon a compound in which all carbon atoms are linked by single bonds to other carbon atoms or hydrogen atoms

scurvy a deficiency disease caused by a lack of vitamin C

secondary consumer an animal that feeds on a primary consumer, a carnivore

secondary growth in plants, growth of the vascular and cork cambium, which leads to an increase in girth

secondary metabolite a substance produced by an organism only at certain stages of its growth or development, which is not required for normal day-to-day metabolism

seed structure formed from a fertilised ovule and containing an embryo plant together with a food store within a seed coat

selection pressure varying survival or reproductive ability of different organisms in a breeding population due to the influence of the environment

semen thick, whitish liquid produced by male mammals, which contains sperm cells

semi-conservative replication each of two partner strands of DNA in a double helix acts as a template for a new strand; after replication each double helix consists of one old and one new strand

semicircular canals three fluid-filled tubes in the inner ear, connected at right angles to one another, which enable the detection of movements of the head, thereby allowing a sense of position and balance

semilunar valves half-moon shaped valves in the arteries leaving the heart, also found in some large veins

seminal vesicle part of the male reproductive system in which components of semen are produced

seminiferous tubule tubules in the testes in which sperm production takes place

sense strand the coding strand of DNA, which is not transcribed but which has the same base sequence as mRNA produced during transcription (though mRNA has uracil instead of thymine)

sensory areas parts of the cerebrum that receive information from various sense organs, via sensory neurons

sensory neuron nerve cell that carries impulses to the central nervous system

sepal the outermost parts of a flower that protect the bud

seral stage one of the communities that exists during a succession

sere (see seral stage)

Sertoli cell cells in the seminiferous tubules that protect and nurture developing sperm cells

sex chromosomes chromosomes that determine the sex of an individual

sex linkage the pattern of inheritance of genes carried on only one of the sex chromosomes, which therefore show a different pattern of inheritance in crosses where the male carries the gene from those where the female carries the gene

short-day plant plants that only flower when the night is longer than a critical length

sieve tube cell an element of the phloem containing cytoplasm but few organelles and having perforated ends known as sieve plates

simple diffusion (see *diffusion*)

single-stranded binding protein protein which binds to single-stranded regions of DNA to protect them from digestion and remove secondary structure

sinoatrial node (SA node) the pacemaker cells in the wall of the right atrium, which initiate the heart beat

sinusoids small blood vessel found in the liver, similar to a capillary but with fenestrated endothelium cells

sister chromatids vtwo joined copies of a chromosome after it has replicated and before the centromeres separate at anaphase

skeletal (striated) muscle voluntary muscle tissue, which has multinucleated cells with arrangements of actin and myosin microfilaments

smooth muscle sheets of mononucleate cells that are stimulated by the autonomic nervous system

somal migration the way in which a neuron moves in the embryonic brain by extending a long process from its cell body to the outer region that will later become the cortex

somatic cell body cell – not a cell producing gametes (sex cell)

spatial habitat the space in an ecosystem occupied by an organism

speciation the evolution of new species

species a group of individuals of common ancestry that closely resemble each other and are normally capable of interbreeding to produce fertile offspring

spermatogenesis the production of sperm cells

spindle structure formed of microtubules to which centromeres attach during meiosis and mitosis

stabilising selection selection against extreme phenotypes in a population

starvation when an individual simply does not have enough to eat – there is insufficient energy in the diet for respiration (as opposed to malnutrition, which results from a nutritionally unbalanced diet)

stigma part of the female reproductive organs of a plant, which receives pollen grains

stimulus a change in the environment that is detected by a receptor and leads to a response

stoma (stomata) pore in the epidermis of a leaf, surrounded by two guard cells

striated muscle see *skeletal muscle*

stroke damage to the brain caused by a burst or blocked blood vessel, which starves brain cells of oxygen

style female part of a flower, which links the stigma to the ovary

succession the sequence of different communities that appears in a given habitat over a period of time

supercoiling additional coiling of the DNA helix that reduces the space needed for the molecule and allows the chromosomes to move easily during nuclear division

sustainable a resource or system that can continue indefinitely because it is replenished at the same rate as it is used

symbiosis 'living together' – includes commensalism, mutualism and parasitism

sympatric speciation separation of a species into two groups within the same geographic area; caused by some form of isolation, such as a behavioural, ecological or genetic barrier

sympathetic nervous system part of the autonomic nervous system; effects include increasing blood pressure and heart rate

synapse the connection between two nerve cells – a small gap that is bridged by a neurotransmitter

synaptic plasticity the changes that take place in neural pathways as an organism undergoes life experiences, so that some connections are enhanced through regular use and others that are not used are lost through neural pruning

synovial fluid lubricating fluid secreted by the synovial membrane at a joint

synovial joint a joint that allows movement between two bones; it is lined by a synovial membrane, which produces synovial fluid for lubrication

systems approach study of a system as a whole, rather than examining individual parts within it

systole contraction of the chambers of the heart during the cardiac cycle

target cell specific cell upon which a hormone will act and elicit a response; or a cell that is the recipient of recombinant DNA in biotechnology

target DNA DNA sequences in a sample that are identified using DNA microarray tests – the target DNA hybridises with probe sequences on the surface of a chip, and may be detected and quantified using luminescent labels

target gene in recombinant DNA technology, the gene for the desired protein, which must be isolated in order for it to be inserted into the cells of the organism to be genetically modified

taxis the movement of an organism in a particular direction in response to a stimulus

taxonomy the science of classification

telomere region of repeated nucleotide sequences at the end of a chromatid which protect the chromosome from damage

telophase the phase of cell division when daughter nuclei form

tertiary consumer consumer that feeds on a secondary consumer, often a top carnivore

test cross testing a suspected heterozygote by crossing with a known homozygous recessive

testa seed coat

testosterone the main sex hormone of male mammals

theory a scientific proposal which is based on observable, repeatable experimentation

threshold potential the potential difference beyond which an action potential must take place

thrombin enzyme that converts fibrinogen to fibrin and triggers blood clotting

thrombosis the blockage of a blood vessel by a blood clot

thylakoid membranes membrane system of a chloroplast where the light dependent reactions take place

thyroxin hormone produced by the thyroid gland which influences the rate of metabolism

tidal volume volume of air normally exchanged in breathing

tight junction junction between epithelial cells where plasma membranes of adjacent cells are bonded together by integral proteins, which prevents molecules passing between the cells

tolerance the desensitisation experienced by frequent users of psychoactive drugs, so that more and more of the drug is required to produce the same feelings of well being (see also neuro-adaptation)

top-down control control of community structure by keystone species that reduce competition in lower trophic levels

total lung capacity volume of air in the lungs after a maximum inhalation

trachea the windpipe

transect a line or belt through a habitat in order to sample the organisms present

transcription copying a sequence of DNA bases to mRNA

transfer RNA (tRNA) short lengths of RNA that carry specific amino acids to ribosomes during protein synthesis

transferrin receptor (TfR) a glycoprotein in cell membranes that is involved in the cellular uptake of iron from transferrin and also with the regulation of cell growth *transferrin* glycoproteins found in the blood, which bind to iron and control the level of free iron in the plasma

transgenic containing recombinant DNA incorporated into an organism's genetic material

translation decoding of mRNA at a ribosome to produce an amino acid sequence

translocation transport of food via the phloem of a plant

transpiration loss of water vapour from the leaves and stem of plants

trisomy containing three rather than two members of a chromosome pair

trophic level a group of organisms that obtain their food from the same part of a food web and which are all the same number of energy transfers from the source of energy (photosynthesis)

tropism a growth response of plants in which the direction of growth is determined by the direction of the stimulus

tropomyosin a protein that, together with actin, makes up the thin filaments in myofibrils

troponin a protein that, together with actin and tropomyosin, makes up the thin filaments in myofibrils

tumour a disorganised mass of cells; a malignant tumour grows out of control

turgid having a high internal pressure; the turgidity of plant cells allows them to remain firm

turgor pressure the internal pressure of fluid inside a plant cell which pushes the membrane against the cell wall

type II diabetes diabetes which results from the body developing an insensitivity to insulin over a long time period

ultrafiltration process that occurs through tiny pores in the capillaries of the glomerulus

umbilical cord tissues derived from the embryo that contain and carry blood vessels between the embryo and the placenta

unsaturated unsaturated fats contain at least one double bond between carbon atoms in their molecules

urea NH_2CONH_2 – a molecule formed of amino groups deaminated from excess amino acids in the liver; the main form in which nitrogen is excreted in humans

vaccination injection of an antigen to induce antibody production before a potential infection

vacuole a liquid-filled cavity in a cell enclosed by a single membrane; usually small in animals

variation differences in the phenotype of organisms of the same species

vascular bundles a length of vascular tissue in plants consisting of xylem and phloem

vector a plasmid or virus that carries a piece of DNA into a bacterium during recombinant DNA technology; or an organism, such as an insect, that transmits a disease-causing organism to another species

vein a vessel that returns blood to the heart

ventricle muscular lower chamber of the heart

vesicle a membrane-bound sac

villus (villi) a fold in the lining of the small intestine where absorption occurs

visual processing the way the brain interprets the impulses it receives from the inverted and reversed images on the retinas of the eyes, via the optic nerve, so that we perceive the world 'the right way up' and 'the right way round'

vital capacity the total possible change in lung volume; the maximum volume of air that can be exhaled after a maximum inhalation

water potential the tendency of water molecules to move

Wernicke's area a part of the left cerebral hemisphere responsible for understanding of language

xerophyte a plant that is adapted to withstand drought conditions

xylem tissue water-carrying vessels of plants that transport up the stem only

Z lines the border between one sarcomere and the next in skeletal muscle

zygote the cell produced by the fusion of two gametes

Index

abiotic factors 137
absorption (digestion) 190–91
acetylcholine (ACh) 217
acidification, ocean 154
acrosome reaction 371
actin 350
action potential 216, 217, 218
activating enzymes 251
activation energy 262–63
activator proteins 246
active immunity 339–340, 341
active sites 62
active transport 23, 26–27, 190, 291
adaptive radiation 162
adenine 235–36
adenosine triphosphate (ATP) 78–80, 269, 273–75
ADH (antidiuretic hormone) 359–360
adhesive forces 49–50
adrenalin (epinephrine) 196
aerobic respiration 78, 209
agouti variable yellow (*Avy*) gene 250
AIDS (acquired immune deficiency syndrome) 206–7
alimentary canal 187
allele frequencies 327–28
alleles 93, 94
allergies 341–42
Allium oleraceum 331
allopatric (geographic) speciation 329–330, 332
allosteric control 265
alveoli 209, 212–13
amino acids 44, 45, 59, 61, 94, 257–58
amniocentesis 109, 375
amniotic fluid 372–73
amphipathic compounds 20, 51–52
amylopectin 55
amylose 55
anabolic reactions 46
anaerobic respiration 79–80
analogous characteristics 184
anaphase 37, 105, 106, 107
Angiospermophyta 173
animal cells 17
animal physiology 337–375
Annelida 174
antagonistic pairs 346
anthers 301
antibiotics 167–68, 205–6, 208
antibodies 203–5, 337–345
anticodons 75–76, 251
antidiuretic hormone (ADH) 359–360
antigen presentation 339
antigens 203, 337

antisense strands 245
apoptosis 38
appetite 225
aquaporins 359–360
Arctic ecosystem 155
arteries 193
arthropod muscles 347–49
Arthropoda 174
artificial selection 160–61
assimilation (digestion) 190
atherosclerosis 198, 199
ATP (adenosine triphosphate) 78–80, 269, 273–75
atria 195, 196
atrioventricular node (AVN) 197
autosomes 102
autotrophs 137, 140
auxin efflux pumps 300
auxins 299–300
AVN (atrioventricular node) 197
Avy (agouti variable yellow) gene 250
axons 215–17
B-cells 338–39
bacteria
  antibiotic resistant 167–68
  transgenic 126–27
baldness 248
ball-and-socket joints 347
base substitution mutation 94
basophils 341
bees 304
binary fusion 13
binding proteins 246–47
binomial classification 169–170
biodiversity 169–178
biotechnology 124–132
bird flu 338
blastocysts 372
blood 193–201
  clotting 202
  glucose levels 220–22
  plasma 200
blood groups 344
blood vessels 193–94
body temperature 224
bones 346
Botox 219
Bowman's capsule 355, 357–58
Bryophyta 172
butterflies, GM crops and 131–32
Cairns, Hugh John Forster 102
calcium-containing organic compounds 43
Calvin cycle 279–282
CAM (crassulacean acid metabolism) 288
cambium 297

cancer 38, 343–44
cannabis 219
capillaries 193–94, 196
carbohydrates 44–45, 54–55
carbon compounds 43–44
carbon cycle 147–48
carbon dioxide 149–150, 153
carbon fixation 85, 280
carbon flux 149
carcinogens 38–39
cardiac cycle 197–98, 199
cardiac muscle 196, 348
carrier proteins 23–24
carriers, genetic 110, 116
cartilage 175
catabolic reactions 46, 48
causal relationships 58
cell biology 1–40
cell cycle 33–37
cell differentiation 7, 8
cell division 33–40
cell respiration 78–81, 268–274
cell size 3–5
cell theory 1, 30
cell walls 12, 17
cellulose 54–55, 192
central nervous system (CNS) 215
centrioles 36
centromere 36, 37
CFCs (chlorofluorocarbons) 152
channel proteins 23–24
charged tRNA 251
Chase, Martha 236–37
CHD (coronary heart disease) 57, 198–99
chemiosmosis 273–75, 279
chi-squared test 323
chiasmata 309
Chlorella 2–3
chlorofluorocarbons (CFCs) 152
chlorophyll 82
chloroplasts 17, 31–32, 83, 276–77
cholesterol 20, 51
cholinergic synapses 217, 219
Chordata 175
chorionic villi 373–74
chorionic villus sampling (CVS) 109, 375
chromatids 36, 308–10
chromatography 83
chromosomes 12, 99–103, 308–11, 315–320
chyme 188
circadian rhythms 225–26
circulation, blood 195–99
cis fatty acids 56–57
cisternae 16
clades 178–79

cladistics 178–184
cladograms 178–79, 182–84
classification of living organisms 169–178
climate change 151–55
clonal selection 338–39
cloning 130–32
Clostridium botulinum 219
Clostridium tetani 219
clotting factors 202
Cnidaria 174
CNS (central nervous system) 215
codominant alleles 110, 115
codons 75–76, 93
cohesion-tension theory 285–87
cohesive forces 49–50
colour blindness 117
communities (ecosystems) 137
companion cells 293
competitive inhibitors 263
complementary base pairing 68
concentration gradients 23
condensation reactions 46–47
Coniferophyta 173
conjugated proteins 256
conservation 304
consumers (food chains) 141
continuous variation 162–63, 320
contraceptives, oral 376
contraction, muscle 350–53
coral reefs 154
coronary heart disease (CHD) 57, 198–99
coronary thrombosis 198
corpus luteum 228–29, 368
correlation 58
cortical reaction 371
cotyledons 302
CpG islands 250
crassulacean acid metabolism (CAM) 288
cristae 16
crops, genetically modified 128–132
cross-pollination 301
CVS (chorionic villus sampling) 109, 375
cyclic phosphorylation 279
cyclins 37–38, 40
cystic fibrosis 119–120
cytochrome c 180–81
cytoplasm 12, 17
cytosine 235–36
Darwin's finches 165–66
Davson-Danielli model 21
ddNTPs (dideoxynucleotides) 240
decarboxylation reactions 272
decomposers 138
deductive reasoning 208
dehydration 362
denaturation 60, 62
deoxynucleoside triphosphates (dNTPs) 238–39, 243
deoxyribonucleic acid (DNA) see DNA
desert animals 364

detritivores 138
diabetes 221–23, 360–61
dialysis 362–63
dialysis tubing 192
diaphragm, thoracic 212
dichotomous keys 175–77
dideoxynucleotides (ddNTPs) 240
differentiation, cell 7, 8
diffusion 23–24
digestion 187–192
dihybrid crosses 313–15, 325
dihydrotestosterone 248
dipeptides 47
diploid cells 100
directional selection 327–28
disaccharides 45, 46, 54
discrete variation 320
dispersal, seed 302
disruptive selection 327–28
distal convoluted tubule 356, 358–59
disulfide bonds 257
disulfiram 265
DNA
methylation 249
profiling 124–26, 242
replication 71–73, 237–241
sequencing 97–98
structure 68–70, 235–37
transcription and translation 74–77
DNA gyrase 240, 243
DNA ligase 240, 243
DNA methyl transferase 250
DNA polymerase 72, 238–39, 243
dNTPs (deoxynucleoside triphosphates) 238–39, 243
Dolly the sheep 131
dominant alleles 110, 112
Down syndrome 107–8
Drosophila 317, 318, 319, 325–26
E. coli 102, 241
ecology 136–156
ecosystems 136–37
egestion 191, 192
elbow joint 346–47
electron carriers 273
electron microscopes 3–4, 18–19
electron tomography 272
electron transport chain 273–74
elongation stage of translation 253
embryo 372
embryonic stem cells 9
emphysema 213
end-product inhibition 265–66
endocrine disrupters 376
endocrine system 220–21
endocytosis 28
endometrium 372
endoplasmic reticulum 16
endosperm 303
endosymbiosis 31–32

energy coupling 270
energy pyramids 144–45
enhancer region 247
Ensatina 329–330
enzymes 62–67, 263–66
epidemiology 58, 214
epididymis 366
epigenetics 249–251
epinephrine (adrenalin) 196
erythrocytes (red blood cells) 200, 202
Escherichia coli 102, 241
estrogen 228, 375, 376
ethical issues 9, 80–81, 121, 129, 230–31, 344–45, 364
Euglena 140
eukaryotic cells 13–18, 31–33, 99, 238
eukaryotic chromosomes 99–100
evolution 159–168, 327–28
excretion 355, 361–62
exhalation 212
exocytosis 28
exons 241, 244, 247
exoskeletons 347–49
experimental design 29, 66, 88, 122–23
expiratory reserve 210
external fertilisation 370
facilitated diffusion 23–24, 190, 290
factor XI deficiency 126
fats 45
fatty acids 44, 56–57
fenestrations 357
fertilisation
in animals 370–71
in plants 301–2
fetus 372–75
fibrin 202, 203
fibrinogen 202, 203
fibrous proteins 60, 257
field garlic 331
filaments 301
Filicinophyta 172–73
filtrate 358
finches, Darwin's 165–66
fish
osmoregulation 354
populations 139
flagella 13
flowering 303–4
flowers 300–301
fluid mosaic model 19–20, 21, 22
follicle-stimulating hormone (FSH) 228, 230, 365
fomepizal (fomepizole) 264
food chains 141–44, 146
food webs 142–43
forest ecosystems 155
fossil fuels 148, 150, 152
fossil record 159–160
'founder effect' 328
Franklin, Rosalind 242–43

freeze-etching 21
FSH (follicle-stimulating hormone) 228, 230, 365
fungi 1
Galton, Francis 248
gametes 112, 226, 318
garlic, field 331
gas exchange 209–13
gated channels 23
gel electrophoresis 124–25
gene pools 327–28
general transcription factors (GTFs) 246–47
genes 77, 93–98, 315–320
  expression 245–48
  mutations 93–94
  transfer 126–27
genetic code 32–33
genetic diseases 119
genetic drift 328
genetic engineering 124–132
genetically modified organisms (GMOs) 128–29
genetics 93–132
genome 94, 95–96, 100
genotypes 110, 113
genus 170
geographic (allopatric) speciation 329–330, 332
germination 302–3
gibberellin 303
glasswort 288, 290
global warming 153
globular proteins 60, 257
glomerulus 355, 357–58
glucagon 220
glucose 44, 78, 360
glycaemic index 223
glycogen 46, 54
glycolysis 78, 79, 269–270
GMOs (genetically modified organisms) 128–132
goitre 224
goldenrod gall fly 330–32
Golgi apparatus 16
gradualism 332
grana 277
greenhouse effect 151–55
growth, plant 297–300
GTFs (general transcription factors) 246–47
guanine 235–36
guard cells 287
halophilic bacteria 170
halophytes 288
haploid cells 100
Harvey, William 200–201, 232
HCG (human chorionic gonadotrophin) 343, 372
heart
  disease 57, 198–99
  physiology 195–98

  transplantation 196
helicase 72, 238, 243
helper T-cells 339
hemodialysis 362–63
hemoglobin 94–95, 256
hemophilia 117
herbicide-tolerant crops 128
Hershey, Alfred 236–37
heterotrophs 137–38, 141
heterozygous alleles 110
HFCs (hydrofluorocarbons) 152
hip joint 347, 348
His fibres 196, 197
histamine 341
histones 236
HIV (human immunodeficiency virus) 206–7
holoparasites 140
homeopathic remedies 52
homeostasis 220
homologous chromosomes 99–100
homologous structures 161–62, 184
homozygous alleles 110
Hooke, Robert 1
hormone replacement therapy (HTR) 225
hormones 220–230
  plant 298–300
human chorionic gonadotrophin (HCG) 343, 372
Human Genome Project 96, 259
human growth hormone 225
human immunodeficiency virus (HIV) 206–7
human physiology 187–232
human skin colour 321–22
Huntington's disease 119
hybridisation 181
hybridoma cells 343
hydrocarbon chains 56
hydrofluorocarbons (HFCs) 152
hydrogen bonds 49, 52–53, 235–36, 257
hydrogenated fats 57
hydrolysis reactions 48
hydrophilic properties 20, 51, 258
hydrophobic properties 20, 51, 257, 258
hydrostatic pressure 295
hyperglycemia 221
hypertonic solutions 29
hyphae 1
hypotheses, scientific 10–11
hypotonic solutions 29
ice sheets 155
imbibition 303
immobilised enzymes 65
immune response 337
immune system 202
immunity 204, 337, 339–341
in vitro fertilisation (IVF) 230–31
induced-fit enzyme action model 263
inductive reasoning 208

industrial enzymes 64–65
industrial melanism 166–67
infectious diseases 202–8
inflammatory response 341
inhalation 212
inheritance 110–18, 313–325
inhibitors, enzyme 263–64
initiation stage of translation 252
inorganic compounds 43
insects
  muscles 347–49
  osmoregulation 361
inspiratory reserve 210
insulin 220–23
integral proteins 20–21
intercostal muscles 212
internal fertilisation 370
interphase 34
intestines 189–192
introns 241, 244, 247
ionic bonds 257
iron-containing organic compounds 43
isoleucine 266
isotonic solutions 25, 29
IVF (in vitro fertilisation) 230–31
Jenner, Edward 344–45
jet lag 226
joints 346–47
'junk DNA' 244
karyograms 101
karyotyping 100–102, 107–8
kidney 353–361
kidney failure 362–63
knee joint 347, 348
Krebs cycle 271–73
LAC operon 241
lactate 79
lactation 375
lactose 66, 67
lagging strands 239–240
large intestine 191–92
leading strands 238–39
leptin 225
leucocytes (white blood cells) 200, 202, 337
leukaemia 10
LH (luteinising hormone) 228, 365
ligaments 346
light-dependent reactions 277–78
light-independent reactions 279–282
light spectrum 82
link reaction 271–73
linkage groups 315
linked genes 315–18
lipids 44, 45, 47, 55–58
lipophilic properties 20
lipoproteins 51–52
Locke, John 248
locus, gene 110
long-day plants 304
loop of Henle 355, 358–59

lung cancer 39–40, 214
lungs 210, 212
luteinising hormone (LH) 228, 365
lymphatic system 194
lymphocytes 203–4, 338–39
lysis 269
lysosomes 16
magnification 4–6
maize, genetically modified 131–32
Malpighian tubules 361
mantle cell lymphoma 344
marram grass 288–89
mast cells 341
mature mRNA 247
mature RNA 241
meiosis 103–9, 164, 308–12
melanin 321–22
melanism, industrial 166–67
melatonin 225–26
membranes 19–22, 23–30, 258
memory cells 204, 339
Mendel, Gregor 111–12, 123, 311–12
Mendelian ratios 316
menstrual cycle 228–29
meristems 297–98
mesocosms 138
messenger RNA (mRNA) 74–76
metabolic pathways 262
metabolic rate 223–24
metabolism 46, 262–67
metaphase 37, 105, 106
metastasis 38
methane 44, 50, 147–48, 152
methanogenic bacteria 170
micrometres 5
micropropagation 298
micropyles 302
microscopes 3–4, 18–19
milk, lactose-free 66
mineral ions 290
Mitchell, Peter 275
mitochondria 16, 31–32, 271–72
mitosis 34–37, 38
mitotic index 37
molecular biology 43–92
Mollusca 174
monoclonal antibodies 342–44
monohybrid crosses 115, 324
monomers 44
monosaccharides 45, 46, 54
monounsaturated fatty acids 56
Morgan, Thomas Hunt 325–26
moths, peppered 166–67
motor neurons 215, 216, 346
movement, animal 346–352
mRNA (messenger RNA) 74–76
multicellular organisms 6–8
multiple alleles 115
muscle fibres 350
muscle tone 350

muscles 346, 349–352
mutagens 38
mutations, gene 93–94, 122
mutualistic relationships 291, 301
myco-heterotrophy 140
myofibrils 350
myogenic contractions 196
myosin 350
NAD^+ 269
nastic responses 300
natural selection 164–68, 327–28
negative feedback 220, 265
neonicotinoids 219
nephrons 355, 356, 358–59
nerve impulses 215–16
nervous system 215–19
neurons 215–19
neurotransmitters 217–18
nicotine 218
nitrogenous waste 361–62
nodes of Ranvier 217
non-competitive inhibitors 264
non-cyclic phosphorylation 279
non-disjunction 107
non-polar amino acids 258
non-specific immunity 203
non-striated muscle 348
normal distribution 320
NTPs (nucleoside triphosphates) 243, 245
nuclear envelope 16
nucleic acids 46, 235–258
nucleolus 16
nucleoside triphosphates (NTPs) 243, 245
nucleosomes 236, 246
nucleotides 46, 68–69
nucleus, cell 16
null hypothesis 323
nutrient recycling 144–45
ocean acidification 154
oils 45
Okazaki fragments 239, 240
omega-3 fatty acids 57
oncogenes 39
oocytes 367
oogenesis 367–69
operons 241
Opuntia 288
organ transplantation 364
organelles 16
organic compounds 43–44
osmolarity 25
osmoregulation 353–361
osmosis 23, 24–26, 29
ovaries 301
overhydration 362
ovulation 228, 367
oxidation reactions 268
oxidative phosphorylation 273–74
oxytocin 375
ozone layer 152

palaeontology 160
pancreas 189
paradigm shifts 333
Paramecium 2–3
parasympathetic nerves 199
parental generation 113
passive immunity 340–41
passive transport 23
pathogens 202, 337, 338
PCR (polymerase chain reaction) 97, 124
peat 147–48
pedigree charts 118, 120–21
penicillin 208
pentadactyl limbs 161–62
pentose 46
peppered moths 166–67
pepsin 188
peptide bonds 75, 253
peripheral nerves 215
peripheral proteins 20–21
peristalsis 188, 189
petals 301
phagocytosis 28, 32, 203
phenotypes 110, 113
phloem 293–97
phospholipids 20
phosphorus-containing organic compounds 43
phosphorylation 269, 270
photoactivation 278
photolysis 85, 278
photoperiodism 303
photophosphorylation 278–79
photosynthesis 82–88, 276–282
phototropism 299
phylogenetics 171, 179–180
physiology 187–232, 337–375
phytochromes 304
pigs, genetically modified 128
pili 13
pineal gland 226
pinocytosis 28, 190
pituitary gland 224, 228
placenta 373–74
plant biology 285–304
plant cells 17
plant growth 297–300
plant tissue culture 298
plaque 198
plasma, blood 200
plasma cells 339
plasma membrane 12, 17
plasmids 12, 99
platelets 200, 202
Platyhelminthes 174
pluripotent cells 9
pneumocytes 213
polar amino acids 257–58
polar molecules 49
pollen 301

pollen tubes 302
pollination 301, 304
polyclonal response 338–39
polygenes 308, 321–22
polygenic inheritance 321–22
polymerase chain reaction (PCR) 97, 124
polymers 44
polypeptides 47, 59, 252–55
polyploidy 331–32, 334
polysaccharides 45, 46, 54–55
polysomes 255
polyspermy 371
polyunsaturated fatty acids 56
populations 136–37
Porifera 173–74
positive feedback 375
post-transcriptional modification 247–48
potometers 292, 293
precautionary principle 156
pregnancy 343, 372–75
prenatal screening 109, 375
prickly pear 288
primary follicles 367
primary tumours 38–39
producers (food chains) 141
progesterone 228, 375
prokaryotic cells 12–13, 17–18, 31–33, 99, 238
prolactin 375
promoter region 245
prophase 35–36, 104–5, 106
prostate gland 366
prosthetic groups 256
protein channels 23–24
proteins 59–61
 digestion 188
 in urine 360
 structure 45, 255–59
prothrombin 202, 203
proximal convoluted tubule 356, 358
punctuated equilibrium 332–33
Punnett grids 113
pure-breeding 110
purines 235–36
Purkinje fibres 196, 197
pyrimidines 235–36
pyruvate 78, 79
radioactive labelling 296, 353
rainforests 152
reabsorption 358
recessive alleles 110, 112
recombinant gametes 318
redox reactions 268
reduction reactions 268
refractory period 217
regulator genes 241
reliable data 66, 122–23, 149–150
replication forks 238, 239
reproduction
 human 226–232

plants 300–304
residual volume 210
resolution, microscope 3
resolving power 3
respiration 209–13
respirometers 80–81
resting potential 216, 217
retinal pigment epithelium (RPE) 10
ribonucleic acid (RNA) 68–77
ribose 44
ribosomes 12, 13, 16–17, 75–76, 251–52
ribulose biphosphate (RuBP) 280
rice genome 100
rigor mortis 352
risk assessment 128–29, 131–32, 156, 208, 376
RNA (ribonucleic acid) 68–77
RNA polymerase 75, 245
RNA primase 238, 239, 243
RNA splicing 241
root pressure 291
roots 290–91
RPE (retinal pigment epithelium) 10
RuBP carboxylase (Rubisco) 280
salamanders 329–330
Salicornia europaea 288, 290
saltatory conduction 217
SAN (sinoatrial node) 196
Sanger, Frederick 240
saprotrophs 138
sarcolemma 350
sarcomeres 350
sarcoplasmic reticulum 350
saturated fatty acids 56
scanning electron microscopes (SEMs) 3–4
secondary growth 297
secondary tumours 38–39
seeds 302
selection pressures 159
selective breeding 160–61
self-pollination 301
semi-conservative replication 72–73, 238
seminiferous tubules 365
SEMs (scanning electron microscopes) 3–4
sense strands 245
sepals 300
serendipity 40
Sertoli cells 365
sex chromosomes 101–2, 116–17
sex linkage 116, 117
sexual reproduction 165, 365–375
short-day plants 304
short tandem repeats 125
SI units 6
sickle-cell anemia 94–95, 97, 119
sieve tube cells 293
simple diffusion 23, 190
single-stranded binding proteins (SSBs) 238, 243
sink cells 293–94

sinoatrial node (SAN) 196, 197
sister chromatids 36, 37, 308–10
skeletal muscle 348, 349, 350
skin colour 321–22
small intestine 189–191
smallpox 344–45
smoking, and cancer 39–40
smooth muscle 348
sodium-containing organic compounds 43
solvents 51–52
somatic cells 100
source cells 293–94
speciation 329–332
species 136, 170
species barrier 338
specific immunity 203–5
sperm cells 366
spermatocytes 365
spermatogenesis 365–66, 369
spermatozoa 365
spindles, chromosome 36
Spirogyra 84
spirometers 210–11
SSBs (single-stranded binding proteins) 238, 243
stabilising selection 327
starch 54–55
Stargardt's disease 10
stem cells 8–10
stems 292
steroids 45
stigma 301
stomach 188
stomata 285–87
striated muscle 348, 350
style 301
sulfur-containing organic compounds 43
supercoiling 35–36
Svedberg unit 13
sympathetic nerves 199
sympatric speciation 329, 330–32
synapses 217–19
synovial joints 346
systems approach 8
T-cells 339
T2 phages 236–37
tandem repeats 242
Taq DNA polymerase 72
taxonomy 170
telomeres 241–42
telophase 37, 105, 106
TEMs (transmission electron microscopes) 3–4
tendons 346
termination stage of translation 255
terminator region 245
test crosses 110
testa 302
testis 365–66
testosterone 228, 248, 365

tetanus 219
tetraploids 331
β thalassemia 119
thalidomide 248
theories, scientific 10–11, 21, 168
thermophilic bacteria 170
threonine 266
threshold potential 217
thrombin 202, 203
thylakoid membranes 277
thymine 235–36
thyroxin 223–24, 225
tidal volume 210
tissue culture, plant 298
Tragopogon 331–32
trans fats 57
trans fatty acids 56–57
transcription 245–47
transcription factors 246–47
transcriptional regulation 246–47
transfer RNA (tRNA) 77, 242, 251
transgenic microorganisms 77, 126–27
translation 251–58
translocation (plants) 293
translocation (stage of translation) 253–54
transmission electron microscopes (TEMs) 3–4
transpiration 285–88
transplantation, organ 364
triglycerides 47, 57
trisomy 107
tRNA (transfer RNA) 77, 242, 251
trophic levels 137, 141–42
tropisms 299–300
tropomyosin 350–51
troponin 350–51
tumours 38–39
tundra 155
turgor 26, 292
ultrafiltration 357
umbilical cord 10, 372
unicellular organisms 1–3
unsaturated fatty acids 56
urea 48, 360
uric acid 361
urine 360–61
vaccination 204, 337–345
vacuoles 17
variation 159, 308–11
continuous 162–63, 320
discrete 320
meiosis and 108
vascular bundles 285
veins 194
ventilation 209–13
ventricles 195–96
vertebrates 175
vesicles 28
villi 190–91
viruses 202

vitalism theory 48
water 49–53
water potential 26
white blood cells (leucocytes) 200, 202, 337
Woese, Carl 171
X-ray crystallography 242–43
xerophytes 288
xylem 285–293, 295
yeast 80
Z lines 350
zygotes 301–2, 370

Acknowledgements

The authors and publishers acknowledge the following sources of copyright material and are grateful for the permissions granted. While every effort has been made, it has not always been possible to identify the sources of all the material used, or to trace all copyright holders. If any omissions are brought to our notice, we will be happy to include the appropriate acknowledgements on reprinting.

Artwork illustrations throughout © Cambridge University Press and © Geoff Jones

- The chapter on Nature of Science was prepared by Dr. Peter Hoeben.
- The publisher would like to thank Caryn Obert of Curtis High School, Staten Island, New York for reviewing the content of this second edition.

Cover image: Tischenko Irina/Shutterstock; p. 1 Dr Kari Lounatmaa/SPL; p. 2r M.I. Walker/SPL; pp. 2l, 138l Sinclair Stammers/SPL; pp. 5, 212, 366 Astrid & Hanns-Freider Michler/SPL; p. 10 Tek Image/SPL; p. 12 BSIP, SERCOMI/SPL; p. 13l Phototake Inc./Alamy; p. 13r Herve Conge/ISM/Phototake Inc.; p. 14 Moran & Rowley/Visualhistology.com; p. 15 Anton Page, Biomedical Imaging Unit, University of Southampton; p. 34 *all* Michael Abbey/SPL; pp. 42, 203, 271 CNRI/SPL; p. 70l A. Barrington Brown/SPL; pp. 70r, 196, 201, 345 Image source/SPL; p. 82 GIPHOTOSTOCK/SPL; p. 85 Nigel Cattlin/Alamy; p. 95 Eye of Science/SPL; pp. 100, 289b Power & Syred/SPL; p. 107t Denys Kuvaiev/Alamy; p. 107bl, br Soverign, ISM/SPL; p. 108 British Library/SPL; p. 125 J.C. Revy/ISM/SPL; p. 130 Jerome Wexler/SPL; p.132l, r Frans Lanting, Mint Images/SPL; pp. 138r, 331 Bob Gibbons/SPL; p. 140 Dr Morley Read/SPL; p. 148 Robert Harding World Imagery/Alamy; p. 154 Georgette Douwma/SPL; p. 160t Image Source/Alamy; p. 160b Chris Hellier/SPL; p. 161 J.S & S. Bottomley/ardea.com; p. 166l Chris Howarth/Galapagos/Alamy; p. 166m Dr P. MarazzI/SPL; p. 166r blickwinkel/Alamy; p. 167tl, tr John Mason/ardea.com; p. 167b John Durham/SPL; p. 169 Sheila Terry/SPL; p. 172 Greg Vaughn/Alamy; p. 172t Bob Gibbons/Alamy; p. 173b Matthew Oldfield/SPL; p. 174t Peter Scoones/SPL; p. 174m David Fleetham, Visuals Unlimited/SPL; p. 174b David Fleetham/Alamy; p. 191l Steve Gschmeissner/SPL; p. 191r Dennis Kunkel Microscopy, Inc./Phototake Inc.; p. 200 Biophoto Associates/SPL; p. 225 Oak Ridge National Laboratory/US Department Of Energy/SPL; p. 250 Randy L' Jirtle at the University of Bedfordshire and the University of Wisconsin; p. 255 Dr Elena Kiseleva/SPL; pp. 277, 290b Dr Jeremy Burgess/SPL; p. 282 Roy Kaltschmidt/Lawrence Berkeley National Lab Photographer; p. 288 Dirk Wiersma/SPL; p. 289tl Science Pictures Limited/SPL; p. 289m Sidney Moulds/SPL; p. 290t Adrian Bicker/SPL; pp. 296tl, 370 Dr Keith Wheeler/SPL; pp. 296bl, 368 Garry Delong/SPL; p. 298 Rosenfeld Images Ltd/SPL; p. 316 Hermann Eisenbeiss/SPL; p. 330 Valerie Giles/SPL; p. 350 Manfred Kage/SPL; p. 352 Dr Rosalind King/SPL; p. 362 Purestock/Alamy

Key
l = left, r = right, t = top, b = bottom, c = centre
SPL = Science Photo Library